TRIASSIC LIFE ON LAND

Critical Moments and Perspectives in Earth History and Paleobiology

CRITICAL MOMENTS AND PERSPECTIVES IN EARTH HISTORY
AND PALEOBIOLOGY

David J. Bottjer, Richard K. Bambach, and Hans-Dieter Sues, Editors

Mark A. S. McMenamin and Dianna L. S. McMenamin,
The Emergence of Animals: The Cambrian Breakthrough

Anthony Hallam, *Phanerozoic Sea-Level Changes*

Douglas H. Erwin, *The Great Paleozoic Crisis: Life and Death in the Permian*

Betsey Dexter Dyer and Robert Alan Obar, *Tracing the History
of Eukaryotic Cells: The Enigmatic Smile*

Donald R. Prothero, *The Eocene-Oligocene Transition: Paradise Lost*

George R. McGhee Jr., *The Late Devonian Mass Extinction:
The Frasnian/Famennian Crisis*

J. David Archibald, *Dinosaur Extinction and the End of an Era:
What the Fossils Say*

Ronald E. Martin, *One Long Experiment: Scale and Process in Earth History*

Judith Totman Parrish, *Interpreting Pre-Quaternary Climate
from the Geologic Record*

George R. McGhee Jr., *Theoretical Morphology: The Concept
and Its Applications*

Thomas M. Cronin, *Principles of Paleoclimatology*

Andrey Yu. Zhuravlev and Robert Riding, Editors, *The Ecology
of the Cambrian Radiation*

Patricia G. Gensel and Dianne Edwards, Editors, *Plants Invade the Land:
Evolutionary and Environmental Perspectives*

David J. Bottjer, Walter Etter, James W. Hagadorn,
and Carol M. Tang, Editors, *Exceptional Fossil Preservation:
A Unique View on the Evolution of Marine Life*

Barry D. Webby, Florentin Paris, Mary L. Droser, and Ian G. Percival, Editors,
The Great Ordovician Biodiversification Event

Frank K. McKinney, *The Northern Adriatic Ecosystem:
Deep Time in a Shallow Sea*

TRIASSIC LIFE ON LAND

The Great Transition

**HANS-DIETER SUES
AND NICHOLAS C. FRASER**

Columbia University Press
NEW YORK

Columbia University Press
Publishers Since 1893
New York   Chichester, West Sussex
Copyright © 2010 Columbia University Press
All rights reserved

Library of Congress Cataloging-in-Publication Data
Sues, Hans-Dieter, 1956–
Triassic life on land : the great transition / Hans-Dieter Sues and
Nicholas C. Fraser.
p. cm.
(Critical moments and perspectives in Earth history and paleobiology)
Includes bibliographical references and index.
ISBN 978-0-231-13522-1 (cloth : alk. paper)
ISBN 978-0-231-50941-1 (e-book)
1. Paleobiology—Triassic.   2. Paleontology—Triassic.   3. Paleobotany—Triassic.
4. Biotic communities.   I. Fraser, Nicholas C.   II. Title.   III. Series.
QE719.8.S84 2010 560′.1762—dc22
2009022848

Columbia University Press books are printed on permanent and durable acid-free paper.
This book is printed on paper with recycled content.
Printed in the United States of America

References to Internet Web sites (URLs) were accurate at the time of writing. Neither the authors nor
Columbia University Press is responsible for URLs that may have expired or changed since the
manuscript was prepared.

CONTENTS

Preface vii

PREFACE

This book provides an overview of the history of life on land during the Triassic Period. Following the most severe biotic crisis of the Phanerozoic, at the end of the Permian, the Triassic represented an extraordinarily important time in the history of terrestrial ecosystems. During some 50 million years, it witnessed not only the recovery of Earth's biotas from the end-Permian extinctions but also the emergence and initial diversification of many of the principal groups of extant animals and plants on land and in the sea.

In view of the fact that the Triassic was a critical period in the evolution of life on land, it is surprising that there have been few reviews of this subject. Padian's (1986a) symposium volume, *The Beginning of the Age of Dinosaurs*, represented the first real attempt to provide a synthesis of terrestrial vertebrate communities and biotic changes during the Late Triassic and remains a valuable reference. Subsequently, another conference volume, *In the Shadow of the Dinosaurs* (Fraser and Sues 1994), updated and further built on this effort. Two geographically more focused collections of papers, *Dawn of the Age of Dinosaurs in the American Southwest* (Lucas and Hunt 1989) and *Trias—Eine ganz andere Welt* (Hauschke and Wilde 1999), significantly expanded the scope of analysis by including plants and invertebrates. The tremendous progress in research in recent years clearly warrants a comprehensive, updated review. Fraser has recently published an introduction to Triassic continental ecosystems for a more general readership, *Dawn of the Dinosaurs* (Fraser 2006). "Critical Moments and Perspectives in Earth History and Paleobiology" provides an appropriate venue to present a more technical treatment of this exciting subject, aimed at advanced students and specialists. It is our hope that this book will kindle further interest in the evolution of life on land during the Triassic and serve as a resource to inform future research.

When we look at the world today we can readily discern distinct faunal and floral provinces, each with their own characteristics and varying degrees of endemism. Thus, the Australian bush, the Canadian tundra, and the East African savannah all have their own distinctive wildlife. It would be exciting to view Triassic terrestrial ecosystems in a similar way, looking at the various floral and faunal provinces and how they changed over this period. Although the existence of Pangaea resulted in cosmopolitan distributions of many animal and plant groups, there also appear to have been areas of endemism.

The last decade has witnessed many important new discoveries, and, as a result, we now have a much clearer picture of the world when dinosaurs and many elements of modern terrestrial ecosystems first appeared. Yet we are still a long way from developing a comprehensive picture of the history of life on land during the Triassic and its distribution across Pangaea. One of the principal obstacles to achieving these goals is our current inability to date most known Triassic terrestrial faunal and floral assemblages with precision. However, much work is now in progress to develop a high-resolution timescale for the Triassic, which is essential for addressing such issues as ecological succession, rates of morphological change, and extinction and speciation rates.

We have opted for a fairly conventional approach and review life on land during the Triassic in terms of the distribution of major fossil sites on contemporary continents. Although oversimplified, Romer's (1966) tripartite division of terrestrial vertebrate communities still proves useful. Shcherbakov (2008b) proposed a similar scheme for the evolutionary history of plants and insects during the Triassic.

Chapter 1 provides a general introduction to the Triassic Period and related topics. Chapters 2 through 9 survey the rich but widely scattered data garnered from the study of major assemblages of Triassic terrestrial animals and plants across the globe, their diversity, and their paleogeographic, paleoenvironmental, and stratigraphic contexts. They trace the succession from the aftermath of the end-Permian extinctions through the interval dominated by characteristic Triassic groups of animals and plants to the emergence of terrestrial biotas of distinctly modern aspect toward the end of the period. We focus on tetrapods, insects, and plants; other groups of terrestrial invertebrates and freshwater animals are still insufficiently known to integrate into a broader narrative. In chapter 10, we review some of the principal biological changes that occurred in terrestrial ecosystems during the Triassic. Finally, chapter 11 discusses the end-Triassic extinction event and looks at the continuing debate concerning the patterns and possible causation of this biotic crisis.

Many surprises still await us as long as researchers continue to explore exposures of Triassic strata across the globe and find new fossils. These new data will frequently force us to reassess our ideas. Recent discoveries of dinosauromorphs living alongside dinosaurs for millions of years in what is now the American Southwest (Irmis et al. 2007) and of a large dicynodont therapsid from the latest Triassic of Poland (Dzik, Sulej, and Niedzwiedzki 2008) represent just two examples to underscore that fact. In the years ahead, the development of a well-constrained chronostratigraphic framework for Triassic biotic evolution will be a key research objective. The efforts by Spencer Lucas and his associates to develop global correlation schemes for continental Triassic strata based on the distribution of certain tetrapod groups provide an important starting point. However, previously proposed schemes have been oversimplified and are not necessarily applicable across different paleolatitudes (Irmis et al. 2007; Mundil and Irmis 2008; Rayfield et al. 2005; Rayfield, Barrett, and Milner 2009). Future work on Triassic insects and plants is also likely to generate important new insights. There has been a resurgence of interest in Triassic insects, due in part to the discovery of important new occurrences such as the Solite Quarry in Virginia (chapter 9). In addition, recent studies have begun to elucidate the complex evolutionary history of interactions between insects and plants (e.g., Labandeira 2006). For plants, the emphasis on reporting form taxa has been an obstacle to getting a more complete picture of vegetational history. Many classical Triassic floras are in urgent need of restudy with modern methods (Kerp 2000). Furthermore, fieldwork continues to yield important new finds. For example, recent work in the Upper Triassic Yangcaogou Formation of Liaoning Province, China, has already led to the discovery of exceptionally complete specimens of a diversity of plants, many in growth position and some even with attached reproductive structures. Furthermore, the unexpected discovery of the corystospermalean reproductive structure *Umkomasia* in this unit further demonstrates that this group of plants was present deep in Laurasia and not restricted to Gondwana (Zan et al. 2008).

We suggest that readers who wish to learn more about the anatomy and evolutionary history of particular groups of Triassic animals and plants consult the following reference works: Carroll (1988) and Benton (2005) for all vertebrate groups, Kielan-Jaworowska, Cifelli, and Luo (2004) and Kemp (2005) for mammals and other synapsids, Weishampel, Dodson, and Osmólska (2004) for dinosaurs, Carroll (2009) for amphibians, Grimaldi and Engel (2005) for insects, and Taylor, Taylor, and Krings (2009) for plants. We have prepared an extensive bibliography to provide an introduction to the vast, widely scattered primary literature on the geology and paleontology of the continental Triassic.

We are indebted to Robin Smith for inviting us to write this book and to his successor at Columbia University Press, Patrick Fitzgerald, for his enthusiastic support of the project. At the Press, Irene Pavitt, senior manuscript editor, directed the production process, and valuable assistance was provided by Maria Petrova and Bridget Flannery-McCoy. Michael Haggett, production editor at Westchester Book Services, coordinated the editorial and production functions. Writing a work of this nature is impossible without the generous assistance and support of many friends and colleagues. First and foremost, we would like to acknowledge Paul Olsen. Both of us have had the good fortune to work with Paul in the field and laboratory for many years, and we have learned much from him through countless stimulating conversations. For assistance, discussions, information, and literature we are greatly indebted to Brian Axsmith, Gerhard Bachmann, Don Baird, Mike Benton, Dave Berman, José Bonaparte, Bob Carroll, the late Alan Charig, Sankar Chatterjee, Arthur Cruickshank, Ross Damiani, David Dilkes, Jerzy Dzik, Susan Evans, David Grimaldi, Andy Heckert, Jim Hopson, Randy Irmis, Farish Jenkins, Conrad Labandeira, Cindy Looy, Spencer Lucas, Zhe-Xi Luo, Andrew Milner, Sean Modesto, Sterling Nesbitt, Stefania Nosotti, Igor Novikov, Kevin Padian, Anna Paganoni, Bill Parker, Robert Reisz, Silvio Renesto, Olivier Rieppel, the late Pamela Robinson, Rainer Schoch, Paul Sereno, Bob Sullivan, Sebastian Voigt, Gordon Walkden, the late Alick Walker, Jonathan Weinbaum, the late Sam Welles, Dave Whiteside, Rupert Wild, and Kate Zeigler. Special thanks are due to Kevin Padian and Liz Sues for their careful editorial reviews and to two anonymous reviewers for helpful comments on the manuscript.

We thank Mike Benton, Jerzy Dzik, Jim Hopson, Farish Jenkins, Conrad Labandeira, Cindy Looy, Greg McHone, Sean Modesto, Leonardo Morato, Sterling Nesbitt, Igor Novikov, Francine Papier, Bill Parker, Fritz Pfeil, Robert Reisz, Ray Rogers, Bruce Rubidge, Tim Ryan, Rainer Schoch, and Jonathan Weinbaum for generously making available to us and allowing us to reproduce drawings and photographs. Special thanks are due to Greg Paul for granting us use of a series of his elegant reconstructions of Triassic archosauriform and archosaurian reptiles. We thank Doug Henderson for permission to reproduce two of his evocative scenes of Triassic life in the American Southwest in this book. Sarah Woenne expertly drafted several maps and diagrams. Hannah Fraser assisted with the compilation of the index.

We gratefully acknowledge the National Science Foundation and the National Geographic Society and Sues also acknowledges the Natural Sciences and Engineering Research Council of Canada for support of our research over the years.

Above all, our greatest appreciation goes to our families for their steadfast encouragement, love, and support.

TRIASSIC LIFE ON LAND

A scene from the Late Triassic of western North America. A large phytosaur (*Smilosuchus*) confronts a group of crocodylomorph reptiles. (Courtesy and copyright of Doug Henderson)

CHAPTER 1

Introduction

The Triassic Period marks one of the great transitions in the history of life (figure 1.1). At the end of the preceding period, the Permian, about 250 million years ago, ecosystems on land and especially in the sea had been devastated during the greatest biotic crisis of the Phanerozoic (Raup and Sepkoski 1982; Benton 2006; Erwin 2006). The Triassic was a time for new beginnings. During some 50 million years, it witnessed not only the recovery of ecosystems from the end-Permian extinctions but also the emergence and diversification of many of the principal groups of extant terrestrial and marine animals. Although the Triassic ended with another mass extinction on land and in the sea about 200 million years ago, at that point the structure and dynamics of ecosystems did not substantially differ from those of the present day.

As the first of the three periods constituting the Mesozoic Era, the Triassic is often referred to as the "Dawn of the Age of Dinosaurs." While this term is not inappropriate, the Triassic encompassed much more than this designation would imply. The first dinosaurs did appear during the Late Triassic, as did the first mammaliaforms, lepidosaurs, and turtles. Beetles proliferated, the first flies buzzed through the air, and the first water bugs sculled across lakes and ponds, possibly satisfying the appetites of some of the earliest frogs and salamanders. Among land plants, the first representatives of a number of extant groups of conifers and ferns appeared during the Triassic. These changes in terrestrial ecosystems parallel those in the marine biosphere, where calcareous nannoplankton (Erba 2006) and important extant groups of animals, including scleractinian corals (Stanley 2003) and teleost fishes or their closest relatives (Arratia 2001; Hurley et al. 2007), made their first appearances. Thus, a more appropriate designation for the Triassic would be "Dawn of Modern Ecosystems."

Ward (2006:160) observed that "the Triassic was a huge experiment in animal design." Numerous unusual land animals flourished during the Triassic and then apparently vanished near or at the end of this period. Among insects, the Titanoptera, best known from the Middle Triassic of Australia and the Middle or Late Triassic of Kyrgyzstan, deserve special mention. Close relatives of grasshoppers and their kin (Orthoptera), they could attain a wingspan of at least 40 centimeters (Sharov 1968; Grimaldi and Engel 2005). The forewings of many titanopterans bear prominent stridulatory structures, similar to those in present-day ensiferan orthopterans, and the

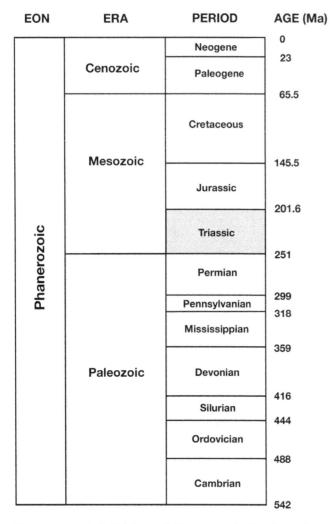

EON	ERA	PERIOD	AGE (Ma)
	Cenozoic	Neogene	0
			23
		Paleogene	
			65.5
	Mesozoic	Cretaceous	
			145.5
		Jurassic	
			201.6
Phanerozoic		Triassic	
			251
	Paleozoic	Permian	
			299
		Pennsylvanian	318
		Mississippian	
			359
		Devonian	
			416
		Silurian	
			444
		Ordovician	
			488
		Cambrian	
			542

Figure 1.1. Subdivision of Phanerozoic Eon based on the timescale by Gradstein, Ogg, and Smith (2004), with the date for the Triassic-Jurassic boundary modified based on Schaltegger et al. (2008).

sounds produced by these giant insects must have filled the forests of their time.

Among reptiles, *Tanystropheus*, from the Middle and Late Triassic of Europe, has an enormously elongated neck that is longer than its trunk and tail combined (Wild 1973; Nosotti 2007). *Longisquama*, from the Middle or Late Triassic of Kyrgyzstan, sports a row of greatly elongated, hockey-stick-shaped integumentary structures on its back (Sharov 1970; Voigt et al. 2009). *Drepanosaurus*, from the Late Tri-

assic of Italy, has a greatly enlarged ungual phalanx on the second digit of each hand and a peculiar, spikelike bone at the end of its apparently prehensile tail (Pinna 1980, 1984).

As its name indicates, the Triassic is divided into three parts. In 1834, the German salt-mining expert Friedrich August von Alberti proposed the name "Trias" (derived from the Greek word *treis* [three]) for a succession of lithostratigraphic units long recognized in southern Germany, which (from oldest to youngest) are the Buntsandstein (colored sandstone), Muschelkalk (clam limestone), and Keuper (derived from a word for the characteristic marls of this unit; figure 1.2). Of these units, the Buntsandstein and Keuper each comprise predominantly continental siliciclastic strata, whereas the intervening Muschelkalk is made up of marine carbonates and evaporites. Alberti noted that similar deposits were widely distributed across Europe and already suspected their presence in India and North America. The threefold rock succession established by Alberti corresponds roughly to the standard division of the Triassic into Lower, Middle, and Upper Triassic series, or, in units of geological time, Early, Middle, and Late Triassic epochs. Later, Alberti (1864) and other researchers employed fossils of marine invertebrates to correlate parts of the Alpenkalk, an old term referring to various carbonate units exposed along the northern and southern flanks of the European Alps, with the Triassic strata in the Germanic basin. During the late nineteenth century, geologists, mostly working in the European Alps, established what would become the standard marine stage–level division for the Triassic (from oldest to youngest): Scythian, Anisian, Ladinian, Carnian, Norian, and Rhaetian (figure 1.3). The Scythian was subsequently further divided into the Induan and Olenekian stages (Kiparisova and Popov 1956). Although this division has been formally adopted by the Subcommission

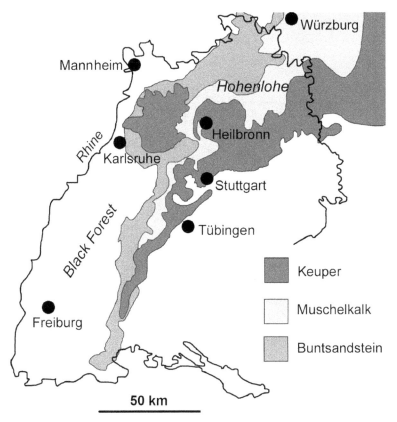

Figure 1.2. Distribution of strata of the classical Germanic Triassic—Buntsandstein, Muschelkalk, and Keuper—in the southern German state of Baden-Württemberg. (Modified from Schoch 2006b)

on Triassic Stratigraphy, some researchers still prefer a four-part subdivision of the Early Triassic (from oldest to youngest: Griesbachian, Dienerian, Smithian, and Spathian). During the second half of the nineteenth century, geologists began to identify and map marine strata of Triassic age in southeastern Europe, Turkey, the Himalayas, the western United States, and British Columbia, Canada. Since then, sedimentary rocks of Triassic age, mostly shallow-water marine carbonates and continental red beds, have been discovered on all continents and on the islands of Greenland, Madagascar, and Svalbard. Gregor (1970) estimated the combined maximum thickness of Triassic-age strata on Earth to be about 9 kilometers and the total volume of Triassic sedimentary rocks to be about 45 million cubic kilometers—substantially less than for either the Jurassic or Cretaceous periods, both of which, as a result, have

historically attracted much more paleontological interest than the Triassic (Sheehan 1977).

STRATIGRAPHIC CORRELATION

Published absolute dates for the Triassic Period and its stages have varied considerably over the years (figure 1.3). More recently, however, refined methods of radiometric dating have led to increasingly better resolution; still, much work remains to be done. Repeated uranium-lead (U-Pb) dating of zircon crystals from an ash bed at the Permian-Triassic boundary at Meishan in Zhejiang Province, China, has generated average ages ranging from 250 million to 251 million years (Ma) (Renne et al. 1995; Bowring et al. 1998). Recent revised processing has yielded a date of

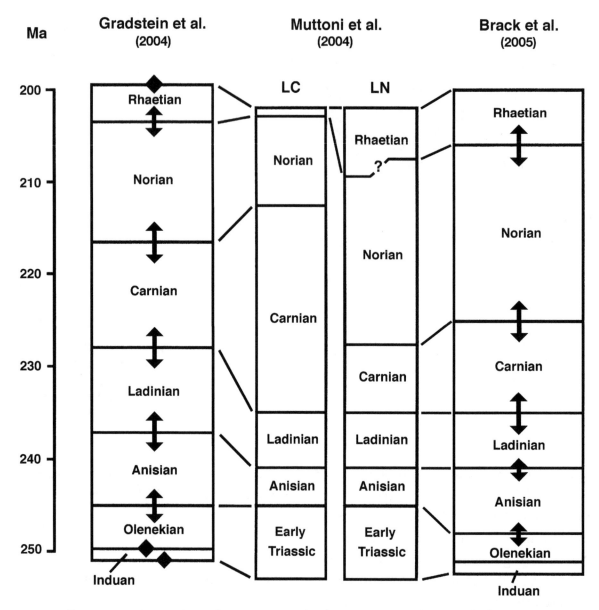

Figure 1.3. Comparison of recent timescales for Triassic Period. *Left to right*: time-scale by Gradstein, Ogg, and Smith (2004), timescale based on Muttoni et al. (2004), with "long Carnian" (LC) and (preferred) "long Norian" (LN) options, and timescale from Brack et al. (2005). (Modified from a diagram by Dickinson and Gehrels [http://gsa.confex.com/gsa/responses/2008CD/283.ppt])

252.6 ± 0.2 Ma for the Permian-Triassic boundary (Mundil et al. 2004). At the top of the period, zircons extracted from a tuff below the Triassic-Jurassic boundary in British Columbia generated a U-Pb date of 199.6 ± 0.3 Ma (Pálfy et al. 2000), but the validity of this result was later questioned (Pálfy and Mundil

2006). Most recently, Schaltegger et al. (2008) obtained new U-Pb dates from chemically abraded zircons from the Pucara Group in northern Peru, which place the Triassic-Jurassic boundary at 201.58 ± 0.17/0.28 Ma. Based on new magnetostratigraphic correlations using an astrochronological polarity timescale based on the Late Triassic formations of the Newark Supergroup in eastern North America, Muttoni et al. (2004) redated the boundary between the Carnian and Norian stages at about 228 Ma. This yields a very long Norian stage, with a duration exceeding 20 Ma (Furin et al. 2006).

Subdivided into Early, Middle, and Late Triassic, the ages for the stages of the Triassic, based on the latest (2009) Geologic Time Scale of the Geological Society of America, result in three very unequal time intervals, with durations of 6, 10, and 33.4 Ma, respectively.

The first problem encountered when trying to get a global overview of the Triassic is the correlation of the suite of sedimentary rocks in the Germanic basin with possibly coeval strata elsewhere. This proves particularly challenging due to the inherent difficulties of correlating continental and marine strata. For example, it has proven difficult to identify terrestrial equivalents of the Middle Triassic marine Muschelkalk. Consequently, it is still impossible to correlate many Triassic continental strata confidently at a level of resolution below that of the stage (Olsen and Sues 1986).

METHODS FOR STRATIGRAPHIC CORRELATION

Biostratigraphy

Although the use of fossils and fossil assemblages has a venerable tradition in the stratigraphic correlation of sedimentary rocks, this process often proves challenging because of the incompleteness of the fossil record. Not only must potential index fossils be common and relatively widespread, and have a short temporal range (in terms of geological time), but they should also include at least a few forms that provide tie-ins with the standard marine sequences.

Pollen and spores have proven particularly useful for the biostratigraphic correlation of nonmarine strata for much of the Phanerozoic. Produced by plants in vast quantities and easily dispersed over great distances by wind or water, they are very abundant in many strata and frequently even find their way into marginal marine environments. Moreover, there are many characteristic forms of pollen and spores whose plant producers were apparently short-lived (in terms of geological time) as well as widely distributed and thus are well suited as index fossils. Perhaps the greatest downside is that pollen and spores are susceptible to destruction under oxidizing conditions, a feature of many terrestrial depositional environments. Nevertheless, palynological zonation and correlation of Triassic continental strata have been widely employed with great success (e.g., Cornet 1993; Visscher and van der Zwan 1980; Litwin, Ash, and Traverse 1991; Heunisch 1999; Deutsche Stratigraphische Kommission 2005).

Dating back at least to the work of Huxley (1869), researchers have used tetrapod fossils for intercontinental correlation of Triassic continental sequences. In recent years, the principal advocates for a tetrapod-based zonation of Triassic continental strata have been Spencer Lucas and his former students Andrew Heckert and Adrian Hunt. Because of the inherent difficulties in correlating continental and marine formations, Lucas proposed a biochronological scheme for continental deposits independent from the Standard Global Chronostratigraphic Scale, which is based on marine strata. He and his associates have published many papers advocating the use of various tetrapod groups, especially phytosaurs and aetosaurs, for regional and even global correlation of continental sequences (e.g., Lucas 1993, 1998, 1999; Lucas and Huber 2003). Lucas (1998) proposed and defined eight successive land-vertebrate faunachrons (LVFs) during the Triassic Period. Each LVF was characterized by the first appearance datum (FAD) in the fossil record of a particular tetrapod taxon. For example, the first (oldest) LVF, the Lootsbergian, was defined on the FAD of the dicynodont therapsid *Lystrosaurus*. Lucas then augmented the definition of each LVF by the occurrence of additional characteristic taxa. Thus, additional characteristic tetrapods for the Lootsbergian are the procolophonid parareptile *Procolophon*, the cynodont therapsid *Thrinaxodon*, and the archosauriform reptile *Proterosuchus* (Lucas 1998).

However, Rayfield et al. (2005, 2009; see also Lucas et al. 2007) and other authors (e.g., Lehman and Chatterjee 2005; Parker 2005, 2007) have shown that several of Lucas's LVFs are problematical because their purported index fossils have longer stratigraphic ranges than originally assumed or have more restricted geographic distribution, or their taxonomic status is uncertain. New radiometric data also underscore the need for calibration of any biostratigraphic zonation scheme against a chronostratigraphic standard (e.g., Mundil and Irmis 2008).

Magnetostratigraphy

Earth has a magnetic field, which is driven by circulation in the planet's molten outer core and flows from pole to pole. At the present day, the north magnetic pole is located close to the north rotational pole (normal polarity). For reasons that are still not fully understood, the magnetic field reverses at irregular intervals, with the north magnetic pole moving close to the south rotational pole (reversed polarity). Such reversals have occurred many times during the Phanerozoic. Certain minerals, such as the iron oxide magnetite, are readily magnetized. Magnetite is common in a variety of rocks including basalts. When basalt cools from molten lava it passes through a threshold termed the Curie point, at which magnetite and other magnetic minerals take up and lock in magnetization from Earth's field at that time. In sedimentary rocks, minute particles of magnetic

minerals orient themselves with Earth's magnetic field at the time of deposition of the sediments. Sophisticated instrumentation now allows researchers to measure this "fossilized" magnetization and put together the complex history of episodic reversals of Earth's magnetic field in deep time, establishing what is called the Global Magnetic Polarity Time Scale. Using radiometric dating of rocks for precise calibration, a regional magnetostratigraphic succession can then be correlated with this global scale (e.g., Kent, Olsen, and Witte 1995; Muttoni et al. 1998, 2004). Magnetostratigraphy has become a critical tool for stratigraphic correlation.

TRIASSIC GEOGRAPHY

Paleogeographic reconstructions of Earth during the Triassic reveal a configuration of continents and seas very different from that of today. During the entire period, there existed a single, vast landmass that encompassed all present-day continents (figure 1.4). The German meteorologist Alfred Wegener, who first formulated the theory of continental drift, named this supercontinent Pangaea (Greek for "all earth") (Wegener 1915). The northern portion of this landmass is termed Laurasia, whereas the southern part of the supercontinent is named Gondwana. Both portions were partially separated by the vast embayment of Tethys in the east. Pangaea formed through a series of collisions between Asia, Laurentia, and Gondwana along the Allegheny, Ural, and Variscian mountain chains during the late Paleozoic. It was surrounded by subduction zones, where subduction of oceanic plate led to intense mountain building as well as the formation and deformation of foreland basins (Catuneanu et al. 2005; Golonka 2007). One interesting phenomenon is the accretion of

220 Ma

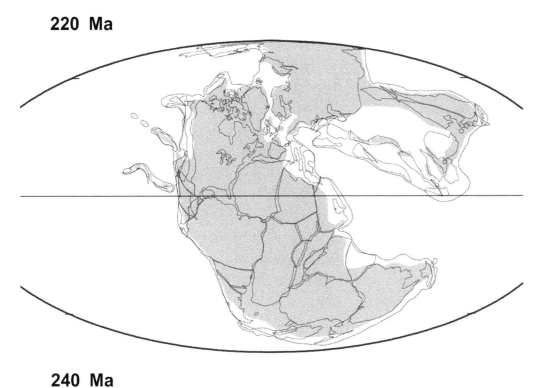

240 Ma

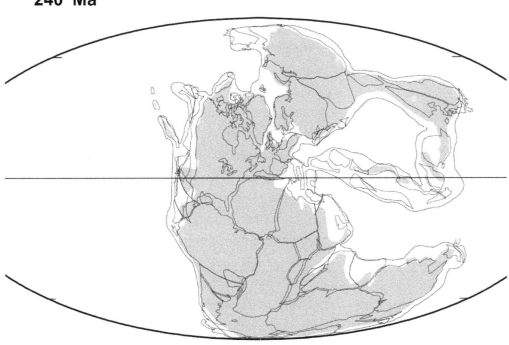

Figure 1.4. Reconstructions of Pangaea at 220 Ma and 240 Ma. Gray areas indicate land. Note the northward drift of Pangaea during the Triassic. (Modified from maps provided by the PALEOMAP Project; copyright of C. R. Scotese 2000)

numerous terranes, or small plates, to Pangaea during the Late Triassic and Jurassic (Golonka 2007). A number of such terranes drifted eastward and ultimately became attached to the western margins of what are now North and South America. In the east, portions of northern Gondwana separated and moved northward, closing Palaeotethys and eventually forming much of what is now eastern and southeastern Asia during the Late Triassic and Jurassic.

An immense ocean, Panthalassa (Greek for "all sea"), surrounded Pangaea. Global sea levels generally rose after a drop at the end of the Permian to a maximum during the Norian and then dropped again close to the end of the Triassic (Hallam 1992). Embry (1988) presented evidence for brief episodes with lower sea levels during the Ladinian and Carnian.

The existence of Pangaea had a profound impact on oceanic currents. Triassic climates, on both a regional and global scale, were markedly different from those of today (figure 1.5). To begin with, there was no significant landmass over much of the globe that would have impeded the exchange of oceanic waters from the equatorial belts to the polar regions. Consequently, the latter were significantly warmer, and there is no evidence for polar ice caps. Global paleoclimatic models suggest that Triassic Earth was a predominantly warm world (Selwood and Valdes 2006). In the north, warm temperate floras occurred as far as about 70° paleolatitude in northeastern Siberia during the Middle Triassic (Ziegler et al. 1993). Second, the physical structure of Pangaea itself, divided into increasingly equal areas of land north and south of the equator (Ziegler et al. 2003), and with the vast embayment of Tethys in the east, would have led to a peculiar climatic regime known as megamonsoon (Dubiel et al. 1991; Parrish 1993, 1999; Wilson et al. 1994). Low atmospheric pressure would

have developed over Pangaea during the cooler winter season, whereas warming during the summer led to the formation of high pressure. Because the supercontinent extended across the northern and southern hemispheres, a pronounced difference in atmospheric pressure existed between the summer and winter hemispheres throughout the year and led to air streaming across the equator and the Tethys for most of the year. The change in seasons caused this airstream to change direction twice a year. Consequently, most regions of the supercontinent would have experienced a monsoon climate with a long dry season and a shorter wet season with abundant rainfall. A pronounced east-west climatic asymmetry must have occurred across Pangaea because the eastern regions of this landmass (at least between paleolatitudes of about 40° north and 40° south) would have been warmer and wetter due to their proximity to the Tethys and the absence of the Atlantic Ocean facilitating oceanic heat exchange (Parrish et al. 1986; Parrish 1993, 1998, 1999). Finally, the northward drift of Pangaea during the Triassic led to increasingly drier conditions in many regions (Parrish et al. 1986; Olsen and Kent 2000; figure 1.4). This trend is clearly evident in sequences of Late Triassic sedimentary rocks in eastern North America and in the American Southwest but also in the basin fills in Argentina, Brazil, and South Africa, which moved from the southern temperate to the southern arid climate belts.

No single location or region of the globe today provides anything remotely approaching a complete picture of terrestrial life and environments during Triassic times. First, the existence of Pangaea does not imply that uniform environmental conditions prevailed across this vast landmass. As today, the distribution of land across such a broad range of latitudes resulted in regional differences in temperature and precipitation, as did the relative proximity to the

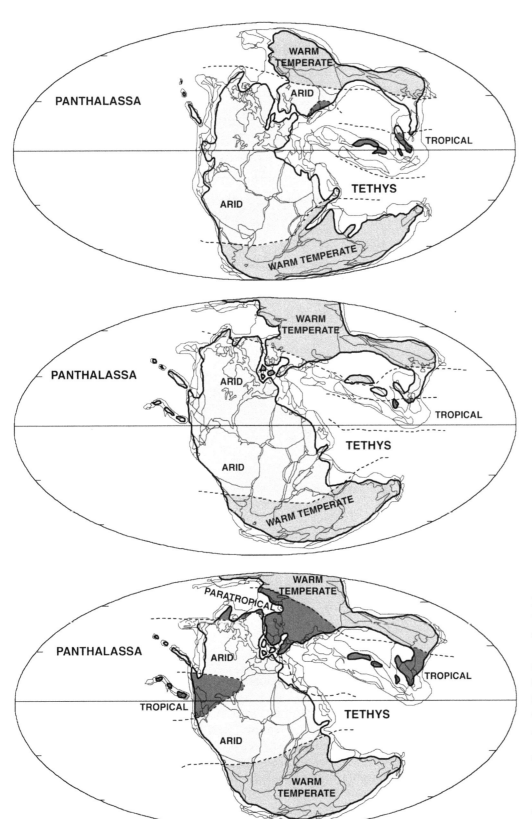

Figure 1.5. Reconstructions of the principal climatic zones of Pangaea based on the distribution of climate-sensitive indicator strata (coals, evaporites, etc.) for (*from top to bottom*) the Early, Middle, and Late Triassic. (Modified from maps provided by the PALEOMAP Project; copyright of C. R. Scotese 2000)

sea and the uneven distribution of mountain chains. Such factors would in turn control the regional distribution of terrestrial plants and animals. Second, the sedimentary rock record is notoriously incomplete, particularly for continental environments. There are no continuous sections documenting even significant intervals of Triassic time, let alone the entire period. Often a single local deposit formed under unusual geological circumstances has yielded fossils of otherwise unknown animals and plants (chapter 9). Thus, the fossil record comprises a series of snapshots that researchers must arrange to trace the complex history of Triassic continental environments and life.

For the Early Triassic, the most extensive and most fossiliferous sections of continental strata are known from the Karoo basin of South Africa as well as from China and Russia (chapter 2). For the Middle Triassic, the Upper Buntsandstein and its equivalents in central and western Europe, the Donguz Svita of southern Russia, and the Moenkopi Formation of the American Southwest (chapter 3) and the Chañares Formation of northwestern Argentina (chapter 4) provide informative glimpses of life on land. The record of continental ecosystems for the Late Triassic (figure 1.6) is much more extensive than that for the earlier parts of this period, which is

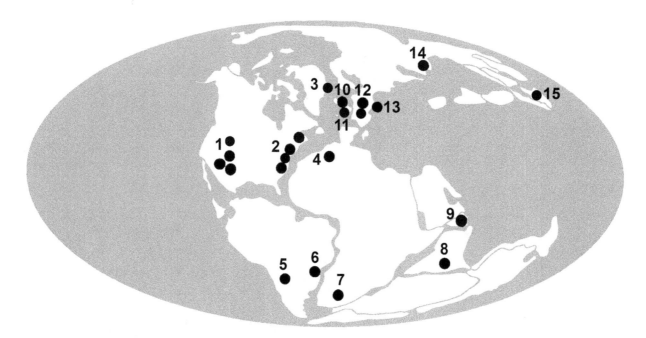

Figure 1.6. Diagrammatic map of Pangaea with important occurrences of Late Triassic continental vertebrates. Localities or locality clusters indicated by Arabic numerals: *1*, Chinle Formation and Dockum Group, western United States; *2*, Newark Supergroup, eastern North America; *3*, Fleming Fjord Formation, East Greenland; *4*, Argana basin, Morocco; *5*, Ischigualasto-Villa Unión basin, northwestern Argentina; *6*, Paraná basin, southern Brazil; *7*, Karoo basin, southern Africa; *8*, Pranhita-Godavari and neighboring rift basins, India; *9*, southwestern Madagascar; *10*, Lossiemouth Sandstone Formation, Scotland; *11*, fissure fillings, southwestern England and southern Wales; *12*, Germanic basin, Germany and eastern France; *13*, northern Italy; *14*, Madygen Formation, Kyrgyzstan; *15*, Khorat Plateau, Thailand. (Modified from Lucas and Huber 2003)

reflected by the amount of space devoted to this long time interval in this book. A series of fossil assemblages, especially from northwestern Argentina and southern Brazil (chapter 4), central and western Europe (chapters 5 and 6), eastern North America (chapter 7), and the American Southwest (chapter 8), offer important examples of Late Triassic continental ecosystems. Although these regions and assemblages form the basis for the discussions, reference is also made to other known occurrences of Triassic land animals and plants where appropriate.

TETRAPOD PHYLOGENY AND DIVERSITY

As noted in the preface, we place particular emphasis on tetrapods in this book. Triassic tetrapods have a rich fossil record, and even fragmentary fossils frequently preserve phylogenetically informative features. Most of the principal groups of extant amphibians and amniotes or their closest relatives made their first appearance in the fossil record during the Triassic (figure 1.7). In addition, the first dinosaurs and pterosaurs originated during this period.

In recent decades, many studies have reviewed the intra- and interrelationships of the major groups of tetrapods in an explicitly phylogenetic manner. It is important to keep in mind that such assessments always represent works in progress because the discovery of new and restudy of known fossils as well as the discovery of new and reassessment of previously published characters continuously generate new phylogenetic hypotheses. In turn, these changing views of relationships are reflected by changes in classification, and frequently disagreements arise concerning the definition of (and taxonomic designation for) a particular clade. For many groups, there now exists a bewildering array of higher-level taxonomic names with multiple, often conflicting definitions. In addition, some familiar names have been imbued with new meanings. Thus, we present here a brief overview of the principal groups of tetrapods known from the Triassic and a series of cladograms

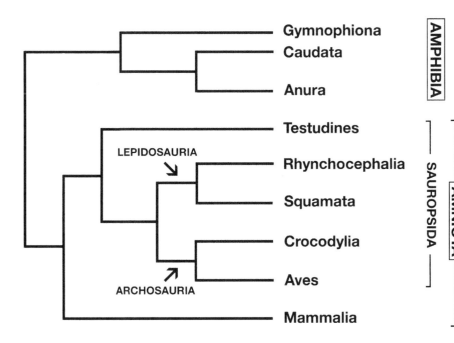

Figure 1.7. Consensus phylogeny of the principal groups of present-day tetrapods (Amphibia and Amniota).

to illustrate our preferred hypotheses of the inter-relationships of these taxa.

AMPHIBIA

Present-day Amphibia comprise Anura (frogs and toads), Caudata (salamanders), and Gymnophiona (caecilians) (figure 1.7). Most are small animals and have moist, naked skin that is rich in glands and forms an important organ for respiration. As they also depend on water for their reproduction, most spend their lives in or at least near water. Most extant adult amphibians have bicuspid, pedicellate teeth. This type of tooth has a weakly mineralized zone between the crown and the base, or pedicel. Extant amphibians also share the reduction or absence of various bones in the skull roof and palate.

There still exists no consensus on whether present-day amphibians form a natural (monophyletic) group. Various authors have argued that at least the limbless Gymnophiona are derived from a different ancestor than Anura and Caudata (which are commonly grouped together as Batrachia) (see Carroll 2009).

Traditionally, the name "Amphibia" has been employed more broadly to refer to all tetrapods that are not amniotes, but this is rather misleading because most of these forms were not similar to extant amphibians and may have even been different biologically. Among the Paleozoic and early Mesozoic "amphibians," most authors have distinguished two principal groups—Lepospondyli and Temnospondyli. The exclusively Paleozoic lepospondyls comprise a heterogeneous assemblage of small forms with holospondylous vertebrae. By contrast, temnospondyls are a remarkably diverse group that ranged in time from the Mississippian to the Early Cretaceous and include the stem forms of at least frogs and salamanders. Temnospondyls typically have solid, completely roofed skulls. The anterior palatal bones border a prominent interpterygoid vacuity, and the parasphenoid bone comprises a broad posterior plate and an anteriorly tapering cultriform process. In the multipartite vertebrae of temnospondyls, the intercentrum forms the largest element and is loosely connected to the neural arch; the small pleurocentra are paired and tend to contact each other dorsally along the midline. During the Triassic, temnospondyls were represented by a variety of lineages and were semi-aquatic or fully aquatic carnivores in many ecosystems. Most Triassic temnospondyls are referred to the Stereospondyli, which are characterized by the possession of a single, large centrum in each vertebra.

Stereospondyls comprise two principal groups (Yates and Warren 2000; figure 1.8). The Capitosauroidea (or Mastodonsauroidea) include often large forms with a dorsoventrally flattened trunk and head and an elongated snout, which lends them a superficially crocodile-like appearance. A characteristic feature of capitosauroids is the presence of a deeply incised otic notch, or opening (which presumably accommodated the tympanic membrane), along the posterior margin of the skull roof.

A second major group, Trematosauria, encompasses a considerable diversity of taxa, ranging from forms with an often greatly elongated snout (Trematosauridae) to ones with short but very broad heads (Plagiosauridae).

ANTHRACOSAUROIDEA

Anthracosauroidea are a still poorly understood assemblage of Paleozoic tetrapods generally considered closely related to amniotes. Characteristic features of anthracosaurs include the contact between the parietal and tabular bones in the skull roof (unlike the

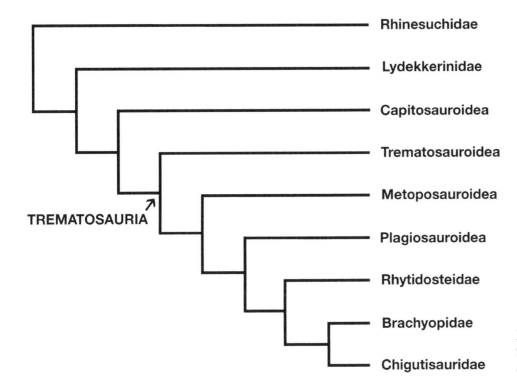

Figure 1.8. Phylogeny of Stereospondyli based on Yates and Warren (2000).

condition in temnospondyls), presence of long tabular horns, absence of posttemporal fossae, and presence of five fingers in the manus (rather than only four as in temnospondyls). Anthracosaurs retain a loose junction between the skull roof and the cheek plate. As in amniotes, the pleurocentrum is the dominant element of the vertebrae and is typically firmly connected to the neural arch. Anthracosaurs were most common during the Pennsylvanian and Early Permian. However, one derived group, the Chroniosuchia, survived well into the Triassic.

AMNIOTA

Among extant tetrapods, reptiles, birds, and mammals are characterized by the shared possession of the amniotic egg, in which a membrane surrounds and encloses the embryo in a liquid-filled sac, and thus are grouped together as Amniota (figure 1.7). Obviously

the structure of the egg is not particularly useful for identifying extinct tetrapods, and various phylogenetic analyses have hypothesized a suite of derived skeletal features that appear diagnostic for amniotes. They include the participation of the frontal in the margin of the orbit, a convex occipital condyle, a scapulocoracoid with (primitively) three separate ossifications, and presence of an astragalus in the ankle (Gauthier, Kluge, and Rowe 1988). Amniotes have an extensive fossil record dating back to the Pennsylvanian; both of the principal lineages of amniotes, Sauropsida and Synapsida, were already present at that time.

Sauropsida comprises reptiles (Reptilia) and their descendants, birds (Aves) (figure 1.7). Some basal forms have skulls with a solid temporal region, but in most sauropsids, one or, more commonly, two openings, or fenestrae, perforate the temporal (cheek) region (diapsid condition). These openings lighten the skull, and their bony margins provide attachment for jaw-closing (adductor) muscles and their overlying connective tissue (fascia). Turtles (Testudines) lack temporal fenestrae and thus were long considered closely related to basal reptiles, which also lack these openings. However, many authors now interpret turtles as diapsid derivatives (Rieppel and Reisz 1999; Lee et al. 2004). A competing hypothesis holds that turtles form part of a clade along with various Permian and Triassic basal amniotes known as Parareptilia (chapter 2).

Most sauropsids have two temporal fenestrae on each side of the skull and are grouped together as Diapsida. In some taxa, one (the lower) or both of those openings may be secondarily absent. An additional diagnostic feature of Diapsida is the presence of a suborbital fenestra bordered by the maxilla, palatine, and ectopterygoid on the palate (Benton 1985). Except for some basal taxa of late Paleozoic age, most diapsids can be referred to one of the two principal groups, Archosauromorpha and Lepidosauromorpha.

Archosauromorpha comprises Archosauria and a number of related clades (figure 1.9). Diagnostic skeletal features for this clade include the presence of a posterodorsal process on the premaxilla, slender, tapering cervical ribs that attached to the vertebrae at a low angle, and the differentiation of the dorsal portion of the ilium into a short anterior and long posterior process (Benton 1985; Gauthier, Kluge, and Rowe 1988; Dilkes 1998). Extant archosaurs comprise only crocodylians and birds, which descended from theropod dinosaurs. Present-day archosaurs share numerous anatomical features, such as pneumatization of numerous cranial bones starting from the middle ear region, presence of a muscular stomach chamber (gizzard) that accommodates ingested stones or sand to break down food, and a completely divided ventricle in the heart. During the Mesozoic, archosaurian reptiles were much more diverse. Using the crown-group approach (Jeffries 1979), Archosauria

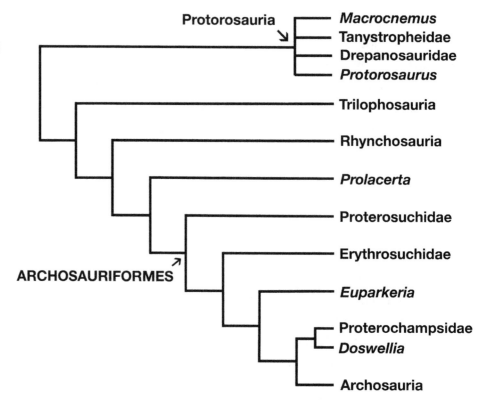

Figure 1.9. Phylogeny of Archosauromorpha based on Dilkes (1998) and Modesto and Sues (2004).

can be defined as comprising the last common ancestor of Crocodylia and Aves and all descendants of that ancestor. This definition of Archosauria includes dinosaurs and pterosaurs as well as a number of extinct groups that can be placed either on the evolutionary lineage leading to crocodylians or on that leading to birds. However, it excludes various taxa of mostly Early and Middle Triassic age, which share with Archosauria the presence of an antorbital fenestra (often surrounded by a distinct depression, or fossa) between the orbit and the external narial opening on the side of the snout as well as the presence of a fenestra at the back of each mandibular ramus (external or lateral mandibular fenestra). These forms are grouped with Archosauria in a clade, Archosauriformes. In turn, Archosauromorpha comprises Archosauriformes and various closely related taxa, which are mostly Triassic in age.

Archosauria is divided into two principal clades, for which we employ the names Crurotarsi (or Suchia) and Ornithodira (figure 1.10). Crurotarsi comprises Crocodylia and a number of related lineages, whereas Ornithodira is made up of Dinosauria (including birds), Pterosauria, and a few related taxa. As their

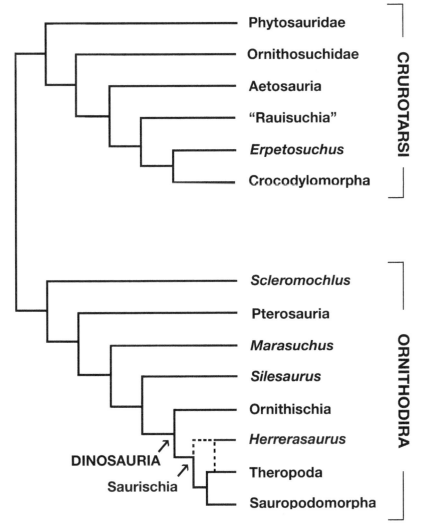

Figure 1.10. Consensus phylogeny of Archosauria. Two alternative positions for *Herrerasaurus* (chapter 4) are indicated by broken lines.

name implies, crurotarsans are characterized by the presence of a mobile joint between the astragalus and calcaneum in the ankle. The astragalus is attached to the tibia and fibula and bears a peg that fits into a socket on the calcaneum. Thus, during locomotion, the body rotates past the foot. By contrast, the body in ornithodirans moves over the foot, with the ankle flexing in a hingelike motion. Furthermore, ornithodirans typically have a long, slightly S-shaped neck, which holds the head higher than the trunk as in birds.

Lepidosauromorpha comprises Lepidosauria and various still poorly known taxa of Permian and Triassic age. The most characteristic feature of this clade is the presence of a large, ossified sternum, which articulates with the coracoids and has been shown to play an important role in locomotion in extant lizards. Lepidosauria comprises Squamata (lizards, amphisbaenians, and snakes) and Rhyncho-

cephalia (or Sphenodontia). Squamates are one of the most diverse groups of present-day tetrapods, whereas rhynchocephalians survive today only as two species of the tuatara (*Sphenodon*) on some isles off the coast of New Zealand. Extant lepidosaurs share a suite of anatomical features including a longitudinally (rather than transversely) oriented cloacal slit, a separate "sexual" segment in the kidney, and skin that contains a distinctive form of keratin and is periodically shed in large pieces (Gauthier, Kluge, and Rowe 1988).

SYNAPSIDA

Mammals and a considerable number of related amniote lineages of late Paleozoic and early Mesozoic age share the derived presence of a single large temporal opening behind the orbit (synapsid condition).

Figure 1.11. Phylogeny of Cynodontia including basal Mammaliaformes based on Hopson and Kitching (2001).

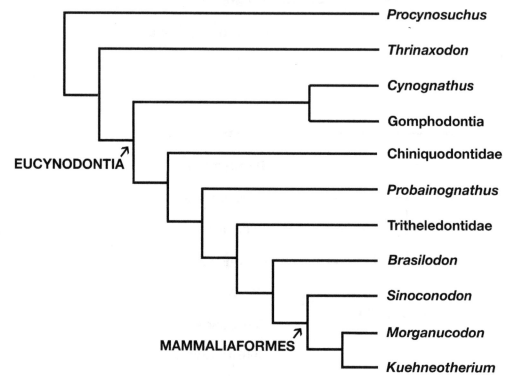

The more derived synapsids of Middle-Late Permian and Triassic age are grouped together as Therapsida (Kemp 2005). Therapsids were the most abundant and diverse terrestrial tetrapods during the Late Permian, but their diversity decreased significantly following the mass extinction at the end of the Permian. A few lineages persisted into the Triassic, during which two groups, Dicynodontia and Cynodontia, underwent diversification and formed important elements of many terrestrial tetrapod communities.

The Dicynodontia have a highly modified skull with a short snout and long temporal region. The premaxilla and usually the dentary lack teeth, and the front of the snout was apparently covered by a keratinous beak in life. Typically each maxilla has a large, caniniform tooth. The jaw joint forms a long sliding contact, which facilitated retraction of the mandible to cut plant fodder between the edges of the beak.

The Theriodontia comprise the morphologically most diverse group of therapsids and include two groups with Triassic taxa, the Therocephalia and the Cynodontia. The most diagnostic feature of theriodonts is the presence of a distinct coronoid process on the posterodorsal portion of the dentary; this process served for insertion of the jaw adductor musculature. The snout is wider than it is tall and has a broadly convex dorsal surface. In both therocephalians and cynodonts, the temporal fossae are wide open dorsally rather than separated by an expanse of skull roof as in other therapsids.

The Cynodontia are of particular interest because they are the stem group of mammals (figure 1.11). We employ here the crown-group definition of Mammalia as comprising the last common ancestor of Monotremata, Marsupialia (Metatheria), and Placentalia (Eutheria) and all of its descendants (Rowe 1988; but see Kielan-Jaworowska, Cifelli, and Luo 2004). This definition excludes most early Mesozoic "mammals," which are classified with Mammalia in a larger grouping, Mammaliaformes.

CHAPTER 2

Early and Early Middle Triassic in Gondwana

THE KAROO BASIN OF SOUTH AFRICA

The Karoo basin covers nearly two-thirds of the present-day territory of South Africa, an area of some 300,000 square kilometers (Smith 1990). Subduction of oceanic plate underneath Gondwana during the late Paleozoic and early Mesozoic led to the formation of the Gondwanides, a continuous mountain range that includes the Cape Fold Belt in South Africa and the Sierra de la Ventana Fold Belt in Argentina, and associated foreland basins (Veevers et al. 2004; figure 2.1). Aside from the Karoo basin, these foreland basins include the Paraná basin in South America, the Beacon basin of Antarctica, and the Bowen basin in Australia (Catuneanu, Hancox, and Rubidge 1998). The term "Karoo" is also frequently applied to the sedimentary fill of similar stratigraphic age in a number of other basins throughout southern and East Africa and even in other regions of Gondwana. Catuneanu et al. (2005) published a detailed review of the genesis and geology of all the South and East African basins, and Rubidge (2005) provided an overview of the Permo-Triassic tetrapod faunas of the Karoo.

In the Karoo basin, sedimentation took place from the late Paleozoic to the Early Jurassic when the onset of the breakup of Africa, Antarctica, and India led to an episode of substantial volcanic activity with emplacement of flood basalts (figure 2.2). The Late Permian and Triassic continental strata of the basin preserve a remarkably rich fossil record of a succession of terrestrial ecosystems, which have been extensively studied ever since the first report of tetrapod remains by Bain (1845). The Karoo basin was located at relatively high paleolatitudes, between 45° and 55° south, during the Early Triassic.

If one could use just two images to illustrate the record of Early Triassic terrestrial environments, they would most likely be of the dicynodont therapsid *Lystrosaurus* and the red mudstones in which it occurs in the Karoo basin. Known from countless specimens, *Lystrosaurus* is by far the most common tetrapod taxon in what Groenewald and Kitching (in Rubidge 1995) have formally defined as the *Lystrosaurus* Assemblage Zone (figure 2.3). This biozone comprises the uppermost part of the Balfour Formation (Palingkloof Member), the Katberg Formation, and the lower third of the Burgersdorp Formation of the Beaufort Group in the Karoo basin. In the field, the base of the *Lystrosaurus* Assemblage Zone strata is marked by a change in the predominant color of the mudstones from the

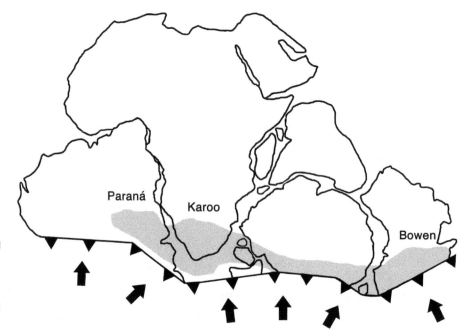

Figure 2.1. Subduction zone along the southern margin of Gondwana and associated foreland basins (shown in gray). Arrows indicate subduction of oceanic crust. (Based on De Wit and Ransome 1992)

greenish gray of the underlying Late Permian strata of the *Dicynodon* Assemblage Zone to red and purplish (Groenewald and Kitching in Rubidge 1995). There is also a notable increase in the sandstone content of the sequence. In the southeastern Karoo basin, the strata of the *Lystrosaurus* Assemblage Zone can reach a thickness of at least 830 meters but decrease in thickness toward both the north and west (figure 2.4). The depositional environments of these sedimentary rocks varied considerably throughout the basin, ranging from predominantly braided rivers in the south to meandering rivers and shallow lacustrine settings in the north. Based on sedimentological features such as large calcareous concretions and pervasive calcitic encrustation of fossil bones, Smith (1995) interpreted the climatic conditions as warm, semiarid to arid with high seasonal precipitation. The *Lystrosaurus* Assemblage Zone is dated as mostly earliest Triassic (Induan), although Hancox (2000) raised the possibility that the uppermost part of this biozone might already be early Olenekian in age. The tetrapod assemblage of the *Lystrosaurus* Assemblage Zone is the type assem-

blage of the Lootsbergian LVF as defined by Lucas (1998) (chapter 1).

Attaining a length of up to 1 meter, *Lystrosaurus* is distinguished from other dicynodont therapsids by its short, distinctly downturned snout and the foreshortened temporal region, both of which lend the skull its distinctive "squashed" appearance (Cluver 1971; figure 2.5*B*). As in other dicynodonts, each maxilla holds a prominent tusk.

Lystrosaurus has a very wide geographic distribution and is known elsewhere from the lower part of the Fremouw Formation of the central Transantarctic Mountains in Antarctica (Colbert 1974), the Jiucaiyuan Formation of the Junggar basin in Xinjiang, China (Young 1935), the Panchet Formation of the Damodar basin in India (Huxley 1865; Tripathi and Satsangi 1963; Ray 2005), and the Vokhmian Gorizont of Nizhnii Novgorod Province in Russia (Kalandadze 1975). Historically, it provided the first compelling paleontological evidence for the existence of Pangaea during the early Mesozoic (Colbert 1974).

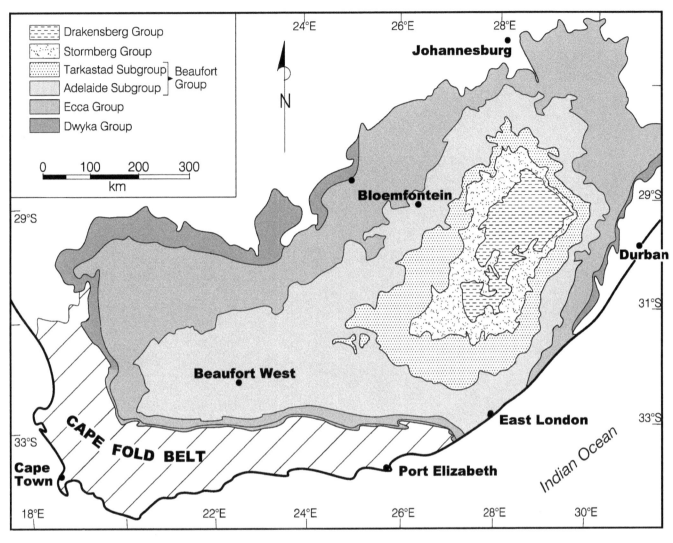

Figure 2.2. Geological map of the Karoo basin in southern Africa. (Map courtesy of B. S. Rubidge)

Period	Formation	Vertebrate biozonation	
JURASSIC	Upper Elliot		*Massospondylus* R.Z.
TRIASSIC	Lower Elliot		*Euskelosaurus* R.Z.
	Molteno		
	Burgersdorp		*Cynognathus* A.Z.
	Katberg		*Lystrosaurus* A.Z.
PERMIAN	Balfour		*Dicynodon* A.Z.

Figure 2.3. Stratigraphic succession and biostratigraphic zonation of the Beaufort Group and "Stormberg Group" in the Karoo basin. (Modified from Hancox 2000)

Figure 2.4.   Outcrop of strata of *Lystrosaurus* Assemblage Zone at Barendskraal, South Africa. (Photograph courtesy of S. P. Modesto)

In recent years, new discoveries have shown that the temporal range of *Lystrosaurus* was not restricted to the Early Triassic. Some species of *Lystrosaurus* already occur in Late Permian strata, and identification of specimens to the species level now becomes critical if *Lystrosaurus* is to be used as a biostratigraphic indicator taxon. Grine et al. (2006) and Botha and Smith (2007) recognized four valid species of *Lystrosaurus* in the Karoo basin. Three of these, *Lystrosaurus curvatus*, *L. maccaigi*, and *L. murrayi*, are first found in Late Permian strata, and *L. mac-*

caigi has not been recorded from Triassic rocks to date. Although *L. curvatus* extends into the earliest Triassic, the uppermost part of the Balfour Formation, it has not yet been found in the overlying Katberg Formation. However, *L. murrayi* extends into the Katberg Formation. The fourth species, *L. declivis*, also ranges into the Katberg but has not been recorded from the Late Permian to date. Thus, it is the individual species of *Lystrosaurus*, rather than the genus itself, that are useful as biostratigraphic indicators.

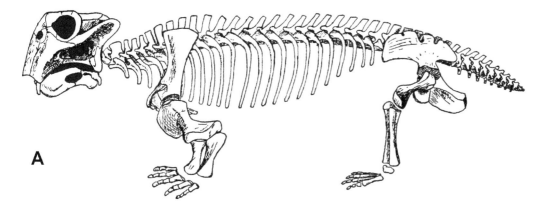

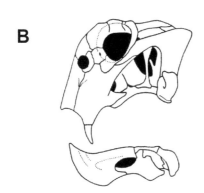

Figure 2.5. (*A*) Lateral view of the reconstructed skeleton of the lystrosaurid dicynodont *Lystrosaurus*. Length about 1 meter. (Modified from Watson 1912) (*B*) Lateral view of the skull of *Lystrosaurus declivis*. (Modified from Cluver 1971)

Traditionally, researchers considered *Lystrosaurus* amphibious in its habits, but this interpretation has been questioned more recently. King (1991) refuted in detail the anatomical features used by earlier authors (e.g., Watson 1912) to argue for an aquatic mode of life. King and Cluver (1991) reconstructed *Lystrosaurus* as a fully terrestrial animal that was possibly capable of excavating burrows for itself. In documenting a find of skeletal remains of a *Lystrosaurus* from the lower part of the Fremouw Formation of Antarctica, Retallack and Hammer (1998) noted that the bones had undergone subaerial exposure and were scattered on a paleosol surface. They accepted the interpretation of *Lystrosaurus* as fully terrestrial and capable of burrowing. Groenewald (1991) reported casts of large burrows from several localities in the *Lystrosaurus* Assemblage Zone. These burrows can reach a length of up to 2 meters and a diameter of up to 45 centimeters and, based on their size, may have been produced or at least occupied by *Lystrosaurus*.

Although it comprises up to 95 percent of identifiable tetrapod fossils recovered at many localities, *Lystrosaurus* is not the only tetrapod known from its eponymous biozone (Rubidge 1995, 2005). The procolophonid *Procolophon* is frequently found with *Lystrosaurus*, and the archosauriform *Proterosuchus* and the cynodont therapsid *Thrinaxodon* also occur in association with *Lystrosaurus*. Other tetrapods from the *Lystrosaurus* Assemblage Zone include a variety of temnospondyls such as *Lydekkerina*, the "protorosaur" *Prolacerta*, and therocephalian therapsids such as *Moschorhinus* and *Regisaurus* (Rubidge 1995, 2005). Associated plant fossils include petrified wood (*Dadoxylon*), the glossopteridalean *Glossopteris*, and the equisetalean *Schizoneura*. Contrary to the impression generated by surveying the literature, some of the smaller tetrapod taxa are fairly common. For example, many as yet unprepared nodules containing skulls and partial skeletons of *Procolophon* and *Prolacerta* are housed in museum collections around the world.

Lydekkerina and its relatives were small (30–35 centimeters long) temnospondyls with narrow, in dorsal view triangular skulls and orbits located at the midlength of the skull (Warren 2000; Warren, Damiani, and Yates 2006). They are considered primarily terrestrial in habits based on the poorly developed lateral-line canal grooves on the skull and the well-ossified postcranial skeleton. The Rhytidosteidae, represented by *Pneumatostega* and *Rhytidosteus*, have triangular or parabolic, flattened, and rather short

skulls, which can reach a length of up to 40 centimeters (Warren 2000). The palate has a dense cover of tiny teeth, and the marginal teeth are small. The *Lystrosaurus* Assemblage Zone is noteworthy for the occurrence of the last known representatives of two temnospondyl clades from the late Paleozoic, the amphibamid dissorophoid *Micropholis* (Schoch and Rubidge 2005) and the tupilakosaurid *Thabanchuia* (Warren 1999).

Procolophonids are small to medium-sized (up to 50 centimeters long) parareptiles, which superficially resemble lizards in appearance (figure 2.6). They ranged in time from the latest Permian to the end of the Triassic Period and had a worldwide distribution. Characterized by a typically broadly triangular head,

Figure 2.6. (*A*) Dorsal view of the reconstructed skeleton of the procolophonid *Procolophon trigoniceps*. (Modified from Watson 1914) (*B*) Dorsal view of the skull of *Procolophon trigoniceps*. (Redrawn from Reisz and Sues 2000a)

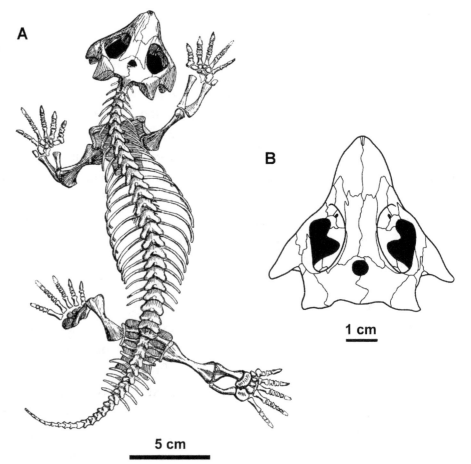

A

B

1 cm

5 cm

many procolophonids have transversely wide marginal teeth in the more posterior portions of the maxillae and dentaries (Ivakhnenko 1979; Carroll and Lindsay 1985). Typically, the large orbit is strongly emarginated posteriorly, imparting a keyhole shape to it. In *Procolophon*, the orbit is only moderately emarginated posteriorly (figure 2.6B). As in turtles, the lack of fenestration in the temporal region of the procolophonid skull was long considered a plesiomorphic feature. However, the posterior embayment of the orbit in procolophonids presumably created bony edges for attachment of jaw adductor muscles, much like the temporal openings in diapsid reptiles and in synapsids.

Another diagnostic feature of *Procolophon* is the prominent lateral bony protuberance on each quadratojugal. This process often bears fine ridging and grooving and may have been covered by a horn in life (Carroll and Lindsay 1985). *Procolophon* is known from many well-preserved, articulated skeletons, some of which appear to have been preserved in burrows (Groenewald 1991). It shows a distinct increase in abundance toward the top of the *Lystrosaurus* Assemblage Zone (Rubidge 2005).

The phylogenetic relationships of the Procolophonidae have been the subject of much debate in recent years. Most authors consider them most closely related to the pareiasaurs, which are large, robustly built, and often armored parareptiles from the Middle and Late Permian (Lee 1995). Laurin and Reisz (1995) hypothesized a sister-group relationship between procolophonids and turtles, but most researchers now consider turtles more closely related either to pareiasaurs (Lee 1995) or to diapsid reptiles (Rieppel and Reisz 1999; Lee et al. 2004).

Ever since its original description (Parrington 1935), the phylogenetic position of *Prolacerta* has been contentious. Early workers interpreted the presence of an incomplete lower temporal bar in this taxon as the condition ancestral to that found in lizards (e.g., Camp 1945). Thus, *Prolacerta* and its close relatives came to assume a key role in discussions concerning the origin of lepidosaurian reptiles (Robinson 1973; Wild 1973). Gow (1975) first challenged this placement and argued for archosauromorph affinities for *Prolacerta*. His hypothesis has since received support from various phylogenetic analyses (Benton 1985; Dilkes 1998; Modesto and Sues 2004). Furthermore, recent analyses placed *Prolacerta* as the sister-taxon to Archosauriformes (Dilkes 1998; Modesto and Sues 2004).

Prolacerta has long been referred, along with other taxa such as *Protorosaurus* from the Upper Permian of Europe and the long-necked *Tanystropheus* from the Middle Triassic of Europe and Israel, to a group variously referred to as Protorosauria, Prolacertilia, or Prolacertiformes. *Prolacerta* is now considered most closely related to Archosauriformes, and *Protorosaurus* and related forms appear to form a distinct, monophyletic group. Protorosaurs have skulls with incomplete lower temporal bars and resemble lizards in this respect. A diagnostic feature of the group is the long neck with elongated cervical vertebrae and cervical ribs that extend back across the intervertebral articulations. The Tanystropheidae have the most pronounced elongation of the cervical vertebrae and ribs (Wild 1973; chapter 3). Unfortunately, a number of small, rather fragmentary specimens with elongate cervical vertebrae have been referred rather arbitrarily to the Protorosauria, and consequently the diversity and interrelationships of this group remain poorly understood.

Among Archosauriformes, the Proterosuchidae are generally considered the most basal of several successive outgroups to the Archosauria (Benton and Clark 1988; Sereno 1991; Gower and Sennikov 2000). They include the oldest known archosauriform, *Archosaurus* from the uppermost Permian Vyatskian Gorizont of Vladimir Province, Russia. *Proterosuchus*, which is also known from China and India, attained a length of up to 1.5 meters and is characterized by conspicuously downturned premaxillae that overhang the mandibular symphysis (figure 2.7). Its skull still retains distinct postparietal and supratemporal bones (Cruickshank 1972). Forms apparently closely related to *Proterosuchus* are known from the Lower Triassic of Russia (Gower and Sennikov 2000).

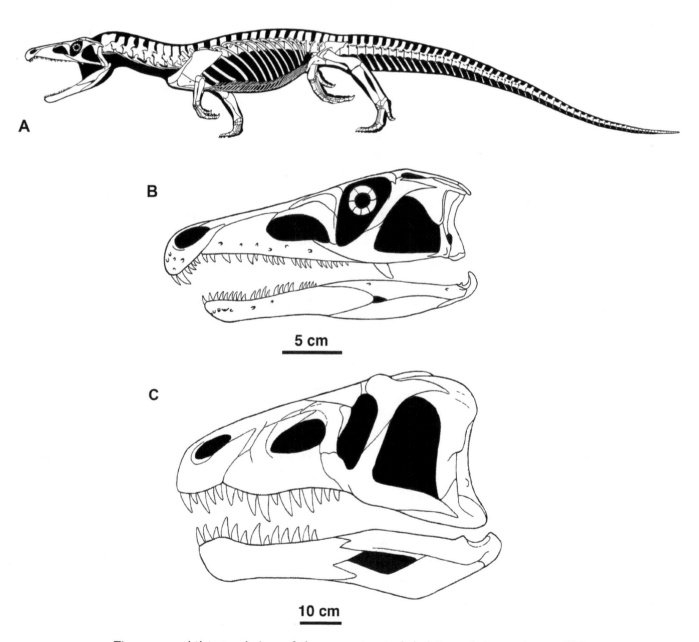

Figure 2.7. (*A*) Lateral view of the reconstructed skeleton of the proterosuchid *Proterosuchus fergusi*. Length up to 1.5 meters. (Drawing courtesy and copyright of G. S. Paul) (*B*) Lateral view of the reconstructed skull of *Proterosuchus fergusi*. (Modified from Cruickshank 1972) (*C*) Lateral view of the skull of the erythrosuchid *Erythrosuchus africanus*. (Modified from Gower 2003)

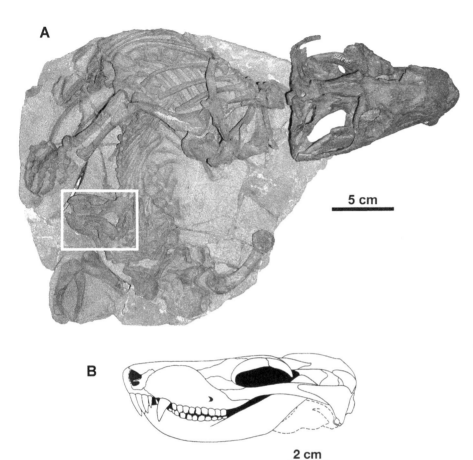

Figure 2.8. (*A*) Skeleton of the whaitsiid therocephalian *Theriognathus* (National Museum Bloemfontein, no. NMQR3375). Note preserved gut contents with the skull of a small dicynodont (white rectangle). (Photograph courtesy of S. P. Modesto) (*B*) Lateral view of the reconstructed skull of the bauriid therocephalian *Bauria cynops*. (Modified from Brink 1963)

The medium-sized therocephalian *Theriognathus* has well-developed incisors and canines but lacks postcanine teeth (figure 2.8*A*). Other therocephalian taxa from the *Lystrosaurus* Assemblage Zone, such as *Regisaurus*, are mainly small and gracile forms, some of which may actually represent juvenile specimens of larger forms.

The small galesaurid cynodonts *Thrinaxodon* (figure 2.9) and *Galesaurus* have been extensively discussed in connection with the origin of mammals. Both represent the earliest well-known representatives of the Epicynodontia (Kemp 1982), in which the coronoid process of the dentary is enlarged and the zygomatic arch has become robust and dorsally arched. The dentition includes four upper and three lower incisors. A distinctive feature of the postcranial skeleton is the presence of overlapping costal plates on the dorsal ribs (Jenkins 1971). Skeletons of *Thrinaxodon* are often found preserved in burrows (Damiani et al. 2003; Sidor, Miller, and Isbell 2008).

In the Karoo basin, the *Cynognathus* Assemblage Zone, which comprises the upper two-thirds of the Burgersdorp Formation, overlies the *Lystrosaurus* Assemblage Zone. There exists some overlap in the stratigraphic ranges of various tetrapod taxa between the two assemblage zones in the southern (but not in the northern) part of the basin (Neveling, Hancox, and Rubidge 2005). The replacement of the fauna from the *Lystrosaurus* Assemblage Zone by that from the *Cynognathus* Assemblage Zone apparently occurred quite rapidly. *Lystrosaurus* was replaced by a

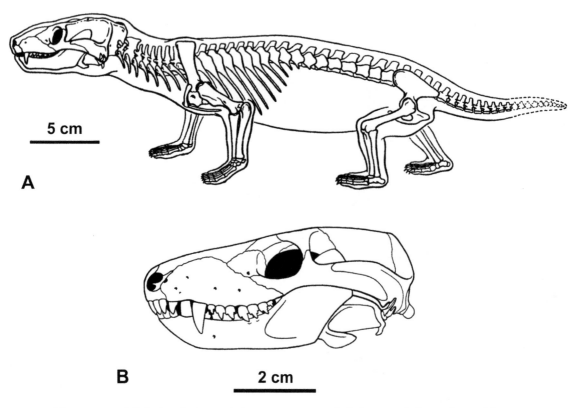

Figure 2.9. (*A*) Lateral view of the reconstructed skeleton of the galesaurid cynodont *Thrinaxodon liorhinus*. (Drawing courtesy of F. A. Jenkins Jr.) (*B*) Lateral view of the skull of *Thrinaxodon liorhinus*. (Modified from Hopson and Kitching 2001)

much larger dicynodont, *Kannemeyeria*, and *Thrinaxodon* was replaced by the large *Cynognathus* as the most common carnivorous cynodont (Kitching in Rubidge 1995). Another common taxon in the *Cynognathus* Assemblage Zone is *Diademodon*, an omnivorous or herbivorous cynodont. The succession of the *Cynognathus* Assemblage Zone reaches a thickness of 600 meters and comprises red, blue-green, and grayish green mudstones with subordinate fine- to medium-grained sandstones, which typically form widely spaced lenses and often yield well-preserved vertebrate remains. The strata were deposited in meandering streams with sandy channels and abundant crevasse splays onto thick, sub-aerially exposed floodplain deposits under semiarid climatic conditions (Johnson and Hiller 1990). There

are also indications for more widespread lacustrine settings in parts of the Karoo basin. The *Cynognathus* Assemblage Zone is considered late Early Triassic (Olenekian) to early Middle Triassic (Anisian) in age, based on correlation with other tetrapod-bearing continental strata elsewhere. Hancox (2000) proposed a tripartite subdivision of this zone (Subzones A–C) based on characteristic temnospondyls and further supported by the distribution of cynodonts and dicynodonts. Lucas (1998) designated the tetrapod assemblage of the *Cynognathus* Assemblage Zone as the type assemblage of the Nonensian LVF. However, Abdala, Hancox, and Neveling (2005) suggested that only the lower two subzones of the *Cynognathus* Assemblage Zone belong to the Nonensian LVF and that the upper subzone should be included in the succeeding Perovkan LVF.

The index taxon for its eponymous biozone, *Cynognathus*, is a large, carnivorous eucynodont (figure 2.10*A*). It has a robust skull, which reached a length of up to at least 40 centimeters (Broili and Schröder 1934). Its dentition includes large canines and postcanines with bladelike crowns that have a recurved main cusp flanked by sharp-edged accessory cusps.

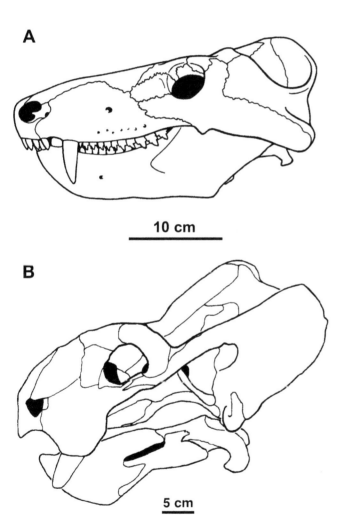

A

10 cm

B

5 cm

Figure 2.10. (*A*) Lateral view of the skull of the cynognathian cynodont *Cynognathus crateronotus*. (Combined from Broili and Schröder 1934 and Hopson and Kitching 2001) (*B*) Lateral view of the skull of the kannemeyeriid dicynodont *Kannemeyeria*. (Modified from Renaut and Hancox 2001)

The temporal region of the skull of *Cynognathus* is broad but short, and the posterior portion of the temporal bar is deep, indicating the presence of a powerful temporalis muscle. *Cynognathus* also occurs in the upper portion of the Puesto Viejo Formation of Mendoza Province, Argentina, where it is found along with the shansiodontid dicynodont *Vinceria* (originally incorrectly referred to *Kannemeyeria*; Renaut and Hancox 2001) and the traversodont cynodont *Pascualgnathus* (Bonaparte 1967, 1970).

Although closely related to *Cynognathus* and very similar in many features, *Diademodon* is distinguished by its postcanine dentition, which indicates possibly omnivorous to herbivorous feeding habits (Grine 1977). The anterior three or four postcanine teeth have simple, conical crowns. They are followed by up to nine molariform (gomphodont) teeth with transversely broad crowns bearing multiple cusps. The lower gomphodont teeth are similar to, but less expanded than, the upper ones. Wear quickly reduced these teeth to mere pegs. Behind the gomphodont teeth, there are up to five sectorial teeth, which are similar to the postcanines in *Cynognathus*. During tooth replacement in *Diademodon*, the number as well as the position of the three types of postcanine teeth continuously changes, with worn teeth being shed at the front of the tooth row and new teeth added at the back (Grine 1977).

Another common gomphodont cynodont from the *Cynognathus* Assemblage Zone is *Trirachodon* (Abdala, Hancox, and Neveling 2005). Its molariform postcanines have three main cusps aligned in a transverse row across the center of the crown and smaller cingular cusps along the anterior and posterior margins of the crown. Two closely related taxa, *Langbergia* and *Cricodon*, are also known from the *Cynognathus* Assemblage Zone (Abdala, Neveling, and Welman 2006). A number of trirachodontid skeletons, probably referable to *Langbergia*, have been discovered in the terminal chambers of complex burrow systems, suggesting communal cohabitation (Groenewald, Welman, and MacEachern 2001).

Cynognathus, Diademodon, and *Trirachodon* belong to the Eucynodontia, which are distinguished by a greatly enlarged dentary with a tall coronoid process and distinct articular process and reduction of the postdentary elements to a bony rod lodged in a medial recess on the dentary (Kemp 1982).

A third key taxon for the *Cynognathus* Assemblage Zone is the dicynodont *Kannemeyeria* (Pearson 1924; Renaut and Hancox 2001). It reached a length of up to 3 meters and was robustly built. Its large skull has a narrow, pointed snout with prominent tusks (figure 2.10*B*). A tall parietal crest separates the temporal openings. The postcranial skeleton of *Kannemeyeria* and related taxa is characterized by a short, barrel-shaped trunk and robust limbs. In the pelvic girdle, the ilium has an anteriorly expanded blade and the pubis is rather small.

Therocephalians include the small *Bauria*, which is distinguished by the possession of transversely broad teeth in the upper and lower jaws, which met in direct occlusion (Brink 1963; figure 2.8*B*). Each tooth has a large labial cusp and several smaller lingual cusps. During occlusion, the anterior margin of an upper tooth contacted the posterior edge of the corresponding lower tooth, generating shearing suitable for processing tough, fibrous material, presumably plants (Gow 1978). Other distinctive cranial features of *Bauria* include the incomplete postorbital bar and the possession of a nearly complete secondary bony palate (Brink 1963).

Several procolophonid taxa have been named from the *Cynognathus* Assemblage Zone (Modesto and Damiani 2003) but still await detailed anatomical study.

Archosauriform reptiles are represented by *Erythrosuchus* and *Euparkeria*. *Erythrosuchus* reached a

length of up to 5 meters and has a proportionately enormous skull (Gower 2003; figure 2.7*C*). Its snout is tall, narrow, and more or less triangular in transverse section. The teeth have tall, recurved crowns with serrated edges and are deeply implanted in the jaws. The pelvis of *Erythrosuchus* has a distinctly triradiate structure, and its femur has a small internal and incipient fourth trochanter, indicating a more upright stance and gait than in *Proterosuchus*.

Euparkeria has long been interpreted as representing the basic body plan of all archosaurian reptiles (Ewer 1965; Gower and Sennikov 2000). It has been reconstructed as a small form with a length of about 60 centimeters (figure 2.11), but it may have attained larger size because most of the known specimens represent immature individuals. Derived features of *Euparkeria* relative to *Proterosuchus* and *Erythrosuchus* include the possession of well-developed dorsal dermal armor, the absence of intercentra in the trunk region, and the presence of a fourth trochanter and absence of an internal trochanter on the femur (Ewer 1965). The hind limb of *Euparkeria* is distinctly longer than the forelimb, suggesting at least facultative bipedality.

The Erythrosuchidae were widespread during the Early Triassic. Although various taxa have been assigned to the Euparkeriidae, Borsuk-Białynicka and Evans (2003) noted that none of these except *Euparkeria* can be confidently assigned to this group. All the other taxa share similar combinations of plesiomorphic and apomorphic character-states, but there are no derived features uniting these various forms. As currently used, "Euparkeriidae" merely denotes a grade of archosauriform reptiles that are more derived than the proterosuchids but lack character-states diagnostic for other archosauriform groups such as Erythrosuchidae (Borsuk-Białynicka and Evans 2003). There still exists no consensus on the phylogenetic positions of the various groups of archosauriform reptiles relative to one another or to

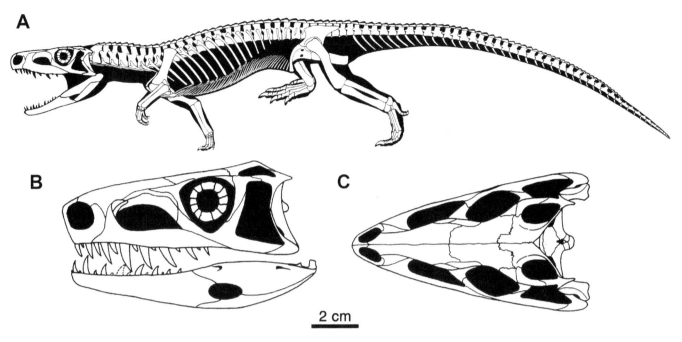

Figure 2.11. (A) Lateral view of the reconstructed skeleton of the euparkeriid archosauriform *Euparkeria capensis*. Length about 50 centimeters. (Drawing courtesy and copyright of G. S. Paul) (B–C) Reconstruction of the skull of *Euparkeria capensis* in (B) lateral and (C) dorsal view. (Combined from Ewer 1965 and a cast of the holotype [South African Museum, SAM 5867])

Archosauria (Benton and Clark 1988; Sereno 1991; Gower and Sennikov 2000; Dilkes and Sues 2009).

The two oldest rhynchosaurs, *Howesia* and *Mesosuchus*, are known from the *Cynognathus* Assemblage Zone (Dilkes 1995, 1998). Rhynchosaurs are a clade of Triassic archosauromorph reptiles characterized by a triangular skull (in dorsal view) with prominent overhanging projections of the paired, robust premaxillae and maxillae that typically have tooth plates with multiple tooth rows that flank a deep longitudinal groove into which the cutting edge of the dentary occluded. In many taxa, the dentary bears additional teeth on its lingual surface. The premaxillary "beak" and the deep, narrow claws on the feet have been interpreted as adaptations for digging up roots and tubers (Huene 1939; Benton 1983b). The larger rhynchosaurs have a barrel-shaped trunk, which suggests the presence of an extensive

gut that could have served as a fermentation chamber for the breakdown of cellulose (Reisz and Sues 2000a). *Howesia* and *Mesosuchus* already share the diagnostic median placement of the external nares and the overhanging premaxillae (Dilkes 1995, 1998). *Mesosuchus* still retains a few teeth on the premaxilla and has single rows of marginal teeth in the dentary and maxilla. However, *Howesia* has additional rows of marginal teeth in the upper and lower jaws, much as in more derived rhynchosaurs.

The tetrapod assemblage from the *Cynognathus* Assemblage Zone includes a variety of temnospondyls. Certain temnospondyl taxa are particularly abundant in portions of the *Cynognathus* Assemblage Zone (Hancox 2000). The capitosauroid temnospondyls such as *Kestrosaurus* and *Watsonisuchus* have large skulls with broad, rounded snouts. The pectoral girdle is robust, but the remainder of the postcranial skeleton is weakly ossified, suggesting predominantly aquatic habits. Trematosauroids include smaller forms with triangular skulls and larger taxa (with skull lengths of up to 50 centimeters) characterized by narrow, long snouts, which are consistent with a predominantly piscivorous mode of life. The brachyopoids, such as *Bathignathus*, have short, parabolic skulls with deep cheek regions and deeply incised lateral-line canals on the skull roof (figure 2.12).

Plant fossils from the *Cynognathus* Assemblage Zone include the equisetalean *Schizoneura*, the corystospermalean *Dicroidium*, and the peltaspermacean *Lepidopteris*. *Schizoneura* has slender stems that reached a height of 2 meters; at each node, it bears opposing sets of flat leaves (Kelber 1999). *Dicroidium* represents a very common form taxon for sterile foliage of various corystospermalean gymnosperms, which have pinnate fronds.

An important Early to early Middle Triassic tetrapod assemblage occurs in the Omingonde Formation of central Namibia (Smith and Swart 2002). This unit represents a succession of red beds in a half-graben that developed during the Early and early Middle Triassic. Massive maroon-colored siltstones in the upper part of the Omingonde Formation have yielded numerous vertebrate remains, representing a variety of cynodonts including *Cynognathus*, the bauriamorph therocephalian *Herpetogale*, the kannemeyeriid dicynodonts *Dolichuranus* and *Kannemeyeria*, and *Erythrosuchus* (Keyser 1973; Pickford 1995; Damiani et al. 2007). Thus, the Omingonde Formation can be correlated with the *Cynognathus* Assemblage Zone. Smith and Swart (2002) interpreted the upper Omingonde deposits as having formed on a well-vegetated floodplain in a semiarid, seasonal climate.

Figure 2.12. Dorsal view of the holotype skull of the brachyopid temnospondyl *Bathignathus watsoni* (Natural History Museum, London, BMNH R3589). (From a drawing by Bystrov 1935 based on a photograph in Watson 1919)

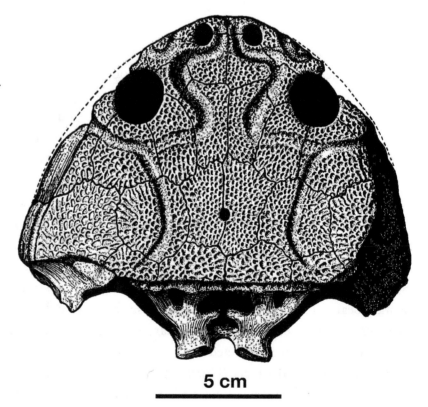

5 cm

TANZANIA

A diverse early Middle Triassic tetrapod assemblage occurs in the Lifua Member (Markwort 1991; part of the "Manda Beds") of the Ruhuhu basin in southwestern Tanzania. The northeast-southwest–trending Ruhuhu basin is 110 kilometers long and forms a half-graben that has been further divided by tectonic activity following deposition (Kreuser 1990; Wopfner 2002; Catuneanu et al. 2005). The Lifua Member reaches a thickness of at least 1000 meters and comprises numerous cyclical sequences of sandstones, siltstones, and mudstones, which were deposited under warm, semihumid conditions (Kreuser 1990; Wopfner 2002). Each cycle begins with medium- to coarse-grained sandstone, which fines upward and is overlain by red and green mudstones and siltstones that, in turn, are followed by coarse-grained material. The basal sandstones are locally interbedded with lenses of greenish gray, laminated mud- and siltstones, which Kreuser (1990) interpreted as deposits of ephemeral lakes. Skeletal remains of tetrapods are most common in the upper portion of the Lifua Member. The tetrapod assemblage is dominated by kannemeyeriid and shansiodontid dicynodonts (*Angonisaurus, Rechnisaurus, Shansiodon,* and *Tetragonias;* Cruickshank 1967; Cox and Li 1983; Cox 1991; Surkov and Benton 2008) and diverse gomphodont and chiniquodontoid cynodonts (Crompton 1955, 1972). Other tetrapods include the rhynchosaur *Stenaulorhynchus* (Huene 1938a), a variety of archosaurs including the large "rauisuchian" *Stagonosuchus* (Huene 1938b) and a smaller, as yet undescribed "rauisuchian" ("*Mandasuchus*"), and a large capitosauroid temnospondyl referable to either *Eryosuchus* (Damiani 2001) or *Stanocephalosaurus* (Schoch and Milner 2000). Closely related, possibly even congeneric kannemeyeriid and shansiodontid dicynodonts are also known from the Middle Triassic

of Russia and India and from the upper member of the Ermaying Formation of China. The trirachodontid cynodont *Cricodon* has recently been identified from the uppermost portion (Subzone C) of the *Cynognathus* Assemblage Zone of the Karoo basin (Abdala, Hancox, and Neveling 2005). Thus, we concur with Cox (1991) and Lucas (1998) in dating the tetrapod assemblage from the Lifua Member as early Middle Triassic (Anisian) in age.

ZAMBIA

The N'tawere Formation in the Luangwa Valley of northeastern Zambia unconformably overlies the Upper Permian Madumabisa Mudstones. Its upper levels have yielded a diverse tetrapod assemblage, including the capitosauroids *Eryosuchus* and *Cherninia* (Damiani 2001), the brachyopid *Batrachosuchus,* possible archosauriforms, the kannemeyeriid dicynodonts *Angonisaurus, Sangusaurus,* and *Zambiasaurus,* and the traversodont cynodont *Luangwa* (Cox 1969, 1991). This assemblage supports correlation of the upper part of the N'tawere Formation with the *Cynognathus* Assemblage Zone of South Africa, especially Subzones B and C, and indicates that this unit is Anisian in age (Catuneanu et al. 2005; Rubidge 2005).

INDIA

Since 1860 the Panchet Formation of the Raniganj coal field in the Damodar basin northwest of Kolkata has yielded numerous tetrapod remains representing an assemblage very similar to that from the *Lystrosaurus* Assemblage Zone of South Africa. The tetrapod assemblage includes several species of *Lystrosaurus*, the proterosuchid *Proterosuchus*, and various temnospondyls including trematosaurids (Huxley 1865; Tripathi and Satsangi 1963; Bandyopadhyay 1999).

In the Pranhita-Godavari basin in central India (chapter 4), the Yerrapalli Formation reaches a thickness of 600 meters and comprises red to purplish mudstones with thin beds of sandstone and caliche deposits (Bandyopadhyay, Roy Chowdhury, and Sengupta 2002). The mudstones are interpreted as deposits of interchannel floodplains, whereas small lenticular sandstone bodies probably represent fills of ephemeral channels. The Yerrapalli Formation has produced a varied tetrapod assemblage including a capitosauroid referable to *Eryosuchus* (Damiani 2001) or *Stanocephalosaurus* (Schoch and Milner 2000), the rhynchosaur *Mesodapedon* (Chatterjee 1980a), the protorosaur *Pamelaria* (Sen 2003), the "rauisuchian" *Yarasuchus* (Sen 2005), and the kannemeyeriid dicynodonts *Wadiasaurus* and *Rechnisaurus* (Roy Chowdhury 1970; Bandyopadhyay 1988, 1999; Cox 1991). Based on this assemblage, the unit is early Middle Triassic (Anisian) in age (Lucas 1998).

AUSTRALIA

The Early Triassic vertebrate assemblages from Australia differ from most coeval communities elsewhere (other than in southern Russia) in the predominance and great diversity of temnospondyls and the virtual absence of therapsids. This difference in faunal composition hints at rather different paleoenvironmental conditions in this region of Pangaea. To date, tetrapod fossils, mostly temnospondyls, have been reported from the Arcadia Formation of the Bowen basin and the Rewan Formation of the Galilee basin in Queensland, the Blina Shale of the West Kimber-

ley District in Western Australia, the Narrabeen Group of the Sydney basin in New South Wales, and the Knocklofty Formation of southeastern Tasmania (Cosgriff 1974, 1984; Warren 2000).

The Arcadia and Rewan formations of Queensland can be readily correlated with the *Lystrosaurus* Assemblage Zone in South Africa on the basis of their temnospondyl assemblages. The common South African temnospondyl *Lydekkerina* has recently been identified from the Rewan Formation (Warren, Damiani, and Yates 2006). The Arcadia Formation comprises mainly massive red-brown mudstones, which form overbank deposits and are interbedded with channel sandstones. Its strata were deposited on a well-vegetated floodplain by meandering and anastomosing river systems under warm, semiarid climatic conditions with seasonal rainfall (Cantrill and Webb 1998). The Arcadia Formation has yielded a diverse tetrapod assemblage dominated by temnospondyls, including the capitosauroids *Rewanobatrachus* and *Watsonisuchus*, the stem-group stereospondyl *Lapillopsis*, the brachyopid *Xenobrachyops*, the chigutisaurid *Keratobrachyops*, and the rhytidosteids *Acerastea* and *Arcadia* (Cosgriff 1984; Schoch and Milner 2000; Warren 2000; Warren, Damiani, and Yates 2006). Of these taxa, *Watsonisuchus* also occurs in the *Lystrosaurus* Assemblage Zone of South Africa (Damiani, Neveling, and Hancox 2001). The representatives of the Brachyopidae and Chigutisauridae are the oldest known members of their respective clades. The Chigutisauridae are closely related to and resemble the Brachyopidae in the structure of their skulls but differ in the presence of distinct, hornlike tabular processes and other projections along the posterior margin of the skull roof (Marsicano 1999). Amniotes are uncommon in the Arcadia Formation. They include the poorly known possible protorosaur *Kadimakara* (Bartholomai 1979), an unnamed procolophonoid, and the archosauriform *Kalisuchus* (Thulborn

1979). To date, therapsids are represented by only a few remains of an indeterminate dicynodont (e.g., Thulborn 1983).

Tetrapod assemblages from the Early Triassic of Gondwana show considerable homogeneity in their taxonomic composition. One taxon in particular, the dicynodont *Lystrosaurus*, achieved a very wide geographic distribution. Even if other tetrapod taxa apparently had more restricted ranges, closely related forms were present in the midlatitudes. This general pattern of faunal homogeneity continues into the Middle Triassic, with kannemeyeriid dicynodonts, gomphodont cynodonts, and various groups of archosauriform rep-

tiles having wide distributions. Localities at higher paleolatitudes tend to have a much greater abundance of temnospondyls, especially in Australia, but therapsids are virtually absent. This may indicate cooler and more humid climatic conditions in these regions. Gondwanan plant communities are characterized by the abundance of the corystospermalean gymnosperm *Dicroidium*. There are as yet no known examples of Early Triassic low-latitude faunas, and there may well be biotic differences in the equatorial regions. Nevertheless, as discussed in chapter 3, Laurasia and Gondwana share remarkably similar Early to early Middle Triassic terrestrial tetrapod assemblages, indicating significant faunal interchange across Pangaea.

Early and Early Middle Triassic in Laurasia

THE BUNTSANDSTEIN OF EUROPE

The Buntsandstein of Europe is the lowermost unit of the tripartite Germanic Trias Supergroup (Bachmann et al. 1999; Stollhofen et al. 2008; figure 3.1). It comprises mostly siliciclastic rocks, which are predominantly fluvial in origin and whose mostly red color lends the unit its name ("colored sandstone"). However, the Buntsandstein also includes paleosols, eolian sandstones, evaporites, and playa deposits. Its strata were deposited in a vast subsiding basin, commonly referred to as the Germanic basin, which extended from present-day central England in the west to southern Poland in the east and from southern Germany and northern Switzerland in the south to Denmark and the North Sea in the north. Located at about 25° north paleolatitude, the basin was surrounded on all sides by geologically older highlands, which formed source areas for the sediments transported by rivers and streams into the basin. Although Buntsandstein strata along the margins of the basin are often less than 100 meters thick, they reach up to 1,000 meters in total thickness near the center of the basin. The climatic conditions in the Germanic basin during the deposition of Buntsandstein were warm and arid (Péron et al. 2005).

The Buntsandstein is thought to overlie the Upper Permian Zechstein conformably and is divided into the Lower, Middle, and Upper Buntsandstein. The Lower and Middle Buntsandstein comprise primarily fluvial strata but also include paleosols as well as playa and eolian deposits, whereas parts of the Upper Buntsandstein (Röt Formation) are marine in origin. The latter document the gradual advance of the Muschelkalk Sea, which, arriving from the east, inundated the low-lying Germanic basin during the Middle Triassic. The Anisian- to Ladinian-age Muschelkalk Group, the second major unit of the Germanic Trias Supergroup, comprises predominantly gray and yellowish limestones and marlstones, along with some evaporites, that were deposited in a warm, shallow sea. The name of this unit, translated as "clam limestone," refers to the countless invertebrate fossils preserved in its carbonates.

The fossil record from the Buntsandstein is rather sparse, but enough animal and plant fossils have been collected over two centuries to develop a picture of the terrestrial communities that existed during this interval. The Lower Buntsandstein preserves few traces of animals and plants and apparently represented an interval of fairly inhospitable environmental conditions for life. By contrast, the Middle

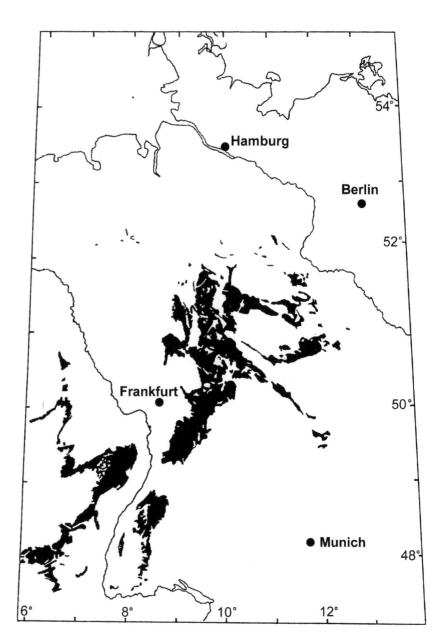

Figure 3.1. Distribution of exposures of Buntsandstein strata (shown in black) in Germany and eastern France. (From Paul 1999; by permission of Verlag Dr. Friedrich Pfeil)

and Upper Buntsandstein have a greater diversity and wider distribution of remains of terrestrial animals and plants. Both the Lower and parts of the Middle Buntsandstein are considered Induan in age, whereas the remainder of the Middle and lower Upper Buntsandstein are dated as Olenekian. Most of the Upper Buntsandstein is now regarded as early Middle Triassic, specifically early Anisian, in age (Bachmann et al. 1999; Bachmann and Kozur 2004).

The lycopsid *Pleuromeia sternbergii* is the most common plant from the Buntsandstein (Grauvogel-Stamm 1993, 1999; figure 3.2*A*). A close relative of extant quillworts (*Isoetes*), it formed unbranched trunks that reached a height of 2 meters and terminated in an apical conelike structure composed of leaves carrying sporangia (sporophylls). Its trunk is covered by short, oval leaves near its base and more elongate, lanceolate leaves toward its apex. Each

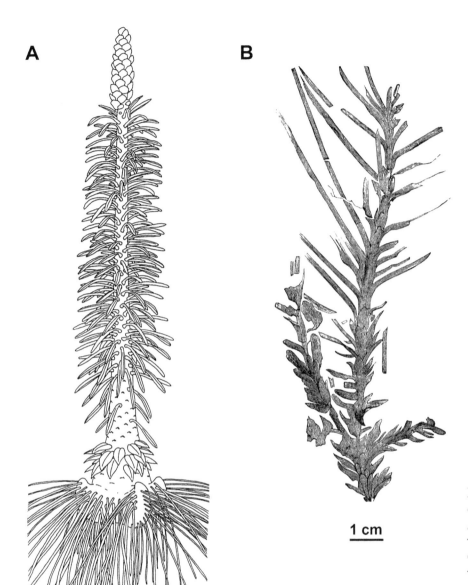

A

B

1 cm

Figure 3.2. (*A*) Reconstruction of a mature plant of the lycopsid *Pleuromeia sternbergii*. Height up to 2 meters. (Drawing courtesy of C. V. Looy) (*B*) Sterile branch of the conifer *Voltzia heterophylla*. (From Schimper 1880)

plant has a basal, four-lobed rhizophore; each lobe of this rhizophore carries rows of individual roots on its underside. Study of the principal occurrences of *P. sternbergii* suggests that this species preferred the periphery of standing water on the floodplains but also grew on sandy areas near river channels. When these channels moved, entire stands of *Pleuromeia* were often buried in place with the stems still in an upright position. Grauvogel-Stamm (1999) surmised that younger plants of this species required moist

substrates, perhaps growing even in very shallow water, whereas older, mature plants could subsist in drier habitats. *P. sternbergii* often formed dense thickets and is the only plant species found at many Buntsandstein localities. *Pleuromeia* had a worldwide distribution, especially in coastal and lacustrine environments throughout Eurasia and ranging into Australia and China (Dobruskina 1994; Retallack 1997).

Grauvogel-Stamm (1993) considered *Pleuromeia* an opportunist that could rapidly colonize pioneer niches. Retallack (1997) and Looy et al. (1999) argued that *Pleuromeia* was the dominant plant following the devastation of the conifer-dominated terrestrial vegetation at the end of the Permian. Palynological data from various regions of Europe confirm the pioneering role for this lycopsid (Looy et al. 1999). Through the development of soil resources, the *Pleuromeia*-dominated plant communities may have contributed to the restoration of habitats necessary for the successional replacement by conifer-dominated vegetation, of which *Voltzia* is the best-known representative (see "The Advent of the Muschelkalk Sea" later in this chapter; figure 3.2*B*).

Associated with *Pleuromeia* and a few less common plant taxa are various temnospondyls including the trematosaurid *Trematosaurus* and the capitosauroids *Eocyclotosaurus* and *Parotosuchus*, which were the principal predators in aquatic communities

(Ortlam 1970; Kamphausen 1989). The skull of *Trematosaurus* forms an elongate triangle in dorsal view and has a slender snout (Burmeister 1849; figure 3.3*A*). It reached a length of 30 centimeters. The orbits are small and placed near the lateral margins of the skull roof. The skull of *Parotosuchus* was up to 40 centimeters long and has a broad, fairly long snout and foreshortened postorbital region (Schroeder 1913; figure 3.4). At the posterior margin of the skull roof, there is a wide embayment of the squamosal, which may have supported the tympanic membrane. The skull of *Eocyclotosaurus* (figure 3.3*B*) attained a length of 30 centimeters. It has an elongated, narrow snout and a completely closed squamosal embayment (Ortlam 1970; Kamphausen 1989). All Buntsandstein temnospondyls have distinctly incised lateral-line canal grooves on their skulls, indicating predominantly aquatic habits.

Figure 3.3. (*A*) Dorsal view of the skull of the trematosauroid temnospondyl *Trematosaurus brauni*. (Combined from Burmeister 1849 and Bystrov 1935) (*B*) Dorsal view of a skull of the capitosauroid temnospondyl *Eocyclotosaurus lehmani* (Staatliches Museum für Naturkunde Stuttgart, SMNS 51562). Broken lines indicate lateral-line canals. (Modified from Kamphausen 1989)

A

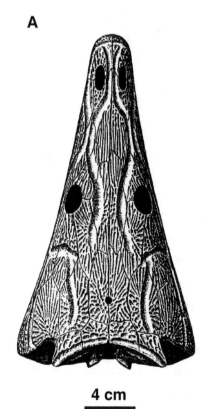

4 cm

B

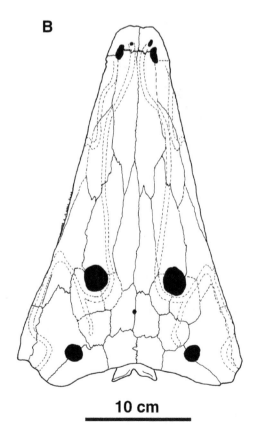

10 cm

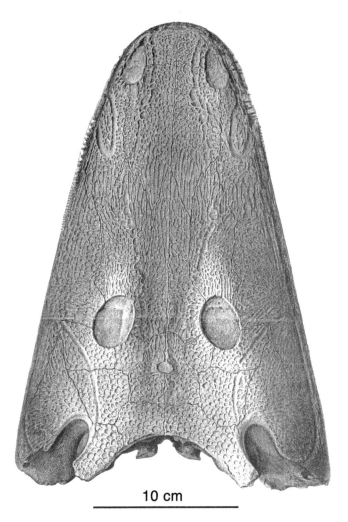

Figure 3.4. Dorsal view of the reconstructed skull of the capitosauroid temnospondyl *Parotosuchus nasutus*. (From Schroeder 1913)

10 cm

Terrestrial tetrapods represented by skeletal remains include procolophonid parareptiles such as the small *Anomoiodon* (Säilä 2008) and *Sclerosaurus*, which attained a length of 50 centimeters (Sues and Reisz 2008). In contrast to the scarcity of skeletal remains, some Buntsandstein horizons contain abundant, often superbly preserved trackways produced by a considerable variety of reptiles and possibly therapsids. The most common ichnotaxon is *Chirotherium* (figure 3.5). Sickler (1834) first reported its trackways with distinctive hand-shaped pes impressions from

the Buntsandstein of Hesse, and this discovery led to an extensive debate concerning the affinities of their producer over the course of the next century. Soergel (1925) first reconstructed the trackmaker as a quadrupedal archosaurian ("thecodontian") reptile. The "thumb" impression was actually left by the fifth digit of the pes. Later, Huene (1935–42) and Krebs (1965) argued that this type of tracks was produced by "rauisuchian" archosaurs. *Chirotherium* and similar "chirotheroid" ichnotaxa such as *Isochirotherium* are particularly common in the Solling Formation of the Middle Buntsandstein, where they occur together with tracks of smaller reptiles (e.g., *Rotodactylus*, *Rhynchosauroides*) and possible therapsids (*Dicynodontipus*) (Haubold 1971). Aside from a few isolated teeth, the only archosaurian skeletal remains reported to date from the Buntsandstein comprise a partial vertebral column and isolated postcranial bones of *Ctenosauriscus* (Krebs 1969; Ebel et al. 1998). *Ctenosauriscus* has greatly elongated, flattened neural spines, especially on its dorsal vertebrae. Recent finds of the apparently closely related *Arizonasaurus* from the Moenkopi Formation in Arizona suggest that *Ctenosauriscus* is a "rauisuchian" (Nesbitt 2003, 2005, 2009).

Early during the Triassic, archosaurian reptiles differentiated into two principal lineages, Crurotarsi and Ornithodira (Sereno 1991; Nesbitt 2003). The Crurotarsi comprise the Crocodylia and a number of related taxa, all of which are characterized by the distinctive structure of their ankles. In the ankle, the plane of bending passes between the two principal bones of the ankle, the astragalus and the calcaneum (crurotarsal condition); typically, the astragalus bears a peg that fits into a socket on the calcaneum (Krebs 1974; Sereno 1991). Presumably this pattern developed in relation to the adoption of a more upright posture. Crurotarsan archosaurs include several groups that flourished during the Middle and Late Triassic: phytosaurs (but see Nesbitt 2009), Ornithosuchidae, and

A

B

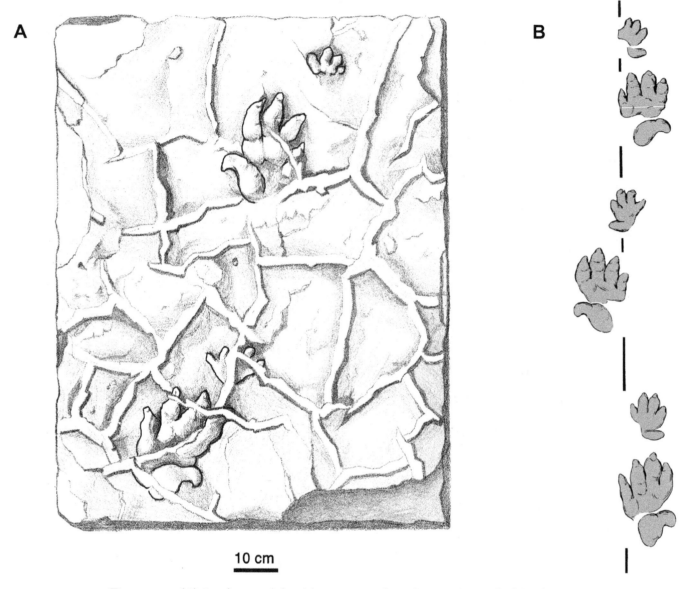

10 cm

Figure 3.5. (*A*) Sandstone slab with manus and pes impressions of *Chirotherium barthii* and desiccation cracks (both preserved as casts) from Hildburghausen, Hesse (Germany). (From Winkler 1886) (*B*) Reconstructed trackway of *Chirotherium barthii*; median indicated by black line. (Modified from Soergel 1925)

Suchia (Krebs 1974), which comprise aetosaurs, "rauisuchians," and crocodylomorphs. With the exception of crocodylomorphs, all crurotarsans became extinct near or at the end of the Triassic. The second major lineage, the Ornithodira, comprises dinosaurs, pterosaurs, and a few related taxa. Ornithodirans retain the typical bending plane between the lower leg (crus) and ankle (mesotarsal condition). The diversity of archosaurian trackways from the Buntsandstein, along with that of the archosaurian assemblage from the Anisian-age Lifua Member of Tanzania (chapter 2), supports the hypothesis that the evolutionary diversification of archosaurian reptiles already took place during the Early Triassic (Demathieu 1989; Nesbitt 2009).

A tetrapod assemblage similar to that from the Upper Buntsandstein is known from the Otter Sandstone Formation on the coast of Devon, England (Milner et al. 1990; Benton et al. 1994). It shares the presence of the capitosauroid *Eocyclotosaurus* and small procolophonids, but also includes the rhynchosaur *Fodonyx* (Hone and Benton 2008). *Fodonyx* reached a length of 1.3 meters and was more derived than *Rhynchosaurus*, which is well-known from possibly slightly older Middle Triassic strata in the English Midlands (Benton 1990), where it occurs along with temnospondyls and "rauisuchian" archosaurs, most of which are known only from incomplete skeletal remains (Benton et al. 1994). Based on paleomagnetic data, the Otter Sandstone Formation is Anisian in age (extending close to the Anisian-Ladinian boundary) and thus equivalent to part of the Muschelkalk succession in the Germanic basin (Hounslow and McIntosh 2003).

THE ADVENT OF THE MUSCHELKALK SEA

The sedimentary rocks of the early Middle Triassic (Anisian) Grès à Voltzia (Voltzia sandstone) in the northern Vosges of eastern France, which is equivalent to the uppermost Buntsandstein in Germany, document the transition from an inland braidplain fluvial environment to a coastal deltaic setting along the edge of the gradually advancing Muschelkalk Sea and thus provide a link between the marine and terrestrial realms (Gall 1971, 1985; Gall and Grauvogel-Stamm 1999).

The name Grès à Voltzia refers to the common occurrence of conifer shoots of *Voltzia* in these clastic strata. The name *Voltzia* was based on, and should be restricted to, a particular type of sterile conifer foliage with falcate to needlelike, rather small leaves (Grauvogel-Stamm 1978; figure 3.2*B*). *Voltzia* shoots are often found associated with female strobili with imbricate, five-lobed scales, which each carry three ovules (Miller 1977). These fossils support the scenario of a coastline with a relatively open forest of conifers. The size and structure of the *Voltzia* shoots suggests that they belonged to trees that reached a height of a few meters.

In addition to *Voltzia*, there are two other common taxa of conifers, *Aethophyllum* and *Pelourdea* ("*Yuccites*"). *Aethophyllum* was a herbaceous form, which reached a height of up to 2 meters and has a very thin "stem" devoid of secondary xylem. The leaves attained a length of 30 centimeters and have parallel venation. It was an ecological opportunist that apparently could rapidly colonize disturbed wetland habitats (Rothwell, Grauvogel-Stamm, and Mapes 2000). *Pelourdea* apparently was shrublike in growth habit and formed dense stands (Grauvogel-Stamm 1978). It has up to 1-meter-long and 20-centimeter-wide leaves with parallel venation.

Additional plants from the Grès à Voltzia include ferns, including *Anomopteris*, with bipinnate fronds that attained a length of 2 meters and a width of 50 centimeters, and the equisetaleans *Equisetites* and *Schizoneura* (Gall and Grauvogel-Stamm 1999; Grauvogel-Stamm 1978).

The Grès à Voltzia comprises two members, the lower Grès à meules, which represents predominantly deltaic sedimentation, and the upper Grès argileux, which reflects the initial transgression of the Muschelkalk Sea. The fossils also reflect this environmental transition and represent a combination of terrestrial and marine forms. Preservation of individual fossils, including soft tissues, can be exquisite in the silt-clay layers of the Grès à meules, and the diversity of preserved organisms is remarkable. Marine forms include jellyfish, worms, mollusks, echinoderms, crustaceans, and fishes (Gall 1971; Gall and

Grauvogel-Stamm 1999). Terrestrial animals include insects (figure 3.6), arachnids (figure 3.7), and myriapods. In addition, trace fossils include egg clusters of insects (figure 3.6B) as well as various examples of plant damage caused by herbivorous insects.

Grauvogel (1947) first reported the occurrence of insects in the Grès à Voltzia. Since that time a great variety of taxa has been recorded, with more than 5,300 insect fossils identified to date (Marchal-Papier 1998). The Blattodea (roaches), Odonata (dragon-

Figure 3.6. Insects from the Grès à Voltzia, northern Vosges Mountains, eastern France. (*A*) Wing of the dipteran *Grauvogelia arzvilleriana*. (*B*) Strings of presumed insect eggs, *Monilipartus tenuis*. (*C*) Forewing of the orthopteran *Triassophyllum leopardii*. (*D*) Forewing of the blattoid *Voltziablatta intercalata*. Scale divisions each equal 1 millimeter. (Photographs courtesy and copyright of F. Papier)

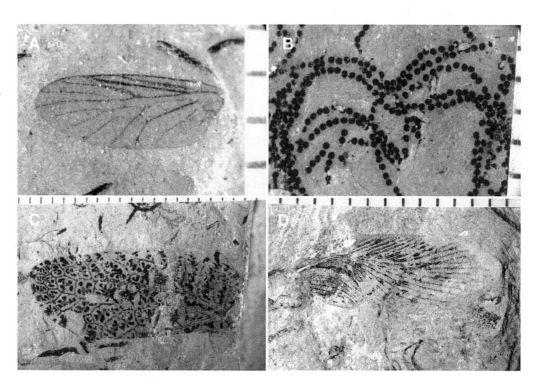

Figure 3.7. Arachnida from the Grès à Voltzia, northern Vosges Mountains, eastern France. (*A*) Mygalomorph spider *Rosamygale grauvogeli*. (*B*) Scorpion *Gallioscorpio voltzi*. Scale divisions each equal 1 millimeter. (Photographs courtesy and copyright of F. Papier)

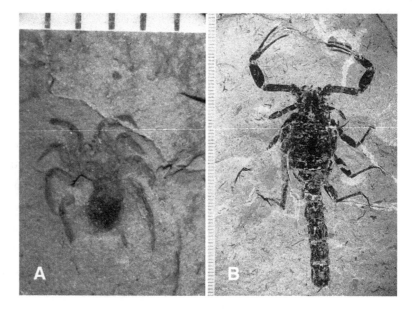

flies and mayflies), Mecoptera (scorpion flies), and Orthoptera (grasshoppers and their relatives) have been most extensively studied (Papier, Grauvogel-Stamm, and Nel 1994, 1996a, 1996b; Papier et al. 1997; Papier and Grauvogel-Stamm 1995; Marchal-Papier, Nel, and Grauvogel-Stamm 2000), but there are also reports of many other insect groups including Coleoptera (beetles; Papier et al. 2005), Diptera (true flies; Krzemiński, Krzemińska, and Papier 1994; Krzemiński and Krzemińska 2003), Hemiptera (sucking bugs including aphids and leaf hoppers; Lefebvre et al. 1998), Heteroptera (true bugs), and Plecoptera (stoneflies).

All of the known insect fossils come from the laminated silty-clayey layers in the Grès à meules. The majority of insect remains comprise isolated wings, but occasionally complete insects have been recovered. Typically, the venation patterns of the wings are clearly visible, and sometimes even color patterns are preserved: as its specific epithet implies, the wings of the orthopteran *Triassophyllum leopardii* bear a leopardlike pattern of spots (figure 3.6*C*).

With more than 2,000 known specimens representing nine different species (Papier, Grauvogel-Stamm, and Nel 1994, 1996b; Papier and Grauvogel-Stamm 1995), cockroaches are by far the most common group. Beetles are second in abundance, with some 500 recorded specimens, but they are much more diverse, with at least 40 species (Papier et al. 2005; figure 3.6*D*).

In many mecopterans, the pointed abdomen is held upward; this feature has given this insect group its vernacular name "scorpion flies." Mecopterans are viewed as central to the evolution of the Endopterygota, the group of insects in which the wings develop internally until the final molt, when they undergo a pupal stage and complete metamorphosis. Trichopterans (caddis flies), lepidopterans (butterflies and moths), and dipterans all are endoptery-

gotes, and the origins of these three major extant groups are thought to lie close to basal Mecoptera or even among the Mecoptera as traditionally defined. From the Permian to the Jurassic, mecopterans (and stem mecopterans) were diverse, but today only about 600 species are known (Grimaldi et al. 2005).

Dipterans account for about 5 percent of all the insects collected from the Grès à Voltzia. Many of the specimens are as yet undescribed aquatic larvae, and some of these may ultimately prove referable to other insect groups. The specimens from the Grès à Voltzia constitute the oldest known record of true flies.

Extant flies have traditionally been divided into two main groups, Nematocera and Brachycera. Nematocerans are characterized by long filiform antennae composed of at least six segments and also have multisegmented palps. Grimaldi and Engel (2005) noted that they probably do not represent a natural group. The more derived brachycerans typically have shorter antennae with fewer than six segments and extended palps with one or two segments. They are the ecologically most diverse group of dipterans today and include the typical, stout-bodied flies. As the oldest dipterans are known only from isolated wings, the diagnostic features of the two major dipteran groups cannot be readily assessed. Fortunately, there exists a subtle difference in the wing-venation pattern between the two groups, so that in theory the two should be identifiable even from wing remains.

Grauvogelia is the most common dipteran in the Grès à Voltzia (Krzemiński, Krzemińska, and Papier 1994; figure 3.6*A*), but Krzemiński and Krzemińska (2003) reported five additional taxa based on a variety of specimens that include some fairly complete individuals. A rather poorly known form, *Gallia*, appears referable to the Brachycera on the basis of the pattern of its wing venation (Marchal-Papier 1998). Thus, this derived dipteran group was already present during the early Middle Triassic, indicating an even earlier origin for Diptera.

A number of specimens of a myriapod have been recovered from the Grès à Voltzia. The presence of two pairs of jointed legs per segment demonstrates that these fossils represent diplopods (millipeds). The material has been assigned to the new genus *Hannibaliulus*, which possibly belongs to the Callipodida (Shear, Selden, and Gall 2009).

Aside from insects and millipeds, the Grès à Voltzia has yielded a mygalomorph spider and two taxa of scorpions (figure 3.7).

Extant spiders (Araneae) are divided into two principal groups, Mygalomorpha and Araneomorpha. The less diverse mygalomorphs include the trapdoor spiders and the large bird-eating spiders. They have chelicerae with fangs that extend parallel to each other so that they move in the same plane as the long axis of the body. Mygalomorphs also have relatively stout legs and always possess two pairs of book lungs. *Rosamygale* from the Grès à Voltzia is the oldest known mygalomorph (Selden and Gall 1992). At a total length of only about 5 millimeters, it is quite small for a mygalomorph (figure 3.7A).

Two taxa of scorpions have been reported from the Grès à Voltzia (Lourenço and Gall 2004). One can be classified with extant groups of the Buthoidea, whereas the other exhibits features shared with Paleozoic forms (figure 3.7B).

Mostly fragmentary, dissociated bones and one skull of the capitosauroid temnospondyl *Eocyclotosaurus* have been found along with plant remains in the coarser sandstone lenses, and this large temnospondyl is thought to have inhabited the levees.

Elsewhere, a few, usually fragmentary bones of terrestrial vertebrates, which presumably were introduced into marine sediments from adjoining land areas, have been recorded from the Muschelkalk Group and correlative Anisian-Ladinian strata in the European Alps. Skeletal remains of capitosauroid temnospondyls and the plagiosaurid temnospondyl *Plagiosternum* are known from nearshore Muschelkalk deposits in southern Germany and eastern France (Schoch and Wild 1999a).

Krebs (1965) described a nearly complete but disarticulated skeleton of the about 2.5-meter-long "rauisuchian" archosaur *Ticinosuchus* from the Grenzbitumenzone of Monte San Giorgio in the southern canton of Ticino (Tessin) in Switzerland, near the border with Italy. This formation (equivalent to the Besano Formation of northern Italy) is renowned for its wealth of superbly preserved marine fishes and reptiles (Peyer 1944; Kuhn-Schnyder 1974; Bürgin et al. 1989) and is early Ladinian in age. It comprises a 5–16-meter-thick unit of black bituminous shales and dolomites and is overlain by the San Giorgio Dolomite. The San Giorgio Dolomite itself grades up into the dark gray Meride Limestone, which has a maximum thickness of about 600 meters. In addition to *Ticinosuchus*, the Grenzbitumenzone has also yielded skeletons of the protorosaurian archosauromorphs *Tanystropheus* (Wild 1973; Nosotti 2007) and *Macrocnemus* (Peyer 1937). For both forms, terrestrial habits have been inferred for at least part of their lives. *Tanystropheus* reached a length of 6 meters and is characterized by its enormously elongated neck, which comprised up to 50 percent of total length in adult individuals (figure 3.8). Juveniles have tricuspid teeth in the posterior portions of the jaws, whereas the adults have only simple conical teeth, suggesting an ontogenetic shift in diet (Wild 1973). Nosotti (2007) has argued that adults of *Tanystropheus* probably spent most of their lives in shallow marine settings. The closely related *Macrocnemus* attained a length of only up to 80 centimeters and has a moderately elongated neck. Its forelimb is

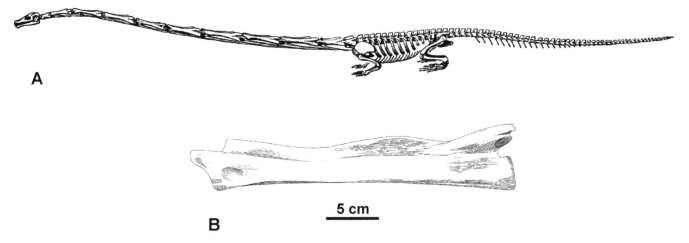

A

B

5 cm

Figure 3.8. (*A*) Lateral view of the reconstructed skeleton of the tanystropheid protorosaur *Tanystropheus longobardicus*. Length up to 6 meters. (From Peyer 1944) (*B*) Lateral view of a tenth cervical vertebra of *Tanystropheus conspicuus*. (From Wild 1973)

distinctly shorter than its hind limb, and it was at least facultatively bipedal (Rieppel 1989). *Macrocnemus* was more terrestrially adapted than *Tanystropheus*. *Tanystropheus* and *Macrocnemus* both also occur in the Muschelkalk of the Germanic basin (Schoch and Wild 1999a).

EARLY TRIASSIC OF EUROPEAN RUSSIA

In the southeastern region of European Russia, also known as the Cis-Urals, fossiliferous continental strata of Early to early Middle Triassic age are widely exposed along the present-day valleys of the Samara, Sakmara, and Ural rivers as well as along the eastern tributaries of the Volga, extending from the foreland fold belt of the Ural Mountains in the east to the Samara median in the west (Tverdokhlebov et al. 2002). These sedimentary rocks have yielded a wealth of fossils representing tetrapods (Shishkin et al. 2000), fishes (of which dipnoans, represented

by their durable tooth plates, are particularly common), conchostracans, ostracodes, and plants (documented by both macrofossils and pollen and spores).

The beginning of the Triassic sedimentation throughout the eastern part of the Eastern European Platform is associated with a period of intensive uplift of the Ural Mountains. The base of the Triassic succession is marked by thick conglomeratic units that formed large alluvial fans as part of a vast terminal fan along the forelands of the Urals. The formation of these deposits probably took place under drier climatic conditions, with greater peak discharges and higher sediment yield in a drainage basin with a reduced vegetation cover due to the end-Permian biotic crisis. This change in sedimentation style compared to that of the underlying Permian horizons is similar to the change reported in the Karoo succession at the Permo-Triassic boundary (Ward, Montgomery, and Smith 2000), which, in turn, has been related to profound environmental changes at the end of the Permian.

Early Triassic continental strata comprise the Vetlugian Supergorizont and the Yarenskian Gorizont (Tverdokhlebov et al. 2002; figure 3.9). Each of these units has its own characteristic suite of temnospondyls: the Vetlugian Supergorizont is associated with the so-called *Benthosuchus-Wetlugasaurus* fauna, and the Yarenskian Gorizont with the *Parotosuchus* fauna (Ochev and Shishkin 1989; Shishkin et al. 2000).

The Vetlugian Supergorizont comprises a series of "svitas" (which are primarily lithostratigraphically delineated units, comparable to formations in Western usage), which (from oldest to youngest) are the Kopanskaya, Staritskaya, Kzylsaiskaya, and Gostevskaya svitas. They, in turn, correspond to the Vokhmian, Rybinskian, Sludkian, and Ust-Mylian "gorizonts" (which are regional stratigraphic units primarily defined by their fossil content; Tverdokhlebov et al. 2002). Each unit has a characteristic tetrapod assemblage. Throughout the Vetlugian succession, skeletal remains of temnospondyls comprise some 90 percent of all identifiable specimens, but reptiles outnumber temnospondyls in terms of recorded generic diversity. There is a distinct increase in both the diversity and size of the temnospondyls from the Kopanskaya to the Gostevskaya Svita (Tverdokhlebov et al. 2002). In contrast to the previously discussed Early Triassic tetrapod assemblages, therapsids are very rare. Among fishes, the dipnoan *Gnathorhiza* is most common.

The Kopanskaya Svita (Vokhmian Gorizont) comprises conglomerates at the tops of the alluvial fans and mixtures of fine pebbles to sand-silt-clay along the periphery of these fans. Eolian sandstones indicate fairly arid conditions, especially during the first half of the deposition of this unit. Aquatic forms dominated the tetrapod assemblage; its low diversity reflects the prevailing climatic conditions (Shishkin et al. 2000). At the base of the Kopanskaya Svita, the small, short-snouted, and long-bodied temnospondyl *Tupilakosaurus* is most common. A surviving representative of the predominantly Permian Trimerorhachoidea, it is characterized by its distinctive, embolomerous vertebrae. Higher up in the section, the capitosauroid *Wetlugasaurus* (Riabinin 1930) is found in habitats away from the alluvial fans (figure 3.10*B*). *Tupilakosaurus* and the associated lydekkerinid *Luzocephalus* also occur together in the Wordy Creek Formation of the Cape Stosch region in East Greenland (Lucas 1998; Shishkin et al. 2000). The latter unit

Stage	Gorizont		Svita
LADINIAN	Bukobay		Bukobayan
ANISIAN	Donguz		Donguz
OLENEKIAN	Yarenskian		Petropavlovskaya
		Ust-Mylian	Gostevskaya
	Vetlugian Supergorizont	Sludkian	Kzylsaiskaya
		Rybinskian	Staritskaya
INDUAN		Vokhmian	Kopanskaya

Figure 3.9. Succession of continental Triassic strata in the Cis-Uralian region of Russia. (Modified from Tverdokhlebov et al. 2002)

A

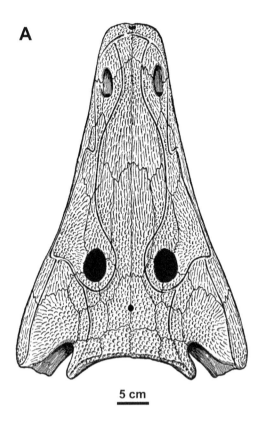

5 cm

B

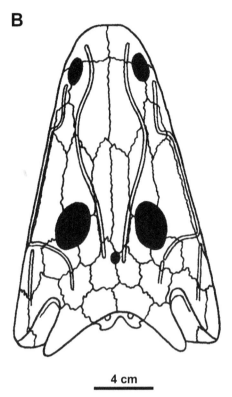

4 cm

Figure 3.10. (*A*) Dorsal view of the reconstructed skull of a large individual of the trematosauroid temnospondyl *Benthosuchus sushkini*. (From Bystrov and Efremov 1940) (*B*) Dorsal view of the reconstructed skull of the capitosauroid temnospondyl *Wetlugasaurus angustifrons*. (From Riabinin 1930)

comprises nearshore marine deposits that contain characteristic ammonoid cephalopods and allow precise dating of the Vokhmian Gorizont as middle to late Induan in age. This date for the Vokhmian Gorizont has since been independently confirmed using palynomorphs and conchostracans (Shishkin et al. 2000).

At some localities, small procolophonids, such as *Contritosaurus* and *Phaantosaurus*, with peglike rather than transversely broad maxillary and posterior dentary teeth, are common (Ivakhnenko 1979). Protorosaurian bones are not uncommon but are generally indeterminate, whereas skeletal remains of archosauriform reptiles are quite rare (Sennikov 1995).

Despite intensive collecting efforts, *Lystrosaurus* is known to date only from a single skeleton and some fragmentary remains from one locality (Kalandadze 1975).

The Staritskaya Svita (Rybinskian Gorizont) mainly comprises thick, poorly sorted fluvial sandstones, which were deposited under fairly arid conditions. The basal trematosauroid *Benthosuchus* (Bystrov and Efremov 1940) is characteristic for this unit (figure 3.10*A*) and is commonly found in association with the closely related *Thoosuchus* (Getmanov 1989). Both forms have long snouts, and deeply incised lateral-line canals on the skulls indicate that these temnospondyls were primarily aquatic.

Among amniotes, the presence of archosauriform and possibly archosaurian reptiles is noteworthy. Procolophonidae are represented by *Tichvinskia*, which differs from earlier taxa in the possession of transversely broad, bicuspid teeth suggestive of herbivorous habits (figure 3.11*A*). Based on various lines of evidence, the Rybinskian Gorizont has been dated as early Olenekian (Shishkin et al. 2000).

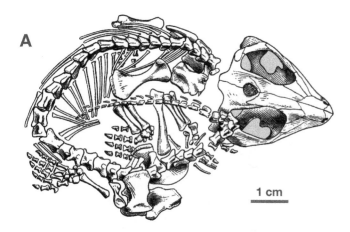

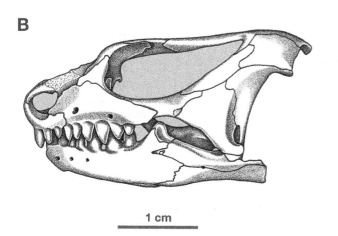

Figure 3.11. (*A*) Dorsal view of a skeleton of the procolophonid *Tichvinskia vjatkensis* as preserved (Paleontological Institute of the Russian Academy of Sciences, PIN 954/1). (From Ivakhnenko 1979) (*B*) Lateral view of the skull of the procolophonid *Kapes* cf. *K. majmesculae* (PIN 4365/40). (From Novikov and Sues 2004)

The Kzylsaiskaya Svita (Sludkian Gorizont) is virtually identical to the Staritskaya Svita in its sedimentary and faunal composition. In addition to fluvial sediments, it also includes deltaic deposits. The capitosauroid *Wetlugasaurus* is the most common tetrapod and occurs together with the trematosauroids *Angusaurus*, *Benthosuchus*, and *Prothoosuchus* (Getmanov 1989).

The Gostevskaya Svita (Ust-Mylian Gorizont) comprises a succession of lacustrine and deltaic strata, pre-

dominantly clays and siltstones. Tetrapods recorded from this unit include *Wetlugasaurus* and *Angusaurus*, procolophonids, the proterosuchid *Chasmatosuchus*, and the possible "rauisuchian" *Tsylmosuchus*.

The Petropavlovskaya Svita, equivalent to the Yarenskian Gorizont, represents a period of renewed tectonic activity in the Urals, and there was also a marine transgression from the region of the Caspian Depression. These events placed the depositional basin closer to the mountains and led to a substantial increase in clastic input into the basin (Tverdokhlebov et al. 2002). The Petropavlovskaya Svita comprises alternating units of obliquely laminated, gray or grayish red sandstones of fluvial origin and more or less horizontally bedded, reddish yellow, reddish brown, and gray clays, siltstones, and clay-rich sandstones, which represent delta floodplains or delta fronts (figure 3.9). Along with various sedimentological features, the presence of a flora dominated by *Pleuromeia* and diverse tetrapod communities indicate significant improvement in environmental conditions, especially an increase in humidity.

The dipnoan *Ceratodus* replaced *Gnathorhiza* in the Yarenskian Gorizont. The capitosauroid *Parotosuchus* (with a skull length of up to 45 centimeters) is the most characteristic tetrapod from the Yarenskian Gorizont, replacing its smaller predecessor *Wetlugasaurus* (with a skull length of up to 22 centimeters) (figure 3.10*B*). It is also known from late Olenekian nearshore marine deposits in the Caspian region and thus facilitates chronostratigraphic placement of the Petropavlovskaya Svita (Shishkin et al. 2000).

The diverse temnospondyls from the Yarenskian Gorizont include representatives of two predominantly Gondwanan clades, Brachyopidae (*Batrachosuchoides*) and Rhytidosteidae (*Rhytidosteus*), which

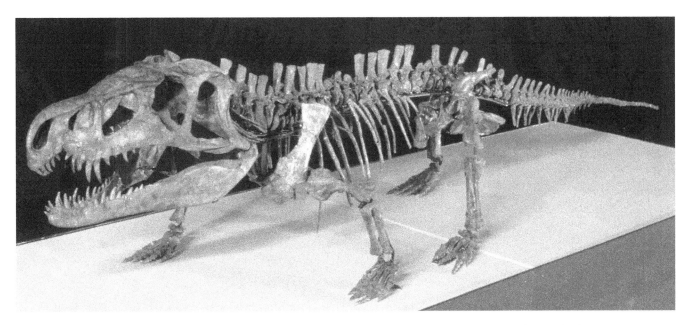

Figure 3.12. Anterolateral view of a reconstructed skeleton of the erythrosuchid archosauriform *Garjainia triplicostata*. Length about 2.8 meters. (Photograph courtesy of Igor V. Novikov)

elsewhere is known from the Karoo basin of South Africa (Shishkin et al. 2000).

Reptiles include a variety of procolophonid parareptiles (figure 3.11*B*), the erythrosuchid *Garjainia* (figure 3.12), and possibly "rauisuchian" archosaurs, which are known only from isolated bones. Ochev (1958) based *Garjainia* on a well-preserved skull and mandible associated with some presacral vertebrae and bones of the pectoral girdle from the Yarenskian Gorizont of Orenburg Province. Huene (1960) named a second taxon, *Vjushkovia*, on the basis of numerous cranial and postcranial bones representing at least two individuals from another locality in the same stratigraphic unit and province. Gower and Sennikov (2000) argued that *Vjushkovia* is indistinguishable from *Garjainia*.

Therapsids are represented to date only by a therocephalian (*Silphedosuchus*) and some undescribed

remains assigned to galesaurid cynodonts (Shishkin et al. 2000).

After a hiatus in sedimentation, the onset of deposition of the Donguz Svita is associated with an abrupt resumption of tectonic activity in the Ural Mountains (figures 3.9 and 3.13). The basal portion of the Donguz Svita comprises coarse-grained fluvial sandstones but, higher in the section, these sandstones are gradually replaced by clays and siltstones deposited in large, shallow lakes, in estuarine ponds, and as deltaic accumulations (Tverdokhlebov et al. 2002). The climate was distinctly seasonal, with alternating wet and dry seasons. The majority of the tetrapod fossils are associated with deltaic alluvial fans.

Temnospondyls are common in the Donguz Svita. They include the large capitosauroid *Eryosuchus* (Ochev 1966, 1972; Damiani 2001) and the plagiosaurids *Plagioscutum* and *Plagiosternum* (Shishkin 1987). The

Figure 3.13. Exposures of fluvial strata of the Donguz Svita at the locality Koltaevo II, Bashkortostan Republic, Russia. (Photograph courtesy of M. J. Benton)

temnospondyls occur together with kannemeyeriid dicynodonts (figure 3.14) and therocephalians (Battail and Surkov 2000) as well as a variety of archosauriforms and archosaurs (Sennikov 1995; Gower and Sennikov 2000). Many of the dicynodonts and reptiles are known only from isolated, incomplete bones; unfortunately, a plethora of taxonomic designations has been applied to these inadequate remains over the years. Lucas (1998) argued that most of the dicynodont taxa from the Donguz Svita are, in fact, referable to *Kannemeyeria,* which is also found in Anisian-age strata in South Africa (chapter 2) and possibly China. This synonymy is certainly

plausible in view of the rather cosmopolitan nature of tetrapod assemblages during the early Middle Triassic.

Archosauriform reptiles from the Donguz Svita comprise the proterosuchid *Sarmatosuchus,* the "euparkeriid" *Dorosuchus,* and the possible erythrosuchid *Uralosaurus.* Archosaurs are represented by *Dongusuchus* and *Vjushkovisaurus,* but the known material is insufficient for more precise phylogenetic assessment of these taxa (Sennikov 1995; Gower and Sennikov 2000).

The Donguz Svita has been dated as Anisian to Ladinian in age (Tverdokhlebov et al. 2002). Lucas

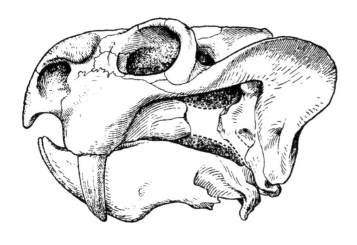

5 cm

Figure 3.14. Lateral view of the holotype skull of the kannemeyeriid dicynodont *Rhinodicynodon gracile* (PIN 1579/50). (From Kalandadze 1970)

(1998) proposed the Perovkan LVF for the global time equivalent of the tetrapod assemblage from the Donguz Svita.

EARLY AND MIDDLE TRIASSIC OF CHINA

Early Triassic communities of archosauriform reptiles, procolophonids, dicynodonts, and cynodonts ranged widely throughout Pangaea. A paleogeographically especially significant tetrapod record is known from the Junggar and Ordos basins in China. Despite the great distances separating these basins from the Karoo basin of South Africa even during the existence of Pangaea, the successive faunal assemblages from the two regions are strikingly similar.

The Jiucaiyuan Formation of the Junggar basin in Xinjiang, northwestern China, comprises an up to 370-meter-thick succession of red and purplish mudstones with numerous calcrete nodules and grayish

green sandstones of fluvial origin (Lucas 2001). It has yielded a tetrapod assemblage dominated by *Lystrosaurus* and including *Proterosuchus* (Young 1936).

The vast Ordos basin covers an area of more than 250,000 square kilometers, extending across the provinces of Nei Mongol, Shanxi, Shaanxi, Ningxia, and Gansu in north-central China, and is framed by mountain ranges. Its massive sedimentary fill includes more than 1,500 meters of Triassic-age strata (Lucas 2001). The Liujiagou Formation has a maximum thickness of 630 meters and is the oldest unit of this succession. It has not yielded tetrapod remains to date but conchostracans, plant macrofossils, and pollen and spores found in this formation in Shaanxi support an earliest Triassic age for this unit. The red and purplish mudstones and sandstones of the overlying Heshanggou Formation reach a thickness of up to 280 meters and have yielded the oldest Triassic vertebrate assemblage from the Ordos basin to date. The upper part of this unit has yielded a vertebrate assemblage comprising dipnoans, capitosauroid temnospondyls, the procolophonid *Eumetabolodon*, the erythrosuchid *Fugusuchus*, the possible "rauisuchian" *Xilousuchus*, and the therocephalian *Hazhenia*. With its large orbits and in dorsal view more or less triangular, short but deep skull, *Eumetabolodon* closely resembles *Procolophon*. It also shares the presence of transversely broad "cheek" teeth in adult individuals, but the teeth in juveniles are conical, reflecting ontogenetic change in feeding habits, with the young presumably having a more omnivorous diet than the presumably herbivorous adults (Li 1983).

The apparent absence of dicynodonts from the Heshanggou Formation suggests that either remains of these therapsids have not yet been found or that the conditions in the depositional setting were inhospitable to dicynodonts.

The Ermaying Formation has yielded two distinct tetrapod assemblages. Reaching a thickness of up to 600 meters, this unit comprises a complex succession of red mudstones intercalated with grayish green and yellow sandstones (Lucas 2001). The Ermaying Formation is usually divided into lower and upper members. The lower member comprises for the most part sandstones and resembles the *Cynognathus* Assemblage Zone of South Africa in the occurrence of large kannemeyeriid dicynodonts, specifically *Shaanbeikannemeyeria* (possibly synonymous with *Kannemeyeria*; Lucas 1998) and *Parakannemeyeria*. It has also yielded a procolophonid (*Paoteodon*), two "euparkeriid" archosauriforms (*Halazhaisuchus* and *Turfanosuchus*; Wu 1981; Wu and Russell 2001), the erythrosuchid *Guchengosuchus* (Peng 1991), and the therocephalian *Ordosiodon* (Sigogneau-Russell and Sun 1979). Most Chinese authors have considered the age of the lower Ermaying Middle Triassic, but Lucas (2001) argued for a late Early Triassic (Olenekian) age. We concur with Lucas's dating because the upper part of the Heshanggou Formation correlates with the lower member of Ermaying Formation and has yielded *Pleuromeia sternbergii*, which is common in Olenekian-age strata elsewhere in Eurasia.

The upper member of the Ermaying Formation comprises intercalated mudstones and sandstones. Its faunal assemblage, like that from the lower member, is dominated by kannemeyeriid dicynodonts. The latter include *Parakannemeyeria* and *Sinokannemeyeria*, both of which are known from numerous skulls and skeletons and differ from *Kannemeyeria* in the possession of broader snouts, smaller temporal openings, and lower intertemporal crests (Sun 1963). Shansiodontid dicynodonts are represented by the small to medium-sized *Shansiodon*, which is characterized by a triangular skull (in dorsal view) with a short, downturned, blunt snout.

Another common tetrapod of the upper Ermaying is the erythrosuchid *Shansisuchus*, which is known from hundreds of dissociated cranial and postcranial bones and reached a length of 3 meters (Young 1964). Associated archosauriforms are represented only by fragmentary skeletal remains and include *Wangisuchus* (Young 1964).

Additional tetrapod taxa recorded from the upper member of the Ermaying include indeterminate capitosauroid temnospondyls, the procolophonid *Neoprocolophon*, the baurioid therocephalian *Traversodontoides*, and the trirachodontid cynodont *Sinognathus*. Based on the composition of its tetrapod assemblage, Lucas (1998, 2001) considered the upper member of the Ermaying Formation early Anisian in age.

In recent years, marine strata in southern China have yielded remarkable assemblages of Middle Triassic tetrapods and fishes that are often closely related to those from Europe (Li, Zhao, and Wang 2007). Of particular significance are occurrences of the protorosaurian archosauromorphs *Macrocnemus* and *Tanystropheus*, which are widely known from the Muschelkalk and correlative strata in Europe. Li, Zhao, and Wang (2007) described *Macrocnemus fuyuanensis* from the Ladinian-age Zhuganpo Member of the Falang Formation of Yunnan Province and also noted that it occurs along with *Tanystropheus*. Furthermore, a tanystropheid apparently virtually identical with *Tanystropheus longobardicus* has recently been found in the same unit in Guizhou Province (Fraser, personal observation, 2008). Although the existence of faunal similarities between marine vertebrates from the western and eastern Tethys is not unexpected, the shared presence of more terrestrially adapted tetrapods such as *Macrocnemus* hints at a possibly widespread coastal or littoral fauna at least along the northern coastline of Tethys.

THE MOENKOPI FORMATION OF
THE AMERICAN SOUTHWEST

Near the western margin of Pangaea, the Moenkopi Formation is widely exposed throughout the western two-thirds of the Colorado Plateau in the western United States, from northern Arizona to northern Utah and from southern Nevada to western New Mexico (Stewart, Poole, and Wilson 1972; Morales 1987). This unit forms a thick sequence of brightly colored marine and continental rocks. In northeastern Arizona, the Moenkopi Formation is divided into three members (from oldest to youngest), Wupatki, Moqui, and Holbrook, which are separated by widespread layers of gypsum (Welles 1947). The strata of the Moenkopi Formation are least thick in the east, and gradually increase in thickness toward the west and into southern Utah. Localities in northeastern Arizona and southeastern Utah have yielded skeletal remains representing a diversity of temnospondyls, reptiles, and therapsids, hybodont sharks, and actinopterygian and coelacanth fishes, along with some plant fossils. Only the Moqui Member has not yielded vertebrate fossils to date. A skull of a capitosauroid temnospondyl from the lower portion of the Wupatki Member appears indistinguishable from that of *Parotosuchus helgolandicus* from the Volpriehausen Formation (lower Middle Buntsandstein) of the Isle of Helgoland in the German sector of the North Sea (Lucas and Schoch 2002). Welles and Cosgriff (1965) and Schoch (2000) described skeletal remains of additional taxa of capitosauroids from the Wupatki and Holbrook members. The presence of *Eocyclotosaurus* in the Holbrook Member (and the correlative Anton Chico Member of the Moenkopi in New Mexico) forms an important biostratigraphic link to the Upper Buntsandstein and Grès à Voltzia (Schoch 2000; Lucas and Schoch 2002). Other temnospondyls include the remarkably slender-snouted trematosaurid *Cosgriffius* (Welles 1993) and the capitosauroid *Stanocephalosaurus* from the Wupatki Member (Welles and Cosgriff 1965; Schoch and Milner 2000), and the brachyopids *Hadrokkosaurus* and *Virgilius* and the capitosauroid *Quasicyclotosaurus* from the Holbrook Member (Welles and Estes 1969; Schoch 2000; Warren and Marsicano 2000; Ruta and Bolt 2008).

Reptiles include *Arizonasaurus* and the poorly known *Anisodontosaurus* (Welles 1947). A recently discovered partial skeleton has established that *Arizonasaurus*, long known only from jaw fragments, is a "rauisuchian" with greatly elongated dorsal neural spines (Nesbitt 2003, 2005). The monophyly of Rauisuchia is questionable, and they probably represent a grade of medium-sized to large crurotarsan archosaurs (Gower 2000; Weinbaum and Hungerbühler 2007; Nesbitt 2009). Other recent finds of tetrapod remains from the Holbrook Member include a rhynchosaur (Nesbitt and Whatley 2004) and large dicynodonts (Nesbitt and Angelczyk 2002).

Peabody (1948, 1956) described tracks and trackways representing a considerable diversity of tetrapods from the Wupatki and Holbrook members. Many of these ichnotaxa are similar to those known from the German Buntsandstein (Haubold 1971). The meager record of fossil plants from the Moenkopi Formation comprises equisetaleans, ferns, and conifers (Morales 1987). Prochnow et al. (2006) inferred semiarid to arid conditions for the upper Moenkopi based on isotopic studies of pedogenic carbonates. Morales (1987) considered the age of the Wupatki Member of the Moenkopi Formation Olenekian (middle Spathian) and that of the Holbrook Member probably early Anisian. Lucas and Schoch (2002) presented additional data in support of the latter date. Thus, the Moenkopi Formation represents the oldest securely dated vertebrate-bearing Triassic continental strata from North America to date.

For the most part, Early Triassic tetrapod assemblages from the midlatitudes of Laurasia, including China and western North America, closely resemble those from Gondwana. Therapsids and archosauriform reptiles are the dominant faunal elements. In regions further to the north, represented by the assemblages from the Cis-Ural region of Russia, temnospondyls predominate and *Lystrosaurus* has been found at only a single locality to date. Continuing into the Middle Triassic, dicynodont therapsids and archosauriforms are the dominant faunal elements. This is also partly true for the Cis-Uralian region, where the two groups occur along with a diversity of temnospondyls. However, the midlatitude tetrapod assemblages from the Buntsandstein Group in the Germanic basin comprise various temnospondyls but few therapsids (inferred only from trackways to date). Interestingly, the latter occurrences would have been in close proximity to the Tethys, which may have resulted in more humid and more equable climatic conditions than further west. Terrestrial taxa dominate the known insect assemblages.

Early Triassic floras from Laurasia are characterized by the absence of corystospermaleans. However, *Pleuromeia* and related isoetaleans not only were abundant throughout much of Laurasia but also ranged widely through the lowland regions in Gondwana, especially Australia.

CHAPTER 4

Late Middle and Late Triassic of Gondwana

SOUTH AMERICA

Middle and Late Triassic continental strata in rift basins in southern South America have yielded a wealth of animal and plant fossils that provides a detailed picture of the composition and succession of terrestrial ecosystems in western Gondwana during this time interval. During the Middle Triassic, Gondwana was still intact, but some regions already showed signs of the crustal extension that subsequently would lead to the breakup of the landmass. The southern margin of Gondwana experienced strong compression caused by the subduction of oceanic crust beneath the supercontinent. As noted in chapter 2, this led to the formation of the Gondwanides, a continuous mountain range that includes the Cape Fold Belt in South Africa and the Sierra de la Ventana Fold Belt in Argentina, and associated foreland basins (Veevers et al. 1994). In South America, Triassic basins represent two kinds of tectonic settings: (1) foreland basins of the Gondwanides, exposed mainly in Argentina and Chile, and (2) intracontinental rift basins in southern Brazil (Zerfass et al. 2003, 2004).

The first phase of the Gondwanides orogeny, Gondwanides I paroxysm of Veevers et al. (1994), led to the formation of the Sanga do Cabral Super-

sequence in the Paraná basin of southern Brazil (figure 4.1) and the Talampaya-Tarjados sequences in northwestern Argentina. These sequences comprise strata that were deposited by braided rivers in a wide alluvial basin (Zerfass et al. 2004). The fluviolacustrine sedimentary rocks of the Sanga do Cabral Formation of southern Brazil have yielded remains of typically Early Triassic tetrapods including *Procolophon* (Dias-da-Silva, Modesto, and Schultz 2007). The Talampaya and Tarjados sequences form the basal strata in the Ischigualasto basin of northwestern Argentina and are probably Early Triassic in age (Zerfass et al. 2004). To date, the Talampaya Formation has produced only chirotheroid tracks (Romer 1960, 1966), and the Tarjados Formation has yielded fragmentary skeletal remains of kannemeyeriid dicynodonts (Bonaparte 1997).

Gondwanides II paroxysm of Veevers et al. (1994) coincided more or less with the early phases of rifting in Gondwana. The Cuyo and Ischigualasto basins of northwestern Argentina and the Santa Maria basin of southernmost Brazil are the principal rift basins that were filled with continental sediments. The two Argentine basins formed in a back-arc setting, whereas the Santa Maria basin represents an intracontinental rift that formed as part of a major

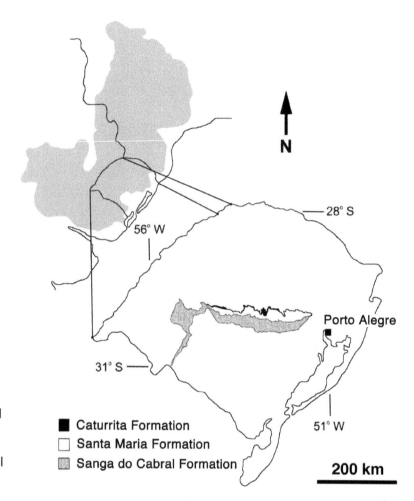

Figure 4.1. Map of the Paraná basin (indicated in gray) in eastern South America with outline map of the Brazilian state of Rio Grande do Sul showing the outcrop belt of Triassic continental strata. (From Langer et al. 2007)

rift system extending all the way to southeastern Africa. The Cuyo and Ischigualasto basins contain much thicker sedimentary fills, presumably reflecting factors such as more rapid subsidence and more elevated source areas (Zerfass et al. 2004).

BRAZIL

The Santa Maria Supersequence of Zerfass et al. (2003) of the Paraná basin in southern Brazil apparently developed as sedimentary fill in a half-graben, with periods of fluvial deposition following episodes of normal fault displacement (figure 4.1). Within the basin, lakes or playa lakes developed in relation to

climatic changes. Langer et al. (2007) united the strata of the Santa Maria and the Sanga do Cabral supersequences as the Rosário do Sul Group.

The Santa Maria Supersequence comprises a succession of Middle to Late Triassic sedimentary rocks that have long been renowned for their wealth of tetrapod fossils. Friedrich von Huene first systematically explored these strata in the late 1920s and described the reptiles and therapsids in much detail (Huene 1935–42). Strata of the Santa Maria Supersequence crop out mainly in a narrow belt that extends for almost 500 kilometers across the territory of Brazil's southernmost state, Rio Grande do Sul. Andreis, Bossi, and Montardo (1980) distinguished the Santa Maria and Caturrita formations and divided the Santa Maria Formation into the Passo das Tropas and Alemoa members. Subsequently, Faccini (1989) identified the Mata Sandstone as a distinct third unit topping the Santa Maria Supersequence.

Based on a succession of characteristic tetrapod assemblages, Brazilian researchers (e.g., Barberena, Araújo, and Lavina 1985) have established a threefold zonation of the Santa Maria strata into a lower "Therapsid" (Ladinian), a middle "Rhynchosaur" (Carnian), and an upper "Ictidosauria" (Norian) association zone. The first two zones make up the Alemoa Member of the Santa Maria Formation, and the third corresponds to the Caturrita Formation. Zerfass et al. (2003) combined this biostratigraphic scheme with other geological information to distinguish three depositional sequences within the Santa Maria Supersequence: Santa Maria 1 (up to 50 meters thick; Ladinian), Santa Maria 2 (up to 130 meters thick; Carnian), and Santa Maria 3 (about 20 meters thick; possibly Rhaetian or younger).

The succession of Santa Maria 1 Sequence begins with fluvial and deltaic sandstones and siltstones, formed by sediments introduced from crystalline basement in the south by a high-energy, low-sinuosity river system. These rocks are overlain by massive or laminated red mudstones, which were deposited in shallow lakes or on the floodplains of an anastomosed river system (Langer et al. 2007). This change in sedimentation is associated with the occurrence of a rich tetrapod assemblage, best known from the region of Chiniquá, west of the town of São Pedro do Sul, and the region of Pinheiros, south of the town of Candelária. This assemblage includes the small owenettid procolophonoid *Candelaria* (Cisneros et al. 2004), up to at least 5 meters long "rauisuchian" archosaurs (*Prestosuchus*), the large dicynodont therapsids *Dinodontosaurus* (figure 4.2) and *Stahleckeria* (the latter with a skull length exceeding 60 centimeters and lacking tusks), and a variety of cynodont therapsids (Huene 1935–42; Cox 1965; Barberena 1978; Barberena, Araújo, and Lavina 1985; Langer et al. 2007).

The cynodonts include chiniquodontoids (*Chiniquodon*) and traversodonts (*Massetognathus, Traversodon*). *Chiniquodon* (including *Belesodon*) and its relatives are distinguished by the sharp angle in the ventral margin of the skull where the zygomatic arch is connected to the maxilla (Abdala and Giannini

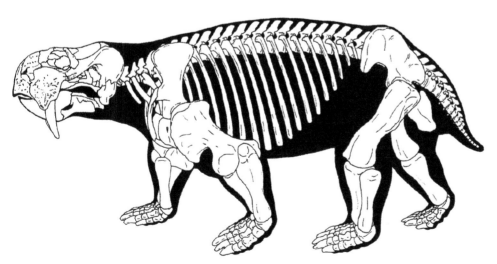

Figure 4.2. Reconstructed adult skeleton of the kannemeyeriid dicynodont *Dinodontosaurus pedroanum*. (Drawing courtesy and copyright of L. Morato)

20 cm

2002). Their sectorial postcanine teeth, indicative of carnivorous habits, have a sharply recurved main cusp. The skull of *Chiniquodon* could reach a length of 25 centimeters. Traversodont cynodonts such as *Traversodon* and the smaller *Massetognathus* are characterized by multicuspid, gomphodont postcanine teeth that met in precise occlusion and appear well-suited for processing high-fiber plant fodder (Crompton 1972). Many of these therapsids also occur in the Chañares Formation of northwestern Argentina, which is considered Ladinian in age (see later in this chapter).

Associated plant fossils include the equisetalean *Neocalamites* and conifers (*Podozamites*). Zerfass et al. (2003) interpreted the Santa Maria 1 Sequence as reflecting a more humid period within an overall semiarid climatic regime. However, they took the apparent absence of temnospondyls as evidence that there were no perennial bodies of water.

The Santa Maria 2 Sequence is separated from the Santa Maria 1 Sequence by a major unconformity. Its fluvial sandstones are interlayered with mudstone lenses, which are interpreted as floodplain deposits and contain a variety of plant fossils, especially the ubiquitous corystospermalean *Dicroidium* (Guerra-Sommer and Klepzig 2000; figure 4.3), along with insect remains and conchostracans (Zerfass et al. 2003). These deposits are overlain by red laminated or massive mudstones, which were laid down in shallow lakes or on the floodplains of an ephemeral, anastomosed river system (Langer et al. 2007). The red mudstones, exposed particularly around the city of Santa Maria, have yielded abundant remains of a tetrapod assemblage that differs from that of the Santa Maria 1 Sequence in the predominance of rhynchosaurs and the absence of dicynodont therapsids (Barberena, Araújo, and Lavina 1985).

The rhynchosaurs from these strata have a confusing taxonomic history. A number of taxa were originally founded on what are now considered nondiagnostic remains (Huene 1935–42), but Langer and Schultz (2000) referred all of them to the widespread

Figure 4.3. Foliage of two species of the corystospermalean *Dicroidium*. (A1–2) *D. intermedium*. (B) *D. lancifolium*. (From Kurtz 1927)

A1

A2

B

2 cm

2 cm

genus *Hyperodapedon* (figure 4.4). Langer and Schultz distinguished three valid species, but, more recently, Langer et al. (2007) suggested greater generic diversity. The Santa Maria rhynchosaurs attained a length of at least 3 meters (Huene 1935–42). *Hyperodapedon* has multiple rows of teeth on the maxillae but only a single tooth row on each dentary. Its skull is transversely wider posteriorly than it is long, indicating substantial development of the jaw adductor musculature (Huene 1935–42; Romer 1962; Benton 1983b).

Cynodont therapsids include the traversodont *Gomphodontosuchus* (known only from a juvenile specimen; Huene 1935–42) and the small, mammal-like *Therioherpeton* (Bonaparte and Barberena 1975, 2001). Archosauriform reptiles include small proterochampsids (*Cerritosaurus*) (figure 4.5C), the rather gracile "rauisuchian" *Rauisuchus* (Huene 1935–42), the

aetosaur *Aetosauroides* (which is closely related to and may even be congeneric with *Stagonolepis*; chapter 6), the 2-meter-long basal saurischian dinosaur *Staurikosaurus* (Colbert 1970b; Bittencourt and Kellner 2009), and the up to 2-meter-long sauropodomorph dinosaur *Saturnalia* (Langer et al. 1999; Langer and Benton 2006). The shared presence of *Hyperodapedon* and *Aetosauroides* supports correlation of the Santa Maria 2 Sequence with the mostly Norian-age Ischigualasto Formation of northwestern Argentina (see "Argentina" later in this chapter). Zerfass et al. (2003) interpreted the strata of the Santa Maria 2 Sequence as showing signs of increased aridity.

The basal sauropodomorph *Saturnalia* has a relatively small skull, the length of which is less than two-thirds that of the femur. The teeth have leaf-shaped crowns with finely serrated margins. The humerus of

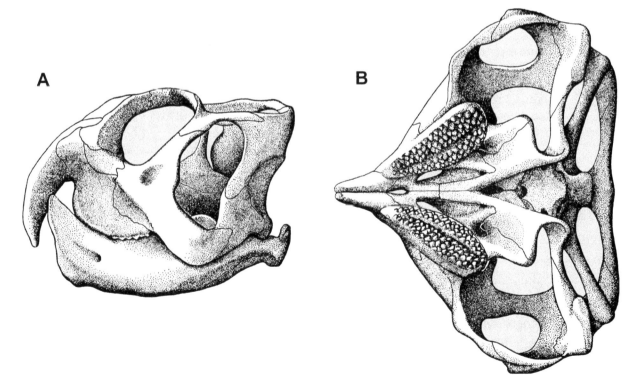

Figure 4.4. Reconstructed skull of the rhynchosaur *Hyperodapedon* sp. in (*A*) lateral and (*B*) palatal views. No scale provided; skulls can reach a length of more than 25 centimeters. (Modified from Romer 1962)

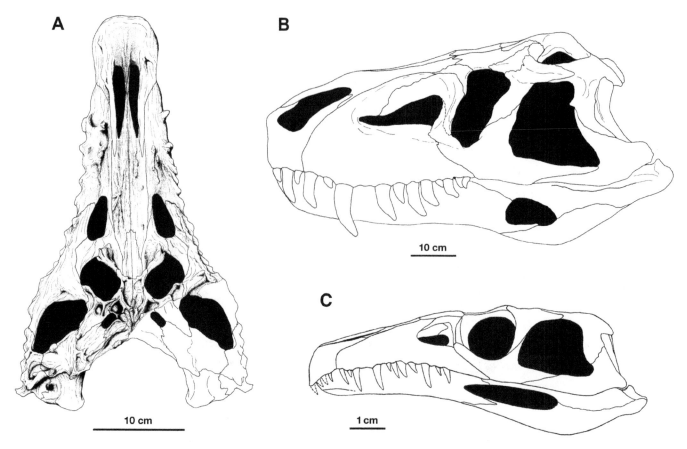

Figure 4.5. (*A*) Dorsal view of the holotype skull of the proterochampsid *Proterochampsa nodosa* (Museu de Ciências e Tecnologia, Pontifícia Universidade Católica do Rio Grande do Sul, Porto Alegre, MCP 1694 PV). (Modified from Barberena 1982) (*B*) Lateral view of a skull referred to the "rauisuchian" *Prestosuchus chiniquiensis* (Universidade Federal do Rio Grande do Sul, UFRGS PV 0156T). (Modified from Barberena 1978) (*C*) Lateral view of the reconstructed holotype skull of the proterochampsid *Cerritosaurus binsfeldi* (Colégio Anchieta, Porto Alegre, no number). (Modified from Barberena and Dornelles 1998)

Saturnalia has a prominent deltopectoral crest. A distinctive feature of the ulna is the presence of a large olecranon process. Noteworthy features of the pelvic girdle include an almost completely closed acetabulum and the large antitrochanter on the ischium. Based on its limb proportions, *Saturnalia* was probably still predominantly bipedal (Langer and Benton 2006).

A coarsening-upward sediment sequence, the Caturrita Formation (Andreis, Bossi, and Montardo 1980), replaces the lacustrine mudstones at the top of the Santa Maria 2 Sequence. It represents primarily lacustrine-deltaic deposits, which are increasingly replaced by fluvial strata. The Caturrita Formation, exposed in the areas of Candelariá and Faxinal do Soturno, has yielded a distinctive tetrapod assemblage that includes the dicynodont *Jachaleria*, the large traversodont cynodont *Exaeretodon*, small, mammal-like cynodonts (*Brasilitherium* and *Brasilodon*), the proterochampsid *Proterochampsa* (figure 4.5*A*), the sphenodontian *Clevosaurus*, the dinosauriform *Sacisaurus*, and several taxa of dinosaurs (Barberena

1982; Barberena, Araújo, and Lavina 1985; Bonaparte, Ferigolo, and Ribeiro 1999; Bonaparte et al. 2003; Bonaparte, Martinelli, and Schultz 2005; Langer et al. 1999; Leal et al. 2004; Bonaparte and Sues 2006; Ferigolo and Langer 2007). Some of these tetrapods also occur in the Carnian- to Norian-age Ischigualasto Formation and the lower part of the Norian-to Rhaetian-age Los Colorados Formation in the Ischigualasto basin of northwestern Argentina and thus provide important biostratigraphic links. The sauropodomorph dinosaur *Unaysaurus* is closely related to *Plateosaurus* from the Norian of Europe and Greenland (Leal et al. 2004). Langer et al. (2007) argued for a Norian age for the tetrapod assemblage from the Caturrita Formation. The vertebrate fossils occur in association with petrified wood and conchostracans.

The brightly colored sedimentary rocks of the Santa Maria 1 and 2 sequences are exposed mostly in small badlands along erosional gullies, which are locally known as *sangas*. Tetrapod bones are typically distorted and often "exploded" by early diagenetic precipitation of calcite and minor amounts of hematite within the pore spaces of the bone (Holz and Schultz 1998). This peculiar postmortem deformation frequently makes it difficult to determine even the overall shape of the affected elements. Holz and Schultz (1998) argued that this diagenetic modification of the bones must have occurred within a few meters of the land surface and required a climate with alternating wet and dry seasons. This inference is consistent with other evidence for the climatic conditions in the Santa Maria depositional environment.

The Santa Maria 3 Sequence consists only of conglomeratic sandstones, which contain abundant petrified wood. Due to the absence of other, biostratigraphically more useful fossils, the age of this unit remains uncertain.

ARGENTINA

Located some 1,300 kilometers to the west of the Santa Maria basin and at paleolatitudes between 40° and 45° south during the Triassic (Colombi and Parrish 2008), the Ischigualasto–Villa Unión basin in the provinces of San Juan and La Rioja in northwestern Argentina can be interpreted as a half-graben (Milana and Alcober 1994; figures 4.6 and 4.7). Its northern depocenter is in the Ischigualasto–Valle de la Luna region. Zerfass et al. (2004) divided the sedimentary fill of the Ischigualasto basin (which can reach a thickness of 4,000 meters) into the Ischichuca–Los Rastros (Anisian to Ladinian) and Ischigualasto–Los Colorados (Carnian to Rhaetian) sequences. The two sequences together form the Agua de la Peña Group.

The Ischichuca–Los Rastros Sequence reaches a thickness of up to 1,750 meters in the Ischigualasto region. Its basal portion contains the most thoroughly documented Middle Triassic tetrapod assemblage to date, from the Chañares Formation in La Rioja Province. Rogers et al. (2001) considered this unit early Ladinian in age. The Chañares Formation is dominated by fluviolacustrine tuffaceous sandstones and siltstones and conglomeratic deposits of alluvial fans (Rogers et al. 2001). Its lower portion comprises white to bluish white, diagenetically altered volcanic ashes that contain bone-bearing carbonate concretions (figure 4.8). Its tetrapod assemblage is notable for its diversity of small archosaurian reptiles and cynodont therapsids and was documented in a long series of papers by Romer (1966–73). Some taxa are represented by fairly complete, at least partially articulated skeletons (figure 4.8), whereas others are known only from fragmentary remains, which are frequently commingled with bones of other animals. In the early 1960s, Romer's field parties collected more than 100 individual vertebrate fossils preserved in concretions at the Los

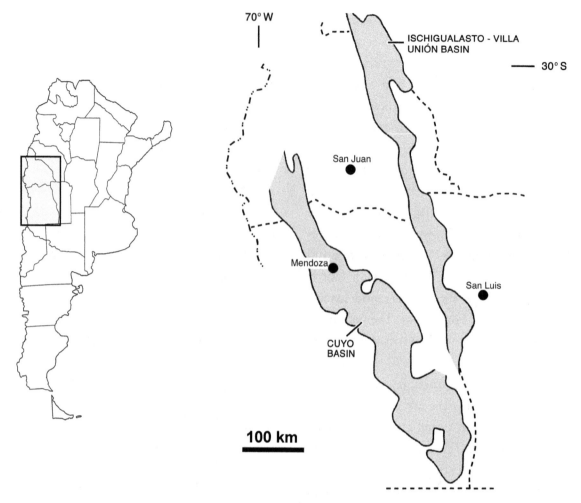

Figure 4.6. Map of Triassic basins in Argentina showing the positions of the Cuyo and Ischigualasto–Villa Unión basins in San Juan, Mendoza, and La Rioja provinces. (Modified from Martins-Neto, Gallego, and Melchor 2003)

Chañares locality. The matrix of these concretions consists of diagenetically altered pieces of volcanic glass that were replaced by calcite. Individual nodules often preserve the skeletal remains of putative predator and prey side by side. The arrangement of the animal carcasses is consistent with their deposition along a strand line. Rogers et al. (2001) argued that local volcanism led to catastrophic flooding in the region due to damming, diversion of local drainages, or both. The lack of compaction of the bones and traces of the outlines of glass shards indicate that the concretions started to form soon after the animal cadavers had been buried in reworked volcanic ash. Decomposition of soft tissues facilitated diagenetic alteration of the concretions, further protecting the entombed skeletons. It is still unclear whether volcanic events caused the death of the animals in the first place.

The Chañares assemblage clearly represents a taphonomically and ecologically highly biased picture of the original biota, with preferential preservation of smaller reptiles and therapsids and complete absence of remains of temnospondyls, terrestrial invertebrates, and plants (Rogers et al. 2001). The few known fossils

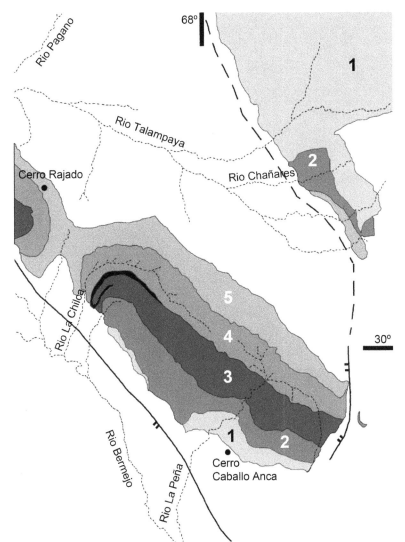

Figure 4.7. Triassic continental strata of the Ischigualasto–Villa Unión basin and adjoining regions in northwestern Argentina. *1*, Talampaya and Tarjados formations; *2*, Chañares Formation; *3*, Los Rastros Formation; *4*, Ischigualasto Formation; *5*, Los Colorados Formation. (Combined from Kokogian et al. 2001 and Rogers et al. 2001)

of large tetrapods, such as the dicynodont *Dinodontosaurus* (with a skull length of possibly up to 50 centimeters) and the "rauisuchian" *Luperosuchus* (with a skull length of about 60 centimeters), tend to be isolated finds outside the concretions. Small archosauriform and archosaurian reptiles are particularly diverse in the Chañares assemblage (figure 4.9). The former are represented by the Proterochampsidae (*Chanaresuchus* [figure 4.9*B*], *Gualosuchus,* and *Tropidosuchus*), the largest of which reached a length of more than 1 meter. Based on their superficially crocodile-like appearance, Romer (1971b, 1972a) inferred an am-

phibious mode of life for these archosauriform reptiles. Among the archosaurian reptiles, the small (up to 70 centimeters long) suchian *Gracilisuchus*, originally considered an ornithosuchid (Romer 1972b), is closely related to crocodylomorphs. *Lagerpeton* and *Marasuchus* ("*Lagosuchus*") (figure 4.9*A*) are small, cursorial ornithodirans with short forelimbs and long, slender hind limbs with digitigrade feet (Romer 1971a; Bonaparte 1975a; Arcucci 1986; Sereno and Arcucci 1994a, 1994b). Numerous features of their postcranial skeletons, such as the sigmoidal curvature of the cervical vertebral column, the inturned head of the femur,

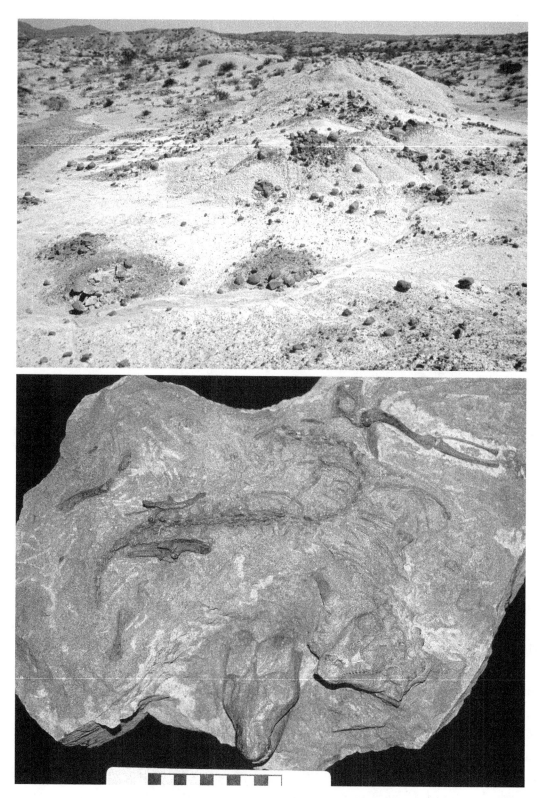

Figure 4.8. Exposures of the Chañares Formation with scattered nodules on the surface (*top*) and a nodule containing two skeletons of traversodont cynodont *Massetognathus pascuali* (*bottom*). (Photographs courtesy of R. R. Rogers)

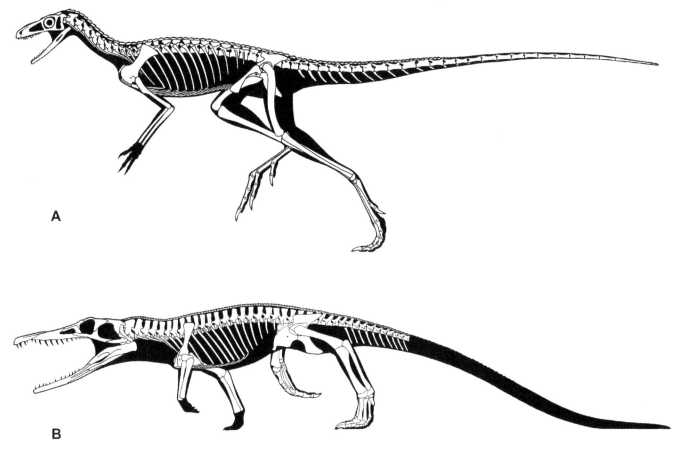

Figure 4.9. (*A*) Lateral view of the reconstructed skeleton of the dinosauriform *Marasuchus lilloensis*. Length up to 50 centimeters. (*B*) Lateral view of the reconstructed skeleton of the proterochampsid *Chanaresuchus bonapartei*. Length about 1 meter. (Drawings courtesy and copyright of G. S. Paul)

and the mesotarsal structure of the ankle, indicate close relationships to dinosaurs (Sereno 1991; Sereno and Arcucci 1994a, 1994b; Benton 1999). *Pseudolagosuchus* (Arcucci 1987) may be even more closely related to Dinosauria (Novas 1996; Irmis et al. 2007).

The Chañares assemblage includes a variety of cynodont therapsids. The most common of these is the traversodont *Massetognathus* (Romer 1967; Jenkins 1970). *Massetognathus* accounts for more than half of all identifiable tetrapod remains from the Chañares Formation and is frequently found in small groups (Bonaparte 1997; Rogers et al. 2001; figure 4.8). Its skull length ranges from 8 centime-

ters to more than 20 centimeters, and its total length could exceed 1 meter (figure 4.10*A*). Carnivorous cynodonts comprise *Chiniquodon* (including *Probelesodon*; Romer 1969; Abdala and Giannini 2002) and *Probainognathus* (Romer 1970); these taxa are considered close to the evolutionary lineage leading to mammals (Kemp 1982). In *Probainognathus*, with a skull length of up to 10 centimeters, the articular process of the dentary extends close to the squamosal (figure 4.11). Romer (1970) argued that a mammal-like contact between these two bones was already present in this form, but, as in many other eucynodonts, the articulation is actually formed between

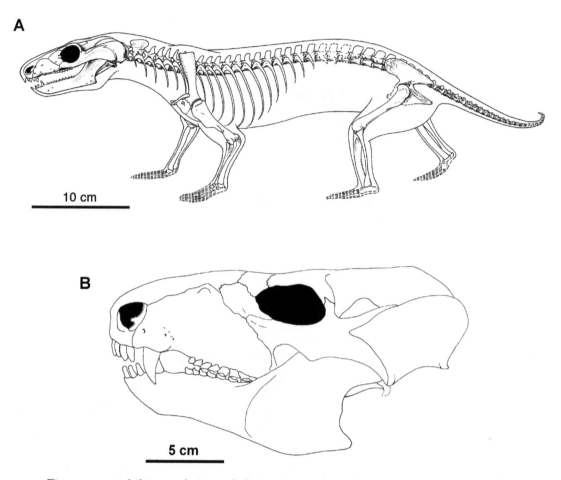

Figure 4.10.   (*A*) Lateral view of the reconstructed skeleton of the traversodont cynodont *Massetognathus pascuali*. (Drawing courtesy of F. A. Jenkins Jr.) (*B*) Lateral view of the reconstructed skull of the traversodont cynodont *Traversodon stahleckeri*. (Combined from Barberena 1981 and specimens)

the surangular and the squamosal. Another noteworthy feature of *Probainognathus* is its long secondary bony palate.

For correlation purposes, Lucas and Huber (2003) proposed the Chañarian LVF ("Chañarense" local age sensu Bonaparte [1973, 1982]) as the time equivalent of the Chañares tetrapod assemblage with *Dinodontosaurus*.

To date, tetrapods from the lacustrine-deltaic strata of the overlying Los Rastros Formation are documented only by large tracks produced by suchian reptiles (*Rigalites*) and temnospondyl bones. This unit comprises carbonaceous shales, siltstones,

and sandstones and has been interpreted as lake-basin fill (Rogers et al. 2001). The Los Rastros Formation has yielded a diversity of insects (Mancuso, Gallego, and Martins-Neto 2007) and actinopterygian fishes, conchostracans, and a varied plant assemblage including *Dicroidium*, ferns, equisetaleans, cycadophytes, and conifers. Mancuso and Marsicano (2008) attempted a detailed paleoenvironmental reconstruction. According to their analysis, the lake margins supported vegetation composed of corystospermaleans, ginkgophytes, and equisetaleans. River margins were covered by thickets of equisetaleans, and the proximal floodplain was home to closed

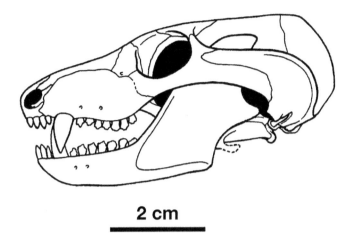

2 cm

Figure 4.11. Lateral view of the reconstructed skull of the probainognathian cynodont *Probainognathus jenseni.* (Combined from Romer 1969 and Hopson and Kitching 2001)

woodlands formed by corystospermaleans, cycadophytes, and ferns. Open conifer forests covered more distal floodplain settings.

The Ischigualasto–Los Colorados Sequence (figure 4.12) is separated from the underlying Ischichuca–Los Rastros Sequence by an unconformity and is up to 1,600 meters in thickness. The Ischigualasto Formation reaches a thickness of 1,000 meters and comprises primarily greenish and brown mudstones and subordinated laminated sandstones and tuffs. Lateritic levels are presumably related to paleosols. Rogers et al. (1993) and Zerfass et al. (2004) interpreted the Ischigualasto strata as having formed on an upland floodplain with shallow lakes and moderate- to high-sinuousity rivers. The Herr Toba bentonite close to the base of the Ischigualasto Formation has yielded an ^{40}Ar/^{39}Ar plateau age of 227.8 + 0.3 Ma (Rogers et al. 1993) or (corrected) 229.2 Ma (Furin et al. 2006). Using the chronostratigraphy proposed by Muttoni et al. (2004), these data indicate an early Norian age for much of the Ischigualasto Formation (figure 4.13). This unit has yielded many well-preserved skeletal remains representing a great variety of tetrapod taxa,

some of which also occur in the Santa Maria 2 Sequence (Bonaparte 1978, 1997).

Skulls of two taxa of temnospondyls, the chigutisaurid *Pelorocephalus* (with a skull length of about 20 centimeters) and the mastodonsaurid *Promastodonsaurus* (with a reconstructed skull length of 45 centimeters), have been reported from the lower portion of the Ischigualasto Formation (Bonaparte 1963, 1975b). Rogers et al. (1993) interpreted the scarcity of temnospondyls and other freshwater animals and sedimentological indicators such as caliches as evidence for a seasonal, rather dry climate during the deposition of this formation. Colombi and Parrish (2008) demonstrated a series of changes in climatic conditions from the base to the top of the Ischigualasto Formation. According to their analysis, deposition began under a moderately humid climate and under unstable tectonic conditions, preventing soil formation. About 40 meters above the base of the formation, tectonic conditions became more stable, and, under dry climatic conditions, calcic paleosols developed. At about 150 meters above the base, climatic conditions became progressively more humid, reaching a peak in humidity between 315 meters and 450 meters. Then they gradually returned to a dry seasonal climate until deposition of the Ischigualasto Formation ceased.

Most tetrapod fossils occur in the lower two-thirds of the Ischigualasto Formation. The most common taxa are the rhynchosaur *Hyperodapedon*, which is particularly abundant in the lower third and then vanishes in the upper portion of the formation, and the large traversodont eucynodont *Exaeretodon* (including *Ischignathus* and *Proexaeretodon*; Hopson and Kitching 1972), which could reach a length of more than 2 meters and is present throughout the succession (Bonaparte 1962, 1997; Rogers et al. 1993). *Exaeretodon* is distinguished from other traversodonts by its massive, bilobed postcanine teeth. The anterior margin of each tooth crown is convex and shoulders

Figure 4.12. Exposures of the Ischigualasto Formation (*lower part of photograph*) and overlying Los Colorados Formation in Ischigualasto National Park, Argentina. (Photograph courtesy of R. R. Rogers)

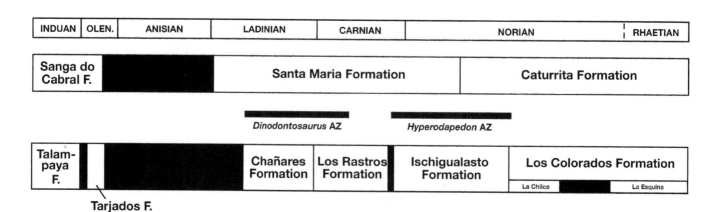

Figure 4.13. Tentative correlation for Triassic tetrapod-bearing continental strata in southern Brazil (*top*) and northwestern Argentina (*bottom*). (Modified from Langer et al. 2007)

tightly into the concave posterior margin of the crown in front of it (Bonaparte 1962; Crompton 1972). Carnivorous eucynodonts are represented by *Chiniquodon* (Martinez and Forster 1996) and *Ecteninion*, which is closely related to *Probainognathus* from the Chañares Formation (Martinez, May, and Forster 1996). The large, robustly built kannemeyeriid dicynodont *Ischigualastia* (with a skull length of up to 60 centimeters; Cox 1965) and a variety of archosaurian reptiles, including three taxa of dinosaurs, are much less common. The top predator was the large "rauisuchian" *Saurosuchus*, with a skull length exceeding 70 centimeters (Reig 1959; Sill 1974; Alcober 2000; figure 4.14*B*). A second form, *Sillosuchus*, appears to be closely related to the North American poposaurid "rauisuchians" (Alcober and Parrish 1997). The dis-

tinctive archosauriform *Proterochampsa* (Reig 1959; Sill 1967) has a dorsoventrally flattened, heavily sculptured skull with dorsally facing orbits and external nares. These features lend its skull (which can attain a length of 35 centimeters) a remarkably crocodile-like appearance, and indeed *Proterochampsa* was initially interpreted as an early crocodylian relative (e.g., Sill 1967). Finally, the aetosaur *Aetosauroides* (including *Argentinosuchus*; Casamiquela 1960, 1962) is closely related to (if not synonymous with) *Stagonolepis* from the Lossiemouth Sandstone Formation of Scotland, and the poorly known ornithosuchid *Venaticosuchus* (Bonaparte 1970) appears to be very similar to *Ornithosuchus* from the same formation (chapter 6).

The Ischigualasto Formation is best known for the presence of various basal dinosaurs. Remains

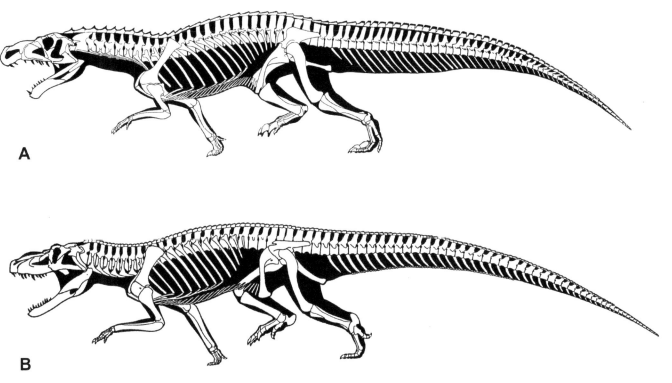

Figure 4.14. (*A*) Lateral view of the reconstructed skeleton of the ornithosuchid *Riojasuchus tenuisceps* from the Los Colorados Formation. Length about 1.5 meters. (*B*) Lateral view of the reconstructed skeleton of the "rauisuchian" *Saurosuchus galilei* from the Ischigualasto Formation. Length up to 6 meters. (Drawings courtesy and copyright of G. S. Paul)

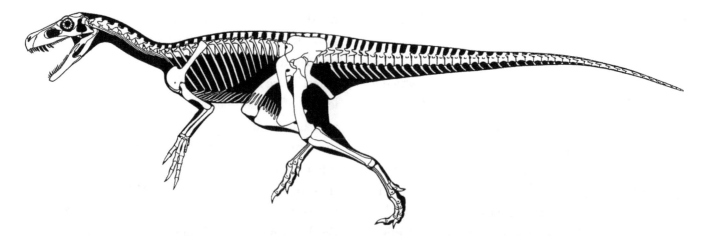

Figure 4.15. Lateral view of the reconstructed skeleton of the basal saurischian dinosaur *Herrerasaurus ischigualastensis.* Length up to 5 meters. (Drawing courtesy and copyright of G. S. Paul)

of basal saurischian dinosaurs, *Herrerasaurus* (Reig 1963; Novas 1994; Sereno 1994; Sereno and Novas 1994; Langer and Benton 2006; figure 4.15) and the smaller *Eoraptor* (Sereno et al. 1993), are known only from the lower third of the Ischigualasto Formation. The single known skeleton of the ornithischian dinosaur *Pisanosaurus* (Casamiquela 1967; Bonaparte 1976) comes from the middle third of the Ischigualasto Formation.

At present, *Herrerasaurus* is the best-known basal saurischian dinosaur (figure 4.15). It reached a length of 5 meters. *Herrerasaurus* has a sliding joint within each mandibular ramus, structurally different from but presumably functionally similar to the condition in many derived theropods (Sereno and Novas 1994). The skull is more or less equal in length to the femur. Its forelimb is less than half the length of the hind limb (Novas 1994). The long manus of *Herrerasaurus* retains the fourth and fifth digits, but the metacarpals of these digits are reduced in length and splintlike (Sereno 1994). Manual digits I–III end in large, trenchant claws. The pelvis of *Herrerasaurus* has a perforated acetabulum, which received the

inturned head of the femur, and the sacrum comprises two primary vertebrae and a dorsosacral. The posteroventrally extending pubis terminates in a robust distal "foot." Although the pes of *Herrerasaurus* still retains five digits, the first and fifth are shorter than the remaining ones.

Eoraptor differs from *Herrerasaurus* in that it is smaller and has a shorter snout with a heterodont dentition (Sereno et al. 1993). It appears to be more derived than *Herrerasaurus* in a number of features, which it shares with other saurischians (Langer and Benton 2006), but the only known specimen is a 1-meter-long juvenile.

The phylogenetic positions of *Herrerasaurus*, *Eoraptor*, and other basal dinosaurs continue to be the subject of lively debate. Some authors have classified these two genera as basal theropods (Sereno and Novas 1994; Sereno 2007), whereas others would place them as basal saurischians (Langer and Benton 2006; Irmis et al. 2007; Martinez and Alcober 2009) or even outside Dinosauria (Fraser et al. 2002).

Most recently, Martinez and Alcober (2009) described another basal saurischian, *Panphagia*, from a level 40 meters above the base of the Ischigualasto Formation. They interpret this taxon as the most basal sauropodomorph. Known from a partial juvenile skeleton with an estimated length of 1.3 meters, *Panphagia* has a dentition that suggests omnivorous habits.

Pisanosaurus is referred to the Ornithischia primarily on the basis of the structure of its jaws and teeth because the only known specimen, a partial skeleton, is too poorly preserved to provide much phylogenetically useful information (Bonaparte 1976).

Based on the characteristic tetrapod assemblage of the Ischigualasto Formation, Bonaparte (1973, 1982) proposed an "Ischigualastense" local age. Lucas and Huber (2003) redefined this local age as the Ischigualastian LVF for correlation of the Argentinian and Brazilian tetrapod assemblages.

Spalletti et al. (1999) interpreted the flora of the Ischigualasto Formation as comprising mixed forests and assigned it to a plant biozone characterized by the corystospermaleans *Rhexoxylon*, *Xylopteris*, and *Zuberia* and the peltaspermaceans *Lepidopteris* and *Scytophyllum*. *Dicroidium* is very abundant.

The contact between the Ischigualasto and Los Colorados formations is marked by a change in color from the drab brown and green sedimentary rocks of the former to the red sandstones and conglomerates of the latter formation, which is up to 1,000 meters thick (Zerfass et al. 2004; figure 4.12). The lower part of the Los Colorados Formation includes lenticular bodies of sandstone and conglomerate, which probably formed in low-sinuosity rivers. The find of a skull of the medium-sized dicynodont *Jachaleria* at the La Chilca locality, in the lower portion of the Los Colorados Formation (Bonaparte 1997), provides an important biostratigraphic link to the uppermost portion of the Santa Maria 2 Sequence. The red beds of the Los Colorados Formation comprise successive

thinning-upward cycles with medium- to coarse-grained sandstones and fine-grained sandstones intercalated with siltstones (Arcucci, Marsicano, and Caselli 2004). Caselli, Marsicano, and Arcucci (2001) interpreted this sequence as deposits of moderately sinuous fluvial systems, which laterally contact and grade into horizontally bedded floodplain deposits. Toward the top of the Los Colorados Sequence, thin-bedded sandstones and siltstones, which formed as pond deposits and crevasse splays in overbank settings, are dominant. Climatic conditions were arid during the deposition of the Los Colorados Formation, especially in the upper portion of the unit (Colombi and Parrish 2008).

Bonaparte (1972) reported a diverse tetrapod assemblage mainly comprising archosaurian reptiles from the upper portion of the Los Colorados Formation, the so-called La Esquina fauna. He later proposed a "Coloradense" local age for this assemblage (Bonaparte 1973, 1982), which Lucas and Huber (2003) defined as the Coloradian LVF for the time equivalent of the La Esquina fauna.

The largest member of this fauna is the large sauropodomorph dinosaur *Riojasaurus* (with a length of up to 11 meters; Bonaparte 1972; Bonaparte and Pumares 1995). The La Esquina fauna represents an interesting transitional community in which typically Late Triassic archosaurs, such as the aetosaur *Neoaetosauroides* (Desojo and Báez 2007), the very large "rauisuchian" *Fasolasuchus* (with a skull length of almost 1 meter; Bonaparte 1981), and the ornithosuchid *Riojasuchus* (figure 4.14*A*), are present along with tetrapods more closely related to forms found in Early Jurassic strata elsewhere, such as the protosuchid crocodyliform *Hemiprotosuchus* (Arcucci, Marsicano, and Caselli 2004).

In addition to these taxa, this faunal assemblage includes the medium-sized sauropodomorph

Coloradisaurus (Bonaparte 1978), the sphenosuchian crocodylomorph *Pseudhesperosuchus* (Bonaparte 1972), the basal turtle *Palaeochersis* (with a carapace length of up to 70 centimeters; Rougier, de la Fuente, and Arcucci 1995), the coelophysoid theropod *Zupaysaurus* (Ezcurra and Novas 2007), and the tritheledontid cynodont *Chalimia* (Bonaparte 1980). Based on its tetrapod assemblage, the uppermost portion of the Los Colorados Formation is late Norian to Rhaetian in age. The Los Colorados Formation has also yielded chirotheroid tracks and plant remains, particularly petrified wood (*Rhexoxylon*) (Arcucci, Marsicano, and Caselli 2004). Spalletti et al. (1999) assigned the flora from the Los Colorados Formation to a biozone characterized by the occurrences of the dipteridaceous fern *Dictyophyllum*, the protopinaceous conifer *Protocircoporoxylon*, and *Linguifolium*, a plant of uncertain affinities.

The Cuyo basin of western Argentina is the largest of the early Mesozoic rift basins in Argentina and comprises several subbasins, mainly in the province of Mendoza (Ávila et al. 2006; figure 4.6). The Triassic succession, the Uspallata Group, commences with the coarse-grained clastic strata of fluvial origin of the Río Mendoza and Cerro de Las Cabras formations (Desojo, Arcucci, and Marsicano 2002). The former formation has yielded the kannemeyeriid dicynodont *Vinceria* (with a skull length of 25 centimeters; Bonaparte 1970) and the poorly known eucynodont *Cromptodon*. It grades up to a suite of cross-bedded sandstones, shale, black shale, and tuffs, which have been interpreted as deposits of a moderate- to high-sinuosity river system and constitute the Potrerillos Formation. This unit is conformably overlain by the lacustrine black shales of the Cacheuta Formation, which in turn is succeeded by the fluvial red beds of the Río Blanco Formation. The Potrerillos Formation can

be correlated with the Los Rastros Formation in the Ischigualasto basin and the Cacheuta Formation with the Ischigualasto Formation (Bonaparte 1978). Both units in the Cuyo basin contain fossils of freshwater bivalves, conchostracans, and insects. Actinopterygian fishes occur throughout almost the entire Triassic succession. The strata have also yielded a considerable variety of plants. The flora from the Cacheuta Formation is dominated by the corystospermalean *Dicroidium* and comprises bennettitaleans, ginkgophytes, conifers, and equisetaleans (Spalletti et al. 1999, 2005). The Cacheuta and Río Blanco formations have yielded chigutisaurid temnospondyls (Marsicano 1999), and the former unit has also produced some reptilian remains including an incomplete skeleton of the enigmatic archosauriform *Cuyosuchus* (Desojo, Arcucci, and Marsicano 2002).

Martins-Neto, Gallego, and Melchor (2003) most recently reviewed the fossil record of insects from the Triassic of South America. They reported specimens from six different localities, with the principal assemblages recovered from the Los Rastros and Potrerillos formations. These assemblages comprise blattoids, coleopterans, ensiferan orthopterans, odonates, plecopterans (stoneflies), and auchenorrhynchan hemipterans. In addition, Martins-Neto, Gallego, and Melchor (2003) tentatively identified trichopterans (caddis flies) and dipterans. Cockroaches and beetles are the most common elements in the Triassic insect communities from Argentina: Martins-Neto, Gallego, and Melchor (2003) recorded well over 1,000 specimens referable to these two groups, whereas a mere 33 specimens represent all other insect groups. Most recently, Martins-Neto, Gallego, and Zavattieri (2007) described new material of blattoids and coleopterans from the Potrerillos Formation of Mendoza Province. They proposed three new genera and one new family based on isolated wings (including elytra). Such material leads to significant difficulties in assessing levels of insect diversity and for meaningful comparisons between insect assemblages from different regions of the globe.

SOUTH AFRICA

The Molteno Formation in the Karoo basin of South Africa and Lesotho is separated from the underlying Burgersdorp Formation (chapter 2) by a major unconformity (Hancox 2000). It is up to 650 meters thick and predominantly comprises yellowish gray or brown, bluish and light gray sandstone, with subordinate dark or olive gray, dark reddish brown, and dusky red siltstone, mudstone, and coal. To the north, the Molteno Formation decreases to about 30 meters in thickness and is represented by only two laterally extensive sandstone horizons. Hancox (2000) interpreted its strata as deposits of braided rivers that formed a vast floodplain. He regarded at least the lower half of the Molteno Formation as Carnian in age based on paleobotanical evidence.

Although the Molteno Formation has not yielded skeletal remains of tetrapods to date, a great variety of tracks and trackways, especially in Lesotho, attest to the presence of a diverse assemblage of reptiles (including dinosaurs) and therapsids (Ellenberger 1970, 1972, 1974). In recent decades, the unit has become widely known for its wealth of plant and insect fossils (Riek 1974, 1976; Anderson and Anderson 1993a, 1993b, 2003; Anderson et al. 1996; Scott, Anderson, and Anderson 2004; Labandeira 2006).

The most common plant taxon is the corystospermalean *Dicroidium*. According to Anderson and Anderson (1993a, 1993b), it occurred in a variety of growth habits ranging from shrubs to tall trees and dominated various habitats. The ginkgophyte *Sphenobaiera* apparently formed woodlands surrounding lakes on the floodplain. The long-leaved, shrubby conifer *Heidiphyllum* formed thickets in areas where the water table was high. *Equisetum* lived in marsh settings, and there were communities composed of ferns and *Kannaskopifolia*.

Anderson and Anderson (1993a, 1993b) and Anderson et al. (1996) reported more than 2,000 specimens of insects from the Molteno Formation, representing some 335 identifiable species in 18 orders. Virtually all specimens recovered to date are isolated wings. Cockroaches are most common in terms of number of specimens. Beetles are second in abundance but represent the taxonomically most diverse insect group in the Molteno Formation (Anderson et al. 1996). On the basis of elytron ornamentation alone, Anderson and Anderson (1993a) distinguished at least 135 different kinds of beetles in their collections. Hemiptera, Odonata, and Plecoptera are additional common insect groups. No identifiable dipterans have been recovered from the Molteno Formation to date. Damage on plants caused by herbivorous insects and their larvae is widespread and takes many forms, including the earliest known examples of leaf mining (Scott, Anderson, and Anderson 2004; Labandeira 2006); indeed, all major categories of insect feeding on plants were already present (figure 4.16).

In addition to insect fossils, the strata of the Molteno Formation have also yielded two specimens of an araneomorph spider, *Triassaraneus* (Selden et al. 1999; Selden, Anderson, and Anderson 2009). Certain features indicate that *Triassaraneus* was a web-living rather than ground spider.

The Lower Elliot Formation conformably overlies the Molteno Formation. The basal 2–3 meters of the Lower Elliot are devoid of vertebrate remains. They are overlain by a lenticular, medium-grained

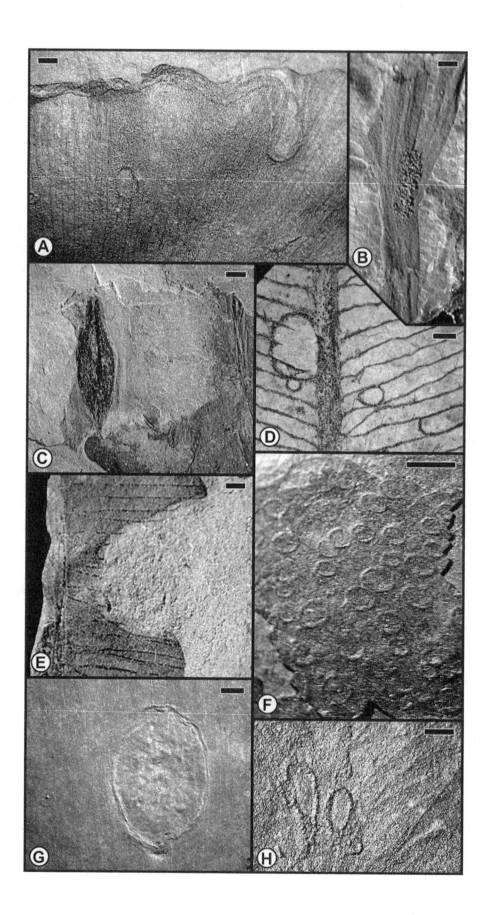

sandstone, above which tetrapod fossils typical of what Kitching and Raath (1984) termed the *Euskelosaurus* Range Zone first occur in red and purple sandstones (Hancox 2000). The diverse tetrapod assemblage comprises sauropodomorph dinosaurs including *Melanorosaurus* and *Antetonitrus*, the basal ornithischian dinosaur *Eocursor*, large chigutisaurid temnospondyls, the large traversodont cynodont *Scalenodontoides*, the tritheledontid cynodont *Elliotherium*, and an unidentified "rauisuchian" (Hopson 1984; Warren and Damiani 1999; Yates and Kitching 2003; Sidor and Hancox 2006; Butler, Smith, and Norman 2007). Although the age of the Lower Elliot is probably Norian, it cannot be constrained more precisely at present (Hancox 2000). The Upper Elliot Formation, which corresponds to the *Massospondylus* Range Zone of Kitching and Raath (1984), is Early Jurassic in age (Kitching and Raath 1984). Bordy, Hancox, and Rubidge (2004) interpreted the sandstones of the Lower Elliot Formation as having been formed by perennial, moderately sinuous fluvial systems under humid conditions, whereas the Upper Elliot Formation comprises mostly tabular sheet sandstones laid down by ephemeral rivers in a semi-arid to arid setting.

MADAGASCAR

Flynn et al. (1999) briefly reported the discovery of a diverse tetrapod assemblage from a unit informally designated as the "Isalo II beds" in the Morondova basin of southwestern Madagascar. These strata can reach a thickness of 1,000 meters and comprise primarily red sandstones. The still mostly unpublished tetrapod remains include a variety of traversodont cynodonts (Flynn et al. 2000), the rhynchosaur *Isalorhynchus*, and an as yet undescribed herbivorous archosauriform reptile (initially identified as a basal sauropodomorph dinosaur). In addition, the presence of dipnoans (*Ceratodus*) is documented by toothplates. Of the traversodont cynodonts, *Dadadon* is similar to *Massetognathus* from the Chañares Formation, and *Menadon* is closely related to *Exaeretodon* from the Ischigualasto Formation (Flynn et al. 2000; Kammerer et al. 2008). Flynn et al. (1999) argued that the Isalo II tetrapod assemblage predates similar Late Triassic faunas elsewhere in Gondwana. However, their assessment was based on assumptions about a relationship between stratigraphic age and the phylogenetic position of the rhynchosaur

Figure 4.16. Types of plant-insect associations from the Molteno Formation of South Africa. Scale bars each represent 1 millimeter. (*A*) Leaf mining on a leaf of *Paraginkgo antarctica* (Ginkgoales) from the Little Switzerland 111 locality. Note the frass trail at the top of the leaf and the terminal chamber at right. (*B*) Gall formation on the petiole of a leaf of *Sphenobaiera schenckii* from the Hlatimbe 213 locality. This feature may represent a mite gall with minuscule nymphal chambers. (*C*) Oviposition on a stem of *Equisetites greenensis* from the Greenville 111-B locality. The lenticular scar is the external expression of egg insertion into the stem of the plant by the ovipositor of a female dragonfly. (*D*) Hole feeding on a leaf of *Yabeiella mareyesiaca* from the Hlatimbe 213 locality. Note the formation of distinctive reaction tissue along the chewed margins of feeding holes. (*E*) Margin feeding on a leaf of *Taeniopteris anavolans* from the Matatiele 111 locality. Resistance of the vascular tissue to insect chewing is provided by the vein stringers along the chewed margin. (*F*) Predation on the seed *Umkoseminites insecta* (of unknown affinities) from the Umkomaas 111 locality. Some of the ellipsoidal punctures, shown here on the surface of the seed test, are overlapping, indicating that they were produced by insects. (*G*) Piercing and sucking on a leaf of *Heidiphyllum elongatum* from the Mazenod 211 locality. Plant damage consists of ovoidal, abraded surface area delimited by the extent of the now absent scale-insect cover. (*H*) Surface feeding on a leaf of *Dicroidium crassinervis* from the Umkomaas 111 locality. Note abrasion of surface tissues, each with a surrounding rim of reaction tissue, on both sides of the midvein of the leaf. (Figure courtesy of C. C. Labandeira)

and traversodont cynodonts, the diversity of traversodonts, and the absence of aetosaurs. Langer (2005) noted problems with these arguments and considered the Isalo II assemblage probably close in age to those from the Ischigualasto Formation and correlative strata.

INDIA

Thick fills of Permo-Triassic continental sedimentary strata were deposited in several intracratonic, typically fault-bounded rift basins on the Indian subcontinent. These basins represent remnants of a single major basin that was tectonically disrupted during and after deposition of these strata (Chakraborty, Mandal, and Ghosh 2003). They form three linear belts along the present-day valleys of the Pranhita-Godavari, Narmada-Son-Damodar, and Mahanadi rivers in central India (figure 4.17). The Pranhita-Godavari and Mahanadi basins extend more or less parallel to each other and meet the east-northeast–west-southwest–extending Narmada-Son-Damodar basin.

A particularly important succession of Middle and Late Triassic tetrapod assemblages from the Pranhita-Godavari basin has been known since the

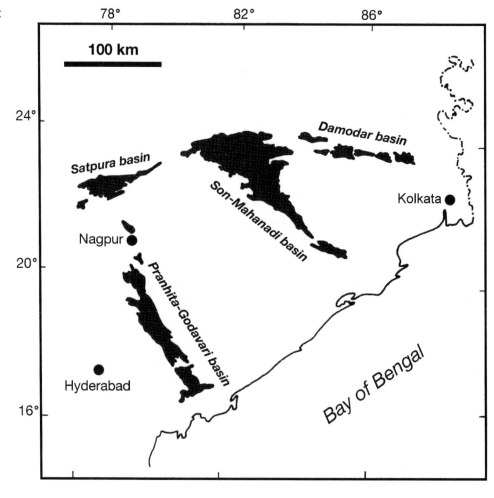

Figure 4.17. Intracratonic rift basins with Triassic continental strata in central India. Outcrop areas are shown in black. The Satpura basin is often called the Narmada basin. (Modified from Bandyopadhyay 1999)

nineteenth century. The north-northwest–south-southeast–trending basin is 400 kilometers long and up to 75 kilometers wide. It is bounded by a major fault in the east and contains some 4,000 meters of Permian and early Mesozoic sedimentary fill.

The Bhimaram Formation overlies the Anisian-age Yerrapalli Formation (chapter 2) and is dominated by coarse-grained, yellowish brown sandstone interbedded with red mudstone. To date it has yielded only fragmentary remains of capitosauroids and dicynodonts (Bandyopadhyay 1999; Bandyopadhyay, Roy Chowdhury, and Sengupta 2002). The Bhimaram Formation, in turn, is overlain successively by the Maleri and Dharmaram formations.

The Upper Triassic Maleri and Dharmaram formations have yielded four stratigraphically distinct tetrapod assemblages. The sequence of the Maleri Formation is up to 600 meters in thickness and begins with a red mudstone, which is up to 150 meters thick and passes upward into a succession of alternating sheetlike sandstones and thicker mudstones (Bandyopadhyay, Roy Chowdhury, and Sengupta 2002). The medium- to coarse-grained sandstones show cross-bedding, with overlapping channel fills forming multistory packages. The mudstones are typically red to brownish red, featureless, and poorly lithified. Most of the vertebrate remains occur in these mudstones, which have been interpreted as floodplain deposits (Bandyopadhyay, Roy Chowdhury, and Sengupta 2002). Distinct lower and upper faunal assemblages have been recognized in the Maleri Formation. The lower assemblage, from the basal portion of the formation, is the best documented to date. It comprises a variety of freshwater fishes, the metoposaurid *Metoposaurus* (Roy Chowdhury 1965), the rhynchosaur *Hyperodapedon* ("*Paradapedon*"; Huene 1940; Chatterjee 1974), the protorosaur *Malerisaurus* (Chatterjee 1980b), the basal phytosaur *Parasuchus* (Chatterjee 1978), the traversodont cynodont *Exaeretodon* (Chat-

terjee 1982), a large dicynodont, and as yet unidentified archosaurian reptiles. A diverse tetrapod assemblage from the Tiki Formation in the west-northwest-east-southeast extending Son-Mahanadi basin is virtually identical with that from the lower Maleri (Datta 2005). The Tiki Formation comprises mottled red mudstones, sandstones, and a lime-pellet horizon. In addition to several tetrapod taxa shared with the lower Maleri, its strata have yielded the "rauisuchian" *Tikisuchus* (Chatterjee and Majumdar 1987), the acrodontan lizard *Tikiguania* (Datta and Ray 2006), the small mammal-like cynodont *Rewaconodon* (Datta, Das, and Luo 2004), the morganucodontid mammaliaform *Gondwanadon* (Datta and Das 1996), and the unusual mammal *Tikitherium* (Datta 2005). A diverse assemblage of pollen and spores from the Tiki Formation has been dated as Carnian (Maheswari and Kumaran 1979).

The upper vertebrate assemblage from the Maleri Formation is placed stratigraphically higher by Indian researchers, but its precise age remains uncertain. It shares some fish taxa with the lower Maleri assemblage but otherwise appears to be different in its faunal composition. Sengupta (1995) described two chigutisaurid temnospondyls, *Cosmocerops* and *Kuttycephalus*, from this assemblage and noted the absence of metoposaurs. Hungerbühler, Chatterjee, and Kutty (2002) compared two phytosaurs from the upper Maleri assemblage to the American taxa *Angistorhinus* and *Leptosuchus*, respectively.

The Dharmaram Formation unconformably overlies the Maleri Formation. It is at least 370 meters thick. The base of this unit is formed by a massive sandstone, followed by an alternating succession of red mudstone and sandstone beds. Compared to the underlying Maleri Formation, the sand-to-mud ratio of the strata increases and the sandstone bodies are thicker (Bandyopadhyay, Roy Chowdhury, and Sengupta 2002). Again, two distinct faunal assemblages have been recorded from the Dharmaram Formation (Kutty and Sengupta 1989), but detailed descriptions of most of the vertebrate fossils have yet to be published. The lower assemblage is reported to include a large, *Nicrosaurus*-like phytosaur, an aetosaur, and a small sauropodomorph dinosaur (Bandyopadhyay 1999) and may be Norian in age. The upper assemblage, from the uppermost mudstone unit of the Dharmaram Formation, is now considered Early Jurassic based on the presence of sauropodomorph dinosaurs and other reptiles (Kutty et al. 2007).

Middle to Late Triassic tetrapod assemblages from Gondwana were much more diverse than those from the Early Triassic. They still show a considerable degree of faunal homogeneity with closely related therapsids, rhynchosaurs, and crurotarsan archosaurs.

However, there are some regional differences in faunal composition, notably the presence of phytosaurs and metoposaurs in the Late Triassic of India and the absence of rhynchosaurs in the Chañares Formation of Argentina. Apart from India, there is a dearth of known freshwater tetrapod communities from mid-latitude Gondwana. Again, there are no known low-latitude assemblages from the Middle to Late Triassic of Gondwana. By contrast, low-latitude tetrapod assemblages are well-known from Laurasia; they are discussed in detail in the following chapters.

Middle to Late Triassic plant communities of Gondwana are dominated by corystospermalean gymnosperms, most commonly represented by the foliage form taxon *Dicroidium*. At the same time they also share many elements of contemporary floras from Laurasia.

CHAPTER 5

Late Middle and Late Triassic of Europe

GERMANY

The Keuper Group represents the upper unit of the Germanic Trias Supergroup. It spans the late Middle Triassic (Ladinian) and the entire Late Triassic (Carnian to Rhaetian) (figure 5.1). Predominantly terrestrial depositional environments rapidly replaced the marine conditions that prevailed during deposition of the Muschelkalk Group (Stollhofen et al. 2008). The name "Keuper" probably derives from the fact that certain Late Triassic marls in southern Germany weather into small pieces, which are variously referred to as "*Kifer*" or "*Kies*" (German for "gravel") (Deutsche Stratigraphische Kommission 2005).

In the nineteenth and early twentieth centuries, strata of the Keuper in the southern part of the Germanic basin (figure 5.2), especially in the conterminous German states of Baden-Württemberg and Bavaria, yielded numerous, often superbly preserved skeletal remains representing a great variety of late Middle and Late Triassic temnospondyls and reptiles (Meyer and Plieninger 1844; Meyer 1847–55, 1861; E. Fraas 1889, 1896, 1913; Huene 1907–8, 1932; Schmidt 1928, 1938; Benton 1993; Schoch and Wild 1999b; Schoch 2006b). Historically, these discoveries provided the first extensive record of late Middle and Late Triassic vertebrates, and the Keuper became the classic succession of continental tetrapod communities from this time interval. After a long hiatus, active collecting and research commenced again in the 1970s and has since led to many important new discoveries.

Over the years, numerous stratigraphic terms have been employed for Keuper strata in the various regions of Germany and adjoining areas. Recently, the German Stratigraphic Commission attempted to sort out this complex terminology and established a new formal lithostratigraphic nomenclature (Deutsche Stratigraphische Kommission 2005), which has been adopted in this chapter (figure 5.2). (The traditionally used stratigraphic designations are added in parentheses.)

The Keuper formations in the Germanic basin can be divided into three principal facies types:

1. Vindelician Keuper: units of the sandy marginal facies of the basin (Gravenwöhr, Benk, Hassberge, and Löwenstein formations, and some parts of the Exter Formation)
2. Nordic Keuper (in the strict sense): units with sandy input from the northern highland of Fennoscandia (Erfurt and Stuttgart formations, and some parts of the Exter Formation)

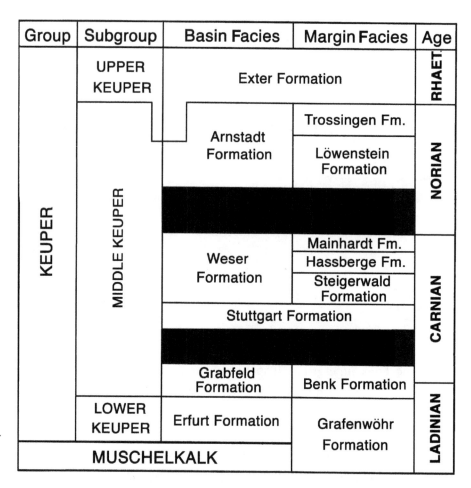

Group	Subgroup	Basin Facies	Margin Facies	Age
KEUPER	UPPER KEUPER	Exter Formation		RHAET.
	MIDDLE KEUPER	Arnstadt Formation	Trossingen Fm.	NORIAN
			Löwenstein Formation	
		■■■■■■■■	■■■■■■■■	
		Weser Formation	Mainhardt Fm.	CARNIAN
			Hassberge Fm.	
			Steigerwald Formation	
		Stuttgart Formation		
		■■■■■■■■	■■■■■■■■	
		Grabfeld Formation	Benk Formation	
	LOWER KEUPER	Erfurt Formation	Grafenwöhr Formation	LADINIAN
MUSCHELKALK				

Figure 5.1. Correlation chart of formations comprising the Keuper Group, using the nomenclature codified by the Deutsche Stratigraphische Kommission (2005). Black indicates major unconformities.

3. Central Basin Keuper: carbonate-clay-evaporite sequences of the central basin (Grabfeld, Weser, Steigenwald, Mainhardt, and Arnstadt formations)

Not all of these formations have yielded significant remains of terrestrial animals and plants, and only the fossiliferous units are reviewed in detail in this chapter.

In southern Germany, the so-called Grenzbonebed marks the transition from the marine Upper Muschelkalk to the Lower Keuper (Erfurt Formation) and overlies a minor unconformity. As its name implies, the Grenzbonebed is made up of countless teeth, scales, and fragmentary, typically water-worn bones representing a diversity of fishes and tetrapods, along with phosphatic coprolites (Reif 1971). It formed

as the result of repeated reworking of bone-bearing marine sediments in shelf areas, with progressive concentration of the heavier skeletal remains and simultaneous removal of lighter sedimentary particles.

The Erfurt Formation (Lower Keuper, or Lettenkeuper) is dated as late Ladinian (Langobardian). It comprises a complex succession of fluvial, lacustrine, and marine sedimentary rocks. During Ladinian times, shallow seas originating from the Tethys repeatedly inundated the Germanic basin, transgressing through the Burgundy Gate and Hessian Depression. Vast swamp and deltaic areas were left behind each time the sea retreated again. Beutler, Hauschke, and Nitsch (1999) described a cyclical pattern of sedimentation for the Erfurt Formation. Limestones and dolomitic deposits formed during the marine incur-

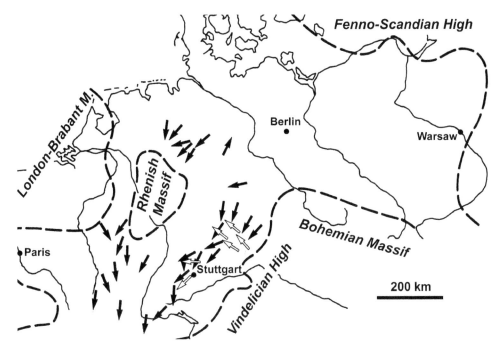

Figure 5.2. Map of central Europe with Late Triassic topographic highs (delimited by broken lines) and directions of sediment transport into the Germanic basin (with black arrows indicating introduction from Fenno-Scandian High to the north and white arrows indicating introduction from Vindelician High to the southeast). (Combined from Schröder 1982 and Geyer and Gwinner 1991)

sions into the basin. They were followed by clays and marls of brackish-water origin, which reflect a decrease in water depth and salinity. Locally, these clays and marls contain coals, which have yielded a rich and varied flora dominated by equisetaleans (Kelber 1999) and indicating humid climatic conditions. Noteworthy plant taxa include the oldest known bennettitaleans (*Pterophyllum*) and caytonialeans (*Sagenopteris*). Finally, there was an increase in more sandy sediments, which are fluvial in origin and contain numerous remains of freshwater organisms, especially ostracodes and the equisetaleans *Equisetites* and *Neocalamites* (Kelber 1999). Shoots of *Equisetites* reached a height of several meters and a diameter of up to 20 centimeters (figure 5.3). *Neocalamites* was up to 2 meters tall; it has well-developed lateral branches with whorls of leaves attached to each node.

Reflecting the depositional setting, the tetrapod fauna from the Erfurt Formation comprises a mixture of marine, brackish-water, and terrestrial elements. The best example of such an assemblage to date came from temporary exposures created during highway construction near the town of Kupferzell-Bauersbach in northern Württemberg. Following the discovery of large temnospondyl bones in a layer of marl in 1977, teams from the Staatliches Museum für Naturkunde in Stuttgart and avocational collectors undertook a salvage excavation. This effort recovered some 30,000 individual bones representing a diverse array of fishes, temnospondyls, and reptiles (Schoch 2006b). The sedimentological context indicates a brackish-water depositional setting.

The capitosauroid *Mastodonsaurus* is the best-documented tetrapod from the Kupferzell locality. The largest known temnospondyl, it has a massive skull reaching a length of 1.25 meters and attained a total length possibly exceeding 6 meters (Schoch 1999, 2006b; figure 5.4). The upper dentition of *Mastodonsaurus* includes robust tusks on the palatine and vomer, which, together with a pair of massive tusks on the dentaries, can be up to 15 centimeters tall and were suitable for impaling struggling prey. The tips of the lower tusks project through paired openings in the roof of the snout. The skull of

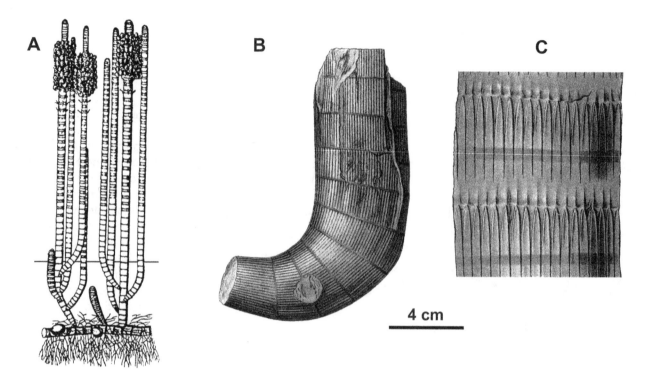

Figure 5.3. Equisetalean *Equisetites arenaceus*. (*A*) Attempted reconstruction of the entire plant. (From Frentzen 1934) (*B*) Fragment of an axis with (*C*) detail showing leaf tips. (From Fraas 1910)

Figure 5.4. Skull of the capitosauroid temnospondyl *Mastodonsaurus giganteus* (Staatliches Museum für Naturkunde Stuttgart, SMNS 54677) in oblique dorsolateral view. (Photograph by NCF)

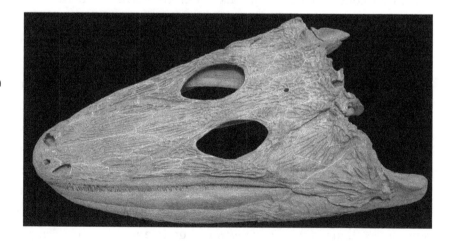

Mastodonsaurus is relatively narrow and has the outline of an elongated triangle. The orbits are proportionately large. The intercentra of *Mastodonsaurus* form complete, massive disks. Its ribs are platelike and bear prominent uncinate processes. Based on the Kupferzell specimens, *Mastodonsaurus* resembled the (much smaller) extant Japanese giant salamander (*Andrias japonicus*) in having a long, anteriorly flattened trunk, a moderately long, laterally flattened tail, and rather small limbs. These body proportions, along with the deeply incised, continuous lateral-line grooves on the skull, suggest that *Mast-*

odonsaurus was adapted to a fully aquatic mode of life.

The second most abundant temnospondyl at Kupferzell is the plagiosaurid *Gerrothorax* (Hellrung 2003; figure 5.5). It attained a length of about 1 meter. A closely related larger form, *Plagiosuchus*, is much less common at this locality; it especially differs from *Gerrothorax* in the enormous size of the orbits and the absence of various circumorbital bones (Damiani et al. 2009). Like other plagiosaurids, *Gerrothorax* has a short but remarkably broad, flattened skull with a convex anterior and nearly straight posterior edge and dense tubercular ornamentation of the bones. Its very large orbits face dorsally and are located directly over huge interpterygoid vacuities. In addition to marginal rows of numerous teeth, the wide mouth contains dense patches of small teeth at the back of the palate. The dorsoventrally flattened trunk of *Gerrothorax* is as wide as its head and covered dorsally and ventrally by overlapping osteoderms, which bear a sculpturing of densely spaced tubercles. On the posterior dorsal vertebrae, each spool-shaped centrum articulates with both the neural arch in front and behind it. The limbs of *Gerrothorax* are small, but its pectoral girdle is robust. Their overall appearance indicates that plagiosaurs were bottom-dwellers. Jenkins et al. (2008) demonstrated that *Gerrothorax* was capable of lifting its

skull through an arc of about 50 degrees, rotating the quadrates forward and thus protruding the mandible. The robust, well-ossified branchial skeleton of *Gerrothorax* indicates that it relied primarily on internal gills for respiration throughout life (Hellrung 2003). The great abundance of remains of this temnospondyl in the marl layer at Kupferzell, along with desiccation cracks at the top of this horizon, suggests that these animals may have congregated in a shrinking body of water and ultimately perished when that water dried up.

Kupferzell and other localities in the Erfurt Formation have yielded fossils representing additional temnospondyl taxa, including the capitosauroid *Kupferzellia* (Schoch 1997), the gracile, long-snouted trematosaurid *Trematolestes* (Schoch 2006a), and a form closely related to metoposaurs, *Callistomordax* (Schoch 2008).

A remarkable find from Kupferzell is referable to a chroniosuchian anthracosaur, *Bystrowiella* (Witzmann, Schoch, and Maisch 2008). Known from vertebrae and osteoderms, the new taxon is referable to the Bystrowianidae, which were previously known only from the Upper Permian to Middle Triassic of Russia (Novikov and Shishkin 2000). Chroniosuchians are distinguished by the fusion of sculptured, butterfly-shaped osteoderms to the apices of

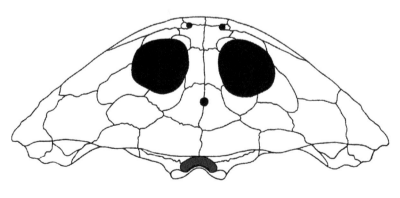

5 cm

Figure 5.5. Dorsal view of the reconstructed skull of the plagiosaurid temnospondyl *Gerrothorax pustuloglomeratus.* (Modified from Hellrung 2003)

the neural spines and by the spherical shape of the intercentra.

The various temnospondyls and *Bystrowiella* presumably subsisted mainly on abundant actinopterygian fishes such as *Gyrolepis* as well as the dipnoans *Ceratodus* and *Ptychoceratodus* (Schultze and Kriwet 1999).

Terrestrial reptiles from the Erfurt Formation include various archosaurs, including an as yet unidentified aetosaur (Schoch 2006b) and the "rauisuchian" *Batrachotomus*. *Batrachotomus* is known from the remains of three associated skeletons from Kupferzell as well as from specimens from other localities in the Erfurt Formation of Württemberg (Gower 1999; Schoch 2006b; Gower and Schoch 2009). It could reach a length of 6 meters. Its skull is up to 50 centimeters long and tall and narrow. The long, rather slender limbs of *Batrachotomus* suggest a rather upright stance and a gait comparable to the "high walk" in crocodylians (Schoch 2006b). As the name of the genus implies, *Batrachotomus* preyed on *Mastodonsaurus* and other temnospondyls. Many limb bones of *Mastodonsaurus* preserve bite marks that can be matched to teeth of *Batrachotomus* (Schoch 2006b).

Therapsids are a still poorly documented element of the tetrapod assemblage from the Erfurt Formation. At present, cynodonts are known only from a few isolated teeth, one of which belongs to a small traversodont, *Nanogomphodon* (Hopson and Sues 2006). Lucas and Wild (1995) also reported a humerus of an indeterminate kannemeyeriid dicynodont from the base of the Erfurt Formation in northern Württemberg.

Finally, marine reptiles are represented at Kupferzell by a well-preserved partial skeleton of the sauropterygian *Nothosaurus*. *Nothosaurus* could attain a length of more than 4 meters; it is common in marine strata of the Muschelkalk and correlative formations in the European Alps and the circum-Mediterranean region (Rieppel 2000). Its skull is long and narrow, with large, procumbent, and interdigitating teeth at the front of the jaws. The neck of *Nothosaurus* is long and flexible, but its dorsal vertebrae have accessory articulations, indicating a fairly rigid trunk region. Although the robust limbs of *Nothosaurus* already show adaptations for rowing, its long, well-developed tail probably still provided the principal means for swimming.

The Middle Keuper succession commences with the Grabfeld Formation (Lower Gipskeuper), which is late Ladinian (late Langobardian) to early Carnian (Julian) in age. Climatic conditions became generally more arid, and fluvial discharge nearly ceased (Stollhofen et al. 2008). The strata of the Grabfeld Formation probably formed under shallow-water, brackish to marine conditions and comprise mudstones, thin limestones, and extensive evaporites, mainly salt (particularly in northern Germany) and gypsum. They have yielded a few skeletal remains of marine reptiles, including the stratigraphically youngest record of *Nothosaurus* (Rieppel 2000).

The onset of the fluviodeltaic sedimentation of the Stuttgart Formation (Schilfsandstein) marks the return of continental conditions over the entire Germanic basin. However, Shukla and Bachmann (2007) noted estuarine, in part even marine-influenced depositional conditions in parts of this unit. The Stuttgart Formation is early Carnian (Julian) in age and separated from the underlying Grabfeld Formation by an unconformity. Rivers introduced large quantities of clastic sediment into the basin from Fennoscandia, a northern highland in the region of present-day Scandinavia and northern Russia (figure 5.2). These deposits formed thick, fine- to medium-grained sandstones that are locally rich in plant fossils (especially *Equisetites*, whose abundant stems and rhizomes gave the unit its original name, which translates as "reed sandstone"; Kelber 1999).

The flora shares numerous taxa with that of the Erfurt Formation, but it also includes conifers and the oldest known dipteridaceous fern (*Dictyophyllum*). A species of the bennettitalean *Pterophyllum* is quite common. Locally extensive claystones and siltstones document areas of flooding. The sandstones occasionally contain vertebrate remains, which represent both terrestrial and freshwater forms. The latter include temnospondyls, which include the largest and stratigraphically youngest known trematosaurid, *Hyperokynodon* (Hellrung 1987), and two capitosauroids, *Cyclotosaurus* and *Metoposaurus* (Schoch and Wild 1999b). *Cyclotosaurus* has a broad skull with a length

of up to 70 centimeters and with a long, broadly rounded snout, terminal nares, far posteriorly placed orbits, and moderately developed lateral-line grooves (Fraas 1889, 1913; Schoch and Milner 2000; Sulej and Majer 2005; figure 5.6). *Metoposaurus* and its relatives are characterized by a large (up to more than 60 centimeters long), unusually flat skull with a rather short snout, an elongated skull table, and conspicuous lateral-line grooves (Fraas 1889; Hunt 1993; Schoch and Milner 2000; Sulej 2007; figure 5.7). The small orbits are placed far forward and face dorsally. The intercentra of metoposaurs are large and spool-shaped. The length of the laterally flattened tail is more or less

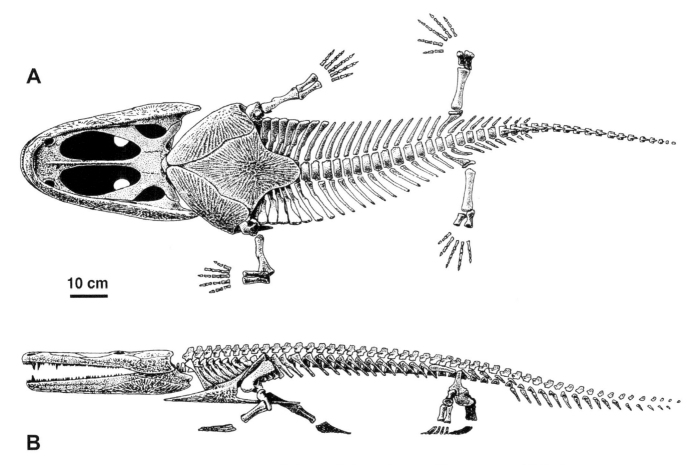

Figure 5.6. Reconstructed skeleton of the capitosauroid temnospondyl *Cyclotosaurus intermedius* in (*A*) ventral and (*B*) lateral views. (From Dzik and Sulej 2007; courtesy of J. Dzik)

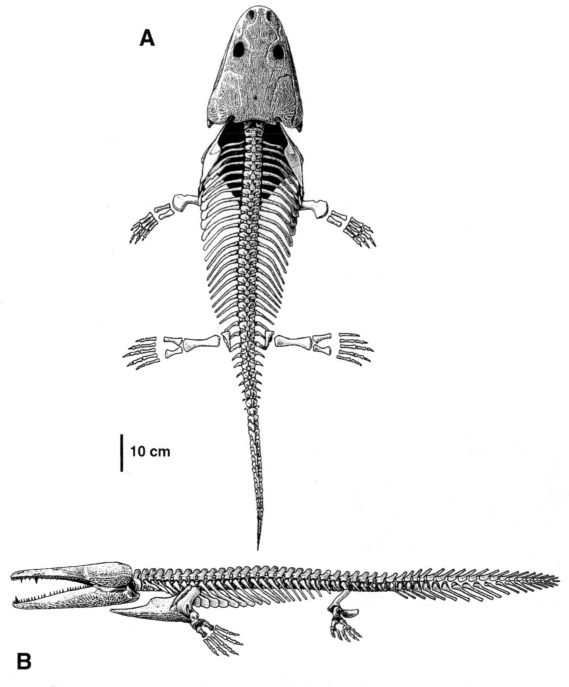

A

10 cm

B

Figure 5.7. Reconstructed skeleton of the metoposaurid temnospondyl *Meto-posaurus diagnosticus* in (*A*) dorsal and (*B*) lateral views. (From Dzik and Sulej 2007; courtesy of J. Dzik)

equal to that of the trunk (Sulej 2007). The limbs are rather small but well ossified. Many of these features indicate that metoposaurs were predominantly aquatic in habits. *Metoposaurus* and its relatives were widely distributed and common throughout Laurasia during the Late Triassic. Among reptiles from the Stuttgart Formation, the small *Dyoplax* is a possible crocodylomorph (Lucas, Wild, and Hunt 1998). Unfortunately, the preservation of the only known specimen as a natural mold makes identification of most anatomical features difficult.

The overlying Weser Formation (Bunte Mergel or Upper Gipskeuper) is Carnian (Julian to Tuvalian) in age and comprises claystones and evaporites, which formed on widespread sabkha mudflats under arid conditions (Stollhofen et al. 2008). The unit has yielded few tetrapod fossils to date. However, the stratigraphically equivalent Blasenstein (which represents the lower part of the Hassberge Formation along the southeastern margin of the depositional basin) at Ebrach in Upper Franconia (Bavaria) contains a tetrapod assemblage including the temnospondyls *Cyclotosaurus*, *Metoposaurus*, and *Plagiosaurus*, the possible aetosaur *Ebrachosaurus*, and the basal phytosaur "*Parasuchus*" (Kuhn 1932, 1936; figure 5.8).

The next higher unit in much of the German portion of the Germanic basin is the Arnstadt Formation (Steinmergelkeuper), which comprises cyclical sequences of mainly siltstones and marly mudstones deposited in a playa setting (Stollhofen et al. 2008). A major unconformity (known as the "Early Cimmerian" unconformity) separates this unit from the underlying Weser Formation. The Arnstadt Formation has yielded relatively few tetrapod fossils. Most noteworthy is an important occurrence of the sauropodomorph dinosaur *Plateosaurus* at Halberstadt in the state of Sachsen-Anhalt. The age of the Arnstadt Formation is middle Norian (Alaunian) to late Norian (Sevatian).

The lower equivalent of the Arnstadt Formation along the southeastern margin of the depositional basin in southern Germany, the Löwenstein Formation (Stubensandstein) of Württemberg, comprises four distinct sandstone members, which are interbedded with siltstones, mudstones, and occasional calcrete horizons. The interbedded mudstones (the so-called Hangendletten) show evidence of paleosol formation (e.g., caliche) and probably formed either on a floodplain or in a playa setting. The sandstone members are fluvial in origin, with their sediments introduced into the basin from highland areas to the south and southeast, specifically the Vindelician High and the Bohemian Massif (figure 5.2). The sandstones are lightly colored and fine- to coarse-grained, occasionally containing conglomeratic layers, and have a relatively high content of feldspar and clay minerals. The old designation "Stubensandstein" for this formation referred to the traditional use of its sandstones in pulverized form for cleaning house floors. The age of the Löwenstein Formation ranges from the middle Norian (Alaunian) to the late Norian (Sevatian). Bachmann and Kozur (2004) argued that a distinct hiatus separates the uppermost part of the Löwenstein Formation (fourth sandstone member) from the remainder of this unit and the former should be correlated with Rhaetian units; however, this claim is based on problematical biostratigraphic comparisons.

The Löwenstein Formation has yielded numerous important tetrapod fossils, the first of which were described by Jaeger (1828).

Temnospondyls are represented by the capitosauroid *Cyclotosaurus* and the plagiosaurid *Gerrothorax*.

The Löwenstein Formation is famous for the occurrence of the oldest known terrestrial turtles, *Proterochersis* and *Proganochelys* (figure 5.9). *Proterochersis*, known only from the lowermost (first) sandstone member of this formation, has a distinctly domed shell (with a length of up to 35 centimeters)

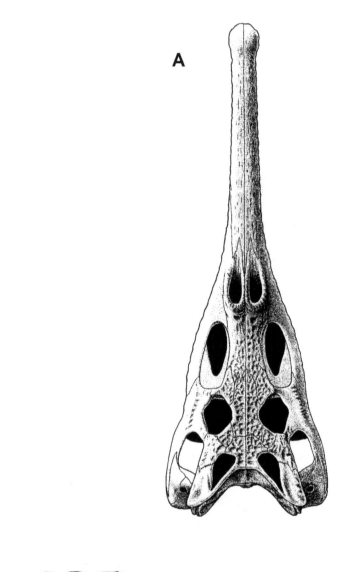

Figure 5.8. Reconstructed skull of the basal phytosaur *"Parasuchus"* sp., based on specimens from the Drawno Beds of Krasiejów (Poland), in (*A*) dorsal and (*B*) lateral views. Note the anterior position of the external nares. (From Dzik and Sulej 2007; courtesy of J. Dzik)

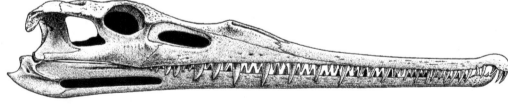

10 cm

(figure 5.9*B*), which, by comparison with extant turtles, suggests terrestrial habits. The stratigraphically slightly younger *Proganochelys* has a more flattened and larger carapace (with a length of up to 60 centimeters) (figure 5.9*A*). It exhibits an interesting mosaic of primitive and derived features (Gaffney 1990; Joyce 2007). The jaws lack marginal teeth, but *Proganochelys* retains rows of small teeth on the palatine and vomer, which are absent in all other known turtles. The articulation between the basicranium

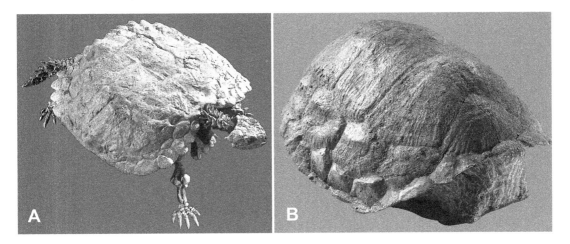

Figure 5.9. (*A*) Oblique anterodorsal view of the reconstructed skeleton of the turtle *Proganochelys quenstedti* (SMNS 16980). Carapace length 70 centimeters. (*B*) Oblique anterodorsal view of a carapace of the turtle *Proterochersis robusta* (SMNS 17561). Carapace length 35 centimeters. (From Schoch and Wild 1999b; courtesy and copyright of Verlag Dr. Friedrich Pfeil)

and palate is open rather than fused as in all other known turtles. A distinctive feature of *Proganochelys* is the presence of dermal ossifications on the neck and tail, the end of which is ensheathed in ossicles. Contrary to earlier interpretations, Joyce (2007) argued that neither it nor the more derived *Proterochersis* can be assigned to either of the two major groups of extant turtles. Neither *Proganochelys* nor *Proterochersis* shed much light on the still controversial phylogenetic position of turtles (chapter 2). Li et al. (2008) recently reported a basal turtle, *Odontochelys*, from early to middle Carnian marine strata of the Wayao Member of the Falang Formation of Guizhou Province, China. *Odontochelys* is more plesiomorphic than *Proganochelys* in retaining marginal teeth in the upper and lower jaws. It also has a fully developed ventral plastron but only expanded dorsal ribs rather than a fully formed dorsal carapace.

Archosaurian reptiles from the Löwenstein Formation include various crurotarsans as well as saurischian dinosaurs. The formation has yielded exquisitely preserved skeletal remains referable to two genera of phytosaurs. Jaeger (1828) coined the misleading name "phytosaur" ("plant reptile") because he mistook the sandstone infillings of the alveoli on poorly preserved jaw fragments for blunt-crowned teeth characteristic of an herbivore. Phytosaurs were, in fact, large predators that resembled crocodylians in overall appearance and presumably mode of life. However, in contrast to the latter, the nostrils in all except basal phytosaurs (*Parasuchus*) are located far back and high up on the skull, just in front of the eyes. The large *Nicrosaurus* has a robust skull with a relatively short snout that typically bears a tall prenarial crest along much of its dorsal surface and has a distinctly heterodont dentition (Meyer 1861, 1863; E. Fraas 1896; figure 5.10*A*). *Mystriosuchus* differs from *Nicrosaurus* in having a more gracile, gharial-like skull with a long, slender snout and homodont dentition (figure 5.10*B*); some specimens bear a premaxillary crest (Huene 1911; Hungerbühler 2002). Its caudal vertebrae have long neural spines and long, distally expanded chevron bones, indicating a deep, laterally flattened tail suitable for sculling through water (Renesto and Lombardo

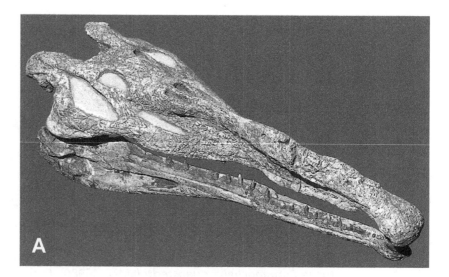

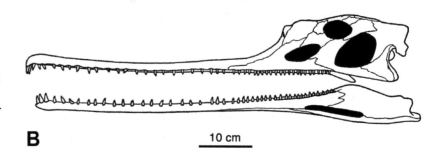

Figure 5.10. (A) Oblique anterodorsal view of a skull (with right mandibular ramus) of the phytosaur *Nicrosaurus kapfii* (SMNS 4379). Skull length 75 centimeters. (From Schoch and Wild 1999b; courtesy and copyright of Verlag Dr. Friedrich Pfeil) (B) Lateral view of the reconstructed skull of the phytosaur *Mystriosuchus planirostris*. (Modified from McGregor 1906)

10 cm

1999). Occurrences of *Mystriosuchus* in the marine Dachsteinkalk in Austria and Calcare di Zorzino of Lombardy, Italy, both of which appear stratigraphically equivalent to the Löwenstein Formation, demonstrate that this phytosaur frequently ventured into shallow marine waters (Buffetaut 1993). Hungerbühler and Hunt (2000) and Hungerbühler (2002) argued that *Mystriosuchus* and *Nicrosaurus* each comprised sympatric gracile, slender-snouted and more robust, broad-snouted morphs, which these authors classified as distinct species.

Aetosaurs (Stagonolepididae) are represented by *Aetosaurus* (with a maximum length of 1.5 meters) and the larger *Paratypothorax*. A cuirass of bony plates encloses most of the body and tail of *Aetosaurus*, like those of other aetosaurs (Walker 1961; Wild 1989; Schoch 2007; figure 5.11). The dorsal armor comprises four rows of slightly sculptured osteoderms, the median ones of which are wider than long. The ventral armor is composed of numerous rows of small plates. Even the limbs bear small osteoderms. The head of *Aetosaurus* is small relative to the body. Based on the small and simple teeth, Walker (1961) suggested that aetosaurs were either herbivores or scavengers. The anterior ends of the upper and lower jaws are devoid of teeth in most aetosaurs, but *Aetosaurus* retains premaxillary teeth. *Aetosaurus* is best known from a group of at least 22 articulated skeletons discovered in a marl lens in the Löwenstein Formation of Stuttgart-Kaltental in 1875 (O. Fraas 1877; Schoch 2007). *Paratypothorax* is known only from large, straplike dorsal osteoderms with a sculpturing of ridges radiating from a posteromedial boss. These plates were initially attributed to phytosaurs (Meyer 1861; Huene 1911), but

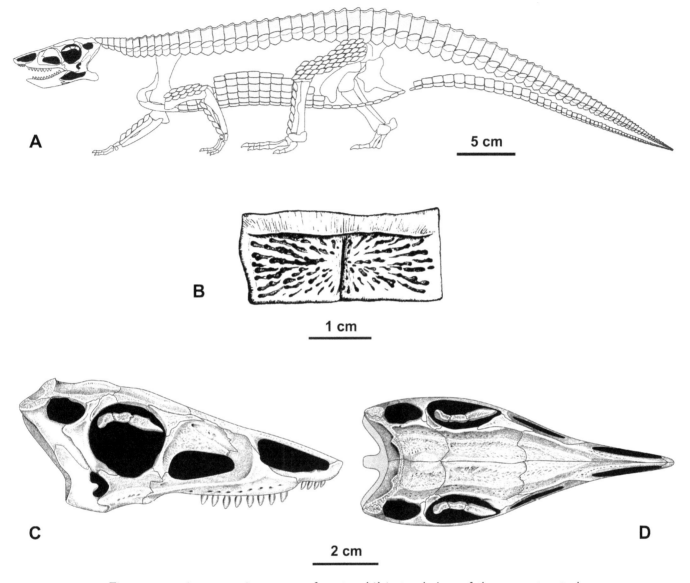

Figure 5.11. Aetosaur *Aetosaurus ferratus*. (*A*) Lateral view of the reconstructed skeleton with complete osteoderm cuirass. (*B*) Dorsal osteoderm. (*C–D*) Reconstructed skull in (*C*) lateral and (*D*) dorsal views. (*A* and *C–D* from drawings courtesy of R. R. Schoch; *B* from Wild 1989)

Long and Ballew (1985) recognized that they were indistinguishable from those of a large aetosaur from the Chinle Formation of the American Southwest (chapter 8).

Teratosaurus, known only from a single maxilla with large, recurved teeth and questionably referred postcranial bones from the Löwenstein Formation (Meyer 1861; Welles 1947; Galton 1985a; Brusatte et al. 2009), was long interpreted as a carnivorous dinosaur (Huene 1907–8, 1932). Bonaparte (1981) first recognized its "rauisuchian" affinities.

Crocodylomorph archosaurs are represented by the sphenosuchian *Saltoposuchus*, a small, lightly built form with hind limbs that are distinctly longer than the forelimbs (Sereno and Wild 1992). Its dorsal dermal armor comprises paired rows of small, sculptured osteoderms.

The lower part of the Löwenstein Formation has yielded skeletal remains referable to several taxa of dinosaurs. Coelophysoid theropods comprise the small *Procompsognathus* (Sereno and Wild 1992) and possibly the poorly known *Halticosaurus* (Huene 1932). A disarticulated skull referred to *Halticosaurus* by Huene (1932) is probably referable to the crocodylomorph *Saltoposuchus*. Sauropodomorph dinosaurs include *Efraasia* (up to 6.5 meters in length) and a small (up to about 5 meters long) species of *Plateosaurus* (Yates 2003b; figure 5.12). *Plateosaurus* and related basal sauropodomorph dinosaurs such as

Efraasia, traditionally grouped together as "Prosauropoda" (Yates 2003a; Galton and Upchurch 2004), have proportionately small skulls with an elongate snout and a dentition comprising simple teeth with labiolingually compressed, leaf-shaped crowns that bear denticulated edges (Huene 1926; Galton 1984, 1985b; figure 5.12*B*). The neck is moderately elongated. Although *Plateosaurus* is frequently reconstructed in a bipedal pose (e.g., Huene 1926), the shape of its vertebral column (Christian and Preuschoft 1996) and its large, robust forelimbs suggest at least facultatively quadrupedal locomotion (figure 5.12*A*). The manus bears a greatly enlarged ungual on its first digit and lacks unguals on its small fourth and fifth digits. *Plateosaurus* and its relatives were the

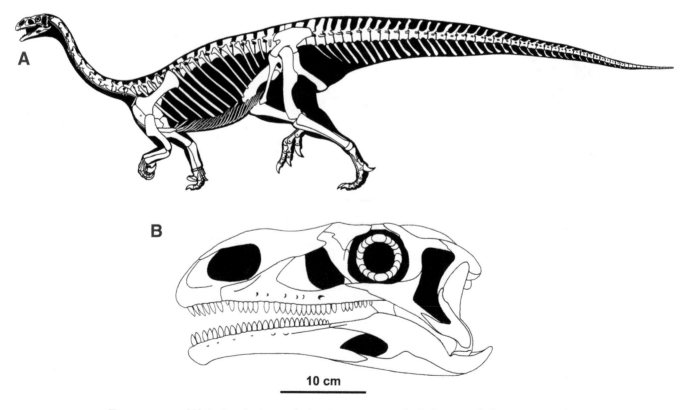

Figure 5.12. (*A*) Lateral view of the reconstructed skeleton of the sauropodomorph dinosaur *Plateosaurus longiceps*. Length up to 10 meters. (Drawing courtesy and copyright of G. S. Paul) (*B*) Lateral view of a skull of *Plateosaurus longiceps* (SMNS 13200). (Modified from Galton 1985b and photographs of specimen)

oldest known large herbivores that could have foraged at levels well above the ground (Galton 1985b). Skeletal remains of *Plateosaurus* were found in the Stuttgart region as early as 1847, but Huene (1907–8) was the first author to recognize their dinosaurian affinities.

The Trossingen Formation (Knollenmergel) is the higher stratigraphic equivalent of the Arnstadt Formation in the marginal facies in southern Germany. It is late Norian (Sevatian) in age and distinguished from the underlying Löwenstein Formation by its predominantly red marls with layers of dolomitic carbonate nodules, which gave the unit its traditional name, Knollenmergel ("nodule marl"). This unit and correlative strata elsewhere in southern Germany and in northern Switzerland (Frick, Aargau) have yielded several accumulations of skeletons of a large species of *Plateosaurus*, which reached a length of about 9 meters and is now referred to as *P. longiceps* (Huene 1926, 1932; Galton 1984, 1985a; Moser 2003; Yates 2003b; Galton and Upchurch 2004; figure 5.12). The *Plateosaurus* "graveyard" near the town of Trossingen in Württemberg has been particularly well studied (Sander 1992). To date, remains of 55 adult individuals of *Plateosaurus* have been collected from this locality, although there are only a few complete skeletons. Sander (1992) argued that these dinosaurs became mired in muddy ground and eventually perished; this scenario would account for the abundance and excellent preservation of the skeletal remains. Other, much less common faunal elements of the Knollenmergel assemblage include capitosauroid and plagiosaurid temnospondyls, the turtle *Proganochelys* (Gaffney 1990), the 5-meter-long coelophysoid theropod *Liliensternus* (Huene 1934; Welles 1984), and the haramiyid ?mammaliaform *Thomasia* (Hahn 1973; figure 5.13).

The Rhaetian-age Exter Formation (Upper Keuper or Rhätkeuper) almost exclusively comprises clastic strata, with only minor carbonate components, and documents a transition from nonmarine to increasingly marine depositional conditions in the Germanic basin. The Rhaeto-Liassic bone bed marks the onset of a marine transgression that inundated the basin from the north at the beginning of the Jurassic Period. Its placer deposits formed in deltaic settings and channels and have yielded isolated teeth and mostly fragmentary bones of tetrapods including a variety of nonmammalian cynodonts and basal mammaliaforms (Clemens 1980). There are no identifiable remains of larger tetrapods, especially crurotarsan archosaurs other than phytosaurs (Buffetaut 1993)

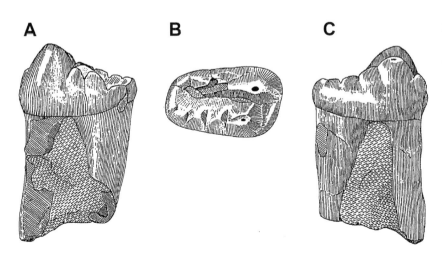

A **B** **C**

Figure 5.13. Lower molariform postcanine tooth of the haramiyid *Thomasia antiqua* (SMNS 90574) in (*A*) buccal, (*B*) occlusal, and (*C*) lingual views. Length of tooth crown 2.1 millimeters. (From Schoch and Wild 1999b; courtesy and copyright of Verlag Dr. Friedrich Pfeil)

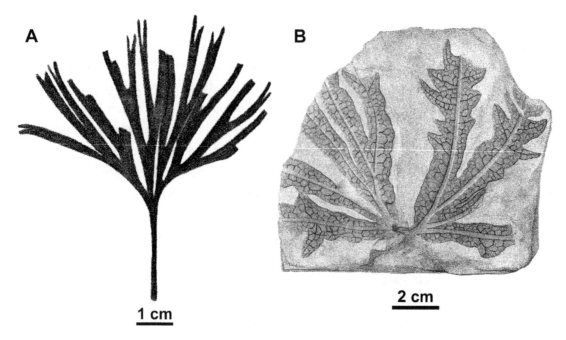

Figure 5.14. (*A*) Leaf of the gingkophyte *Sphenobaiera muensteriana*. (*B*) Leaf of the dipteridaceous fern *Dictyophyllum acutifolium*. (From Schenk 1867)

and dinosaurs, preventing comparisons with potentially coeval tetrapod assemblages elsewhere.

Rhaeto-Liassic continental strata in Franconia, Bavaria, have yielded abundant plant macrofossils as well as pollen and spores, representing diverse plant assemblages composed of ferns, pteridosperms, bennettitaleans, ginkgophytes, and conifers (Schenk 1867; Gothan 1914; Dobruskina 1994; figure 5.14). Gothan (1914, 1935) characterized the Rhaetian flora by the presence of the peltaspermacean *Lepidopteris* and considered the occurrence of the dipteridaceous fern *Thaumatopteris* typical for Early Jurassic (Liassic) plant communities. *Lepidopteris* has a considerable stratigraphic range, extending back to the Permian and disappearing at the end of the Triassic. However, *Thaumatopteris* dates back to at least the Norian (Schweitzer 1978) and thus cannot serve as an index taxon for Jurassic-age strata. Surprisingly, assemblages of pollen and spores from Franconia apparently do

not show the pronounced changes in composition across the Triassic-Jurassic boundary observed, for example, in the Newark Supergroup of eastern North America (Achilles 1981).

FRANCE

An assemblage of mostly small tetrapods has been recovered from the Grès à *Rhaetavicula contorta*, exposed in a now abandoned quarry at Saint-Nicolas-de-Port, near Nancy in eastern France. A Rhaetian age for the Saint-Nicolas-de-Port locality (Sigogneau-Russell 1983a, 1983b, 1989; Sigogneau-Russell and Hahn 1994) is now generally accepted because the Grès à *Rhaetavicula contorta* in the Paris basin can be correlated with Rhaetian strata in neighboring regions of Europe (Deutsche Stratigraphische Kommission 2005).

The vertebrate assemblage from this site is noteworthy for the great diversity of small synapsids, including haramiyid and theroteinid ?mammaliaforms, mammaliaforms (including *Brachyzostrodon*), and small chiniquodontoid (*Meurthodon*) and putative traversodont cynodonts (Godefroit and Battail 1997). More than 1,000 isolated synapsid teeth in various states of preservation have been recovered. Fish remains are very common, and isolated bones and teeth document the presence of various temnospondyls and reptiles (including dinosaurs). The vertebrate-bearing strata were deposited in a shallow, nearshore marine setting (Sigogneau-Russell and Hahn 1994).

The enigmatic Haramiyidae are characterized by the structure of their molariform teeth (Sigogneau-Russell 1989; Butler and MacIntyre 1994). Upper teeth (named *Haramiya*) bear two rows of cusps, one of which has three cusps of subequal size whereas the other has up to five cusps of various sizes. Lower teeth (named *Thomasia*) have rather narrow crowns with two rows of cusps, one of which has two unequal cusps, whereas the other has three to five cusps of various sizes (figure 5.13). A low ridge at one end and the converging rows of cusps at the other end enclose a longitudinal basin on the crown. The molariforms have multiple roots. Based on the similarity of haramiyid teeth to those of paulchoffatiid multituberculate mammals, Hahn (1973) first suggested a phylogenetic relationship between Haramiyidae and Multituberculata. However, there exists as yet no compelling anatomical evidence that haramiyids are even mammaliaforms. Haramiyids were long thought to be restricted to the Late Triassic and Early Jurassic, but recent finds from the Upper Jurassic of Tanzania (Heinrich 1999) and possibly the Upper Cretaceous of India (Anantharaman et al. 2006) have been referred to the Haramiyidae and suggest a possibly much greater temporal range for this group.

Other assemblages with mammaliaforms and nonmammalian cynodonts of Norian and Rhaetian age are known from Switzerland (Hallau: Peyer 1956; Clemens 1980; Achilles and Schlatter 1986), Belgium, and Luxembourg (Hahn, Lepage, and Wouters 1984, 1988; Hahn, Wild, and Wouters 1987; Sigogneau-Russell and Hahn 1994). Discoveries of such occurrences probably reflect the result of employing different collecting techniques because most Late Triassic strata in the Germanic basin and adjoining regions have never been systematically prospected for small vertebrate fossils.

AUSTRIA

Spoil tips from abandoned coal mines in the region of Lunz-am-See, about 100 kilometers west of Vienna, have yielded a remarkably diverse plant assemblage. Together with a coeval but less varied assemblage from Neue Welt near Basel, Switzerland, the Lunz plant assemblage represents one of the most important Late Triassic floras from Eurasia (Stur 1885; Dobruskina 1998). The often superbly preserved plant remains, usually compression fossils that frequently preserve cuticles, occur in the Lunz Formation. This unit consists of sandstones at its base, followed by marine marls that, in turn, grade upward into sands, shales, and coal (Pott et al. 2008). The Lunz Formation is probably equivalent to the upper part of the marine Reingraben Formation and thus late early Carnian (late Julian) in age. The flora is dominated by equisetaleans but also comprises bennettitaleans (especially the foliage taxa *Nilssoniopteris* and *Pterophyllum*), ferns (particularly Dipteridaceae), and ginkgophytes. Pteridosperms are absent and conifers are rare.

POLAND

A remarkable occurrence of excellently preserved tetrapod remains and associated biota has recently been discovered in a commercial clay pit at Krasiejów near Opole in Silesia, Poland (Dzik 2001; Dzik and Sulej 2007). The pit exposes two bone-bearing horizons, the lower of which is lacustrine in origin, whereas the upper appears to be of fluvial derivation. The horizons are separated by only about 7 meters in a succession of mud- and claystones and have a number of tetrapod taxa in common, suggesting that they represent a single biostratigraphic unit. The Krasiejów assemblage can be correlated with the subsurface Drawno Beds, which Dzik and Sulej (2007) and Kozur and Weems (2007) regarded as the stratigraphic equivalent of the Weser Formation (Lehrberg Horizon) in Germany.

Plants are represented by abundant calcified oogonia of charophycean algae (stoneworts) and by remains of conifers and possibly *Pterophyllum* (Dzik and Sulej 2007). The diverse assemblage of freshwater invertebrates comprises unionid bivalves, gastropods, conchostracans, ostracodes, and cycloid crustaceans. Rare beetle elytra record the presence of insects. Fishes include dipnoans and the small actinopterygian "*Dictyopyge*." *Cyclotosaurus* (Sulej and Majer 2005; figure 5.6) and *Metoposaurus* (Sulej 2002, 2007; figure 5.7) represent temnospondyls in the Krasiejów assemblage and are documented by abundant, often excellently preserved skeletal remains. Reptiles include an unidentified sphenodontian, a tanystropheid, the phytosaur "*Parasuchus*" (figure 5.8), the aetosaur *Stagonolepis*, the "rauisuchian" *Polonosuchus*, and the dinosauriform *Silesaurus* (Dzik 2001, 2003; Sulej 2005; Dzik and Sulej 2007; Brusatte et al. 2009; figure 5.15). *Metoposaurus* is the most common faunal element in the lower lacustrine horizon, followed in terms of abundance by *Parasuchus*. The terrestrial reptiles occur with *Cyclotosaurus* (which is also found in the lower bed) in the upper fluvial horizon.

Silesaurus is represented by a considerable quantity of excellently preserved skeletal remains from the upper horizon at Krasiejów (Dzik 2003; Dzik and Sulej 2007) and is a basal dinosauriform (Langer and Benton 2006; figure 5.15). It attained a length of 2 meters. *Silesaurus* differs from other dinosauriform ornithodirans such as *Marasuchus* (chapter 4) in having a dentition indicative of herbivorous rather than carnivorous feeding habits and the presence of long, slender forelimbs, which indicate predominantly quadrupedal locomotion. Another noteworthy feature

Figure 5.15. Lateral view of the reconstructed skeleton of the dinosauriform *Silesaurus opolensis* based on material described by Dzik (2003). Length up to 2.3 meters. (Drawing courtesy and copyright of G. S. Paul)

of *Silesaurus* is the presence of a small beak at the anterior end of the mandible.

Located some 25 kilometers west of Krasiejów, a clay pit at the village of Lisowice, near Lubliniec in southern Poland, has recently yielded a latest Triassic tetrapod assemblage (Dzik, Sulej, and Niedzwiedzki 2008). Bones occur in a sequence of fluvial mudstones and siltstones, which are interbedded with sandstones. A diverse flora, represented by both macrofossils and pollen and spores, comprises cheirolepidaceous conifers, *Lepidopteris*, and *Clathropteris*, which elsewhere occur in Rhaetian and Jurassic strata. A Rhaetian date receives further support from the conchostracans. The tetrapod assemblage from Lisowice is noteworthy for the unexpected co-occurrence of a giant kannemeyeriid dicynodont and a large capitosauroid temnospondyl with a large theropod dinosaur and pterosaurs (Dzik, Sulej, and Niedzwiedzki 2008).

GREENLAND

Faunal assemblages that are virtually identical to that from the lower part of the Löwenstein Formation of southern Germany occur in the Malmros Klint and overlying Ørsted Dal members of the Fleming Fjord Formation in East Greenland (Jenkins et al. 1994). The Malmros Klint Member has yielded plagiosaurid and capitosauroid temnospondyls, possible phytosaurian remains, and *Plateosaurus*. The more diverse tetrapod assemblage from the Ørsted Dal Member comprises a variety of freshwater fish, the temnospondyls *Cyclotosaurus* and *Gerrothorax* (Jenkins et al. 2008), the aetosaurs *Aetosaurus* and *Paratypothorax*, *Plateosaurus*, an indeterminate theropod, and cf. *Proganochelys*, but also sphenodontian and other lepidosaurian jaw fragments, a pterosaur (*Eudimorphodon*), a cynodont of uncertain affinities (*Mitrodon*), the haramiyid ?mam-

maliaform *Haramiyavia* (Jenkins et al. 1997), the mammaliaform ?*Brachyzostrodon*, and the holotherian mammal *Kuehneotherium*. In addition, trackways of theropod dinosaurs have been discovered in the Ørsted Dal Member.

ITALY

In recent decades, two Norian-age formations in northern Italy, the Calcare di Zorzino and the Dolomia di Forni, have yielded many often excellently preserved fossils of a variety of terrestrial tetrapods, including the earliest well-known pterosaurs and some unusual reptiles. Although the two units have various faunal elements in common (and indeed are sometimes confused in the literature), they represent different depositional settings and show differences in the taxonomic composition of their respective tetrapod assemblages (Dalla Vecchia 2006; Renesto 2006).

The Calcare di Zorzino (Zorzino Limestone) in Lombardy comprises dark gray to black limestone or marly limestone, which is typically rather thinly laminated, with a few massive beds. It was deposited in a series of rift basins in a vast carbonate platform that extended from present-day Spain to Greece (Jadoul 1986; Jadoul et al. 1994). These basins had depths of tens to hundreds of meters and could cover an area of many square kilometers. Being surrounded by very shallow water, the basins had only limited water circulation, which resulted in anoxic conditions in deeper water layers and at the bottom (Tintori 1992). Tidal channels formed the sole connection between the basins and the open sea, preventing access by larger marine reptiles such as ichthyosaurs. Patch reefs and organic mounds ringed the basins and occasionally emerged to form small islands. Cheirolepidaceous conifers (*Brachyphyllum*), which apparently could tolerate brackish-water conditions (Francis 1983), colonized these ephemeral islands. The animals and plants preserved in the Calcare di Zorzino lived on the islands or in the well-oxygenated upper water layers in the basins. The faunal assemblage represents a diverse community of marine invertebrates and vertebrates but also includes terrestrial reptiles. Most of the fossils are found in a 3-meter-thick horizon at the transition from the upper part of the unit to the overlying Argillite di Riva di Solto (Riva di Solto Shale; Renesto 2006). The main fossiliferous layer of the Calcare di Zorzino is middle to late Norian in age (Renesto 2006).

The Dolomia di Forni (Forni Dolomite) in northern Friuli lacks the readily accessible, thin vertebrate-bearing layers of the Calcare di Zorzino in Lombardy, and prospecting for fossils is usually restricted to small rock falls (Dalla Vecchia 2006). The basinal facies of Dolomia di Forni consists of a succession of dark, well-bedded, and frequently cherty dolostone (dolomitic limestone), which is up to 800 meters thick and was deposited in a 300–400-meter-deep basin. Graded, thicker layers represent distal turbidites. Closer to the edge of the carbonate platform, the unit also comprises breccia beds and pebbly dolostones, which represent the slope facies (Dalla Vecchia 2006). Like the Calcare di Zorzino, the Dolomia di Forni has yielded a diversity of animal and plant fossils. Conifer foliage (including *Brachyphyllum*) is relatively common, but ferns and equisetaleans are absent. The age of the Dolomia di Forni probably spans the middle to late Norian (Alaunian-Sevatian; Dalla Vecchia 2006).

The Calcare di Zorzino has yielded skeletal remains representing two taxa of marine reptiles: the cyamodontoid placodont *Psephoderma* and the thalattosaur *Endennasaurus*.

Cyamodontoid placodonts superficially resemble turtles in the possession of a rigid carapace of osteo-derms protecting the body. In *Psephoderma*, this carapace is dorsoventrally flat, and a second, smaller armor shield covers the base of the long tail (Pinna and Nosotti 1989). Like all placodonts, cyamodontoids are characterized by the possession of large, massive tooth plates, especially on the palate, that appear suitable for crushing the shells of mollusks and brachiopods (Rieppel 2000). The limbs of *Psephoderma* show adaptations for an aquatic mode of life, such as distal flattening of the humerus and femur and reduced ossification of the articular surfaces (Renesto 2006). *Psephoderma* may have reached a length of more than 2 meters.

By contrast, the thalattosaur *Endennasaurus* attained a length of about 1 meter and has long, well-ossified limbs suitable for terrestrial locomotion. Its long, laterally compressed tail probably provided the principal means for swimming (Renesto 1984, 2006; Müller, Renesto, and Evans 2005). *Endennasaurus* has a long, tapering snout with toothless jaws, and probably fed on soft-bodied benthic organisms. As in other known thalattosaurs, the upper temporal fenestrae form mere slits. *Endennasaurus* has a rigid trunk with a strong ventral "basket" of gastralia, which may have facilitated swimming down to the bottom waters.

Phytosaurs typically thrived in fluviolacustrine settings, but, as noted above, *Mystriosuchus* would at least occasionally venture out into the sea. A nearly complete, 4-meter-long skeleton as well as an isolated large skull document its presence in the Calcare di Zorzino (Renesto and Lombardo 1999; Gozzi and Renesto 2003).

One unusual reptile known from both the Calcare di Zorzino and the Dolomia di Forni is the protorosaur *Langobardisaurus* (Renesto 1994; Renesto and Dalla Vecchia 2000). Like other protorosaurs, it has a distinctly elongated neck. However, its dentition is very different from those of related forms. In each jaw, conical anterior teeth are followed behind by several multicuspid teeth, which, in turn, are followed by a single crushing tooth at the back of the jaw (Renesto and Dalla Vecchia 2000). The pronounced elongation of the first phalanx on the fifth pedal digit of *Langobardisaurus* is a derived feature shared with *Tanystropheus* and *Tanytrachelos.*

In addition to these aquatic or semiaquatic tetrapods, a number of terrestrial tetrapods are known from the Calcare di Zorzino and the Dolomia di Forni. They include the oldest well-known pterosaurs, *Eudimorphodon* (figure 5.16), *Peteinosaurus*, and *Preondactylus* (Wild 1978, 1984; Dalla Vecchia 2003). These three taxa have been referred to the "Rhamphorhynchoidea," a paraphyletic assemblage of long-tailed pterosaurs. Known from fairly complete, articulated skeletons from both units, *Eudimorphodon* has procumbent, conical anterior teeth and more poste-

rior teeth that each bear up to five cusps. Based on the gut contents of one skeleton of *Eudimorphodon* from the Calcare di Zorzino, fish formed at least part of its diet. The still poorly known *Peteinosaurus*, also known from the Calcare di Zorzino, has a similar dentition but differs from *Eudimorphodon* in the presence of bundles of greatly elongated pre- and postzygapophyses on the tail vertebrae, as in the "rhamphorhynchoids" from the Jurassic Period (Wild 1978). *Eudimorphodon* (figure 5.16) and *Peteinosaurus* each probably attained a wingspan of about 1 meter. Finally, the smaller *Preondactylus,* known only from the Dolomia di Forni, was originally reported to have simple conical teeth (Wild 1984), but examination of a second specimen has established that each cutting edge bears small cuspules (Dalla Vecchia 2003).

Other records of fully terrestrial reptiles of the Calcare di Zorzino include a single skeleton of a

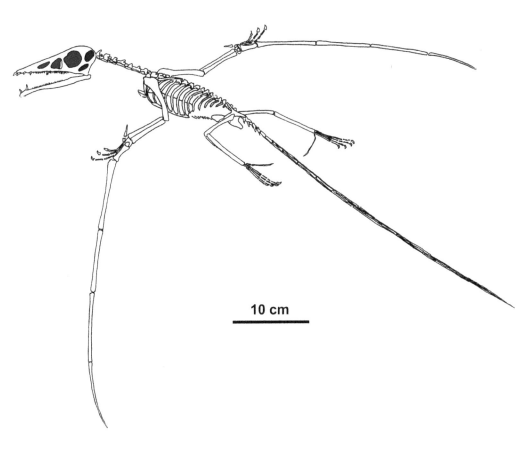

Figure 5.16. Reconstructed skeleton of the "rhamphorhynchoid" pterosaur *Eudimorphodon ranzii* in flying posture. (Modified from Wild 1978 with posture of hind limbs modified based on Unwin 2006)

10 cm

sphenodontian possibly referable to *Diphydontosaurus* from the Upper Triassic of Britain (Renesto 1995; chapter 6) and an isolated fragment of cervical armor referable to the small aetosaur *Aetosaurus* (Wild 1989).

Undoubtedly the most unusual reptiles from the Calcare di Zorzino and the Dolomia di Forni are two small archosauromorphs, *Megalancosaurus* and *Drepanosaurus*. *Megalancosaurus* was first described on the basis of an incomplete but articulated skeleton including the well-preserved skull from the Dolomia di Forni (Calzavara, Muscio, and Wild 1981). The lightly built, triangular skull vaguely resembles that of a bird, with large orbits and nares and a pointed snout. On the basis of these superficial similarities, its original describers suggested possible affinities of *Megalancosaurus* with pterosaurs. Since then additional specimens have established that this taxon is closely related to the equally unusual *Drepanosaurus*. Pinna (1980, 1984) provided a detailed description of *Drepanosaurus* based on a skeleton lacking only the skull and part of the neck from the Calcare di Zorzino (figure 5.17).

He noted the presence of an unusual bony spike (formed by a modified vertebra) at the end of the apparently prehensile tail and a much enlarged ungual phalanx on the second digit of each hand.

More recent finds of *Megalancosaurus* have confirmed Pinna's identifications of these unusual structures and established the presence of additional distinctive features. Reexamination of *Drepanosaurus* revealed that there are three distinct bones between the humerus and the carpus. Two of these bones are relatively slender, as expected for epipodial elements, but the third (ulna?) is flat and platelike. *Megalancosaurus* also has a distinctive forelimb with an elongate ulnare and intermedium aligned parallel to each other in the carpus (Renesto 2000). It is not clear whether one of the additional bones in the forelimb of *Drepanosaurus* represents an elongated ulnare or intermedium. This unusual structure of the forelimb is possibly related to the presence of the greatly enlarged ungual on the second manual digit. Another characteristic feature of *Megalancosaurus* is

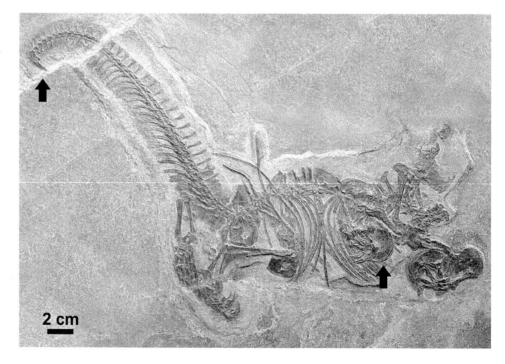

Figure 5.17. Holotype skeleton of the drepanosaurid *Drepanosaurus unguicaudatus* (Museo Civico di Scienze Naturali, Bergamo, MCSNB 5728). Arrows indicate the bony spike at the distal end of the tail and the much enlarged manual ungual, respectively. (Photograph by NCF)

2 cm

the presence of tall, well-developed neural spines on the anterior dorsal vertebrae. The neural spines are expanded distally and fused to each other, superficially resembling the notarium in pterosaurs.

Study of all the known specimens of *Megalancosaurus* revealed some variation in the arrangements and form of the pedal digits. In one specimen, the ungual phalanx of the first digit is a small rounded element with a blunt distal end, rather than a claw. Renesto (2000) suggested that these differences possibly indicate the presence of two species of *Megalancosaurus*. However, it is equally conceivable that this merely represents individual variation.

Renesto and Binelli (2006) recently described a third drepanosaurid taxon, *Vallesaurus,* from the upper portion of the Calcare di Zorzino. The only known specimen is the most complete skeleton of a drepanosaurid recovered to date. Phylogenetic analysis places *Vallesaurus* as the most basal member of Drepanosauridae. *Vallesaurus* lacks both opposable digits and a spine at the tip of the tail and is much smaller than the other drepanosaurs from the Calcare di Zorzino. Renesto and Binelli (2006) further emphasized various features of drepanosaurs that they interpreted as adaptations for an arboreal mode of life for the entire group: modification of the anterior dorsal vertebrae; modifications in the carpus and tarsus to facilitate rotation of the manus and pes; opposable digits in the manus and an opposable hallux; preungual phalanges that are notably longer than the preceding ones; long, mediolaterally compressed ungual phalanges; and a prehensile tail with greatly restricted lateral but considerable dorsoventral flexibility.

Initially, drepanosaurs were considered an endemic element of the Late Triassic tetrapod assemblages from northern Italy, but, more recently, the characteristic vertebrae of these reptiles have been recognized from the Late Triassic fissure fillings in Britain (Renesto and Fraser 2003), and various specimens, including partial skeletons, are now known from the Upper Triassic of the United States (Berman and Reisz 1992; Colbert and Olsen 2001; Harris and Downs 2002). As further discussed in chapter 7, some of these new finds challenge the notion that drepanosaurs were exclusively arboreal in their habits.

An additional record of Norian-age terrestrial tetrapods from northern Italy is provided by numerous tracks of dinosaurs and other reptiles from strata of the Dolomia Principale of the inner carbonate platform (Dalla Vecchia 2006).

Insects from the Calcare di Zorzino include dipterans, beetles, and dragonflies, most of which have yet to be fully described (Whalley 1986).

RUSSIA

The record of late Middle and Late Triassic continental strata in Russia is limited. The Bukobay Gorizont of the southern Cis-Urals has a maximum thickness of 600 meters and is probably late Middle Triassic (Ladinian) in age. Shishkin et al. (2000) correlated this unit with the Lower Keuper in the Germanic basin. The Bukobay Gorizont shows distinctly cyclical deposition. Its basal portion comprises greenish gray medium- to coarse-grained, cross-bedded channel sandstones that contain gravel, clay galls, logs of petrified wood, and disarticulated tetrapod bones. It passes upward into variegated clays of fluviodeltaic origin. Successive cycles of deposition increasingly show a decrease in the thickness of the basal sandstones, and the clays become predominantly gray. The clay layers have yielded remains of a plant assemblage characterized by the presence of the peltaspermacean *Scytophyllum*, which Dobruskina (1994) considered characteristic of Ladinian to Carnian floras in Eurasia, as well as freshwater bivalves and conchostracans.

The Bukobay tetrapod assemblage includes a species of *Mastodonsaurus* (see Damiani 2001) and the possible trematosauroid *Bukobaja* (Ochev 1966). The plagiosaurids *Plagioscutum* and *Plagiosternum* (Shishkin 1987) represent another common faunal element. The reptiles and therapsids from the Bukobay Gorizont are known only from fragmentary skeletal remains. They include the erythrosuchid *Chalishevia* (Gower and Sennikov 2000), possible "rauisuchians," and the protorosaur *Malutinisuchus* (Shishkin et al. 2000). Kannemeyeriid dicynodonts include the apparently very large *Elephantosaurus*. Lucas (1998) proposed the Berdyankian LVF for the time equivalent to the tetrapod assemblage of the Bukobay Gorizont, using the first appearance of *Mastodonsaurus* as the beginning of this faunachron.

The Keuper Group and its equivalents in central Europe have yielded the classic succession of Late Triassic terrestrial tetrapod assemblages. However, because of the complex depositional history of the Keuper strata, with repeated marine incursions into the Germanic basin, some of the communities are not readily comparable to probably coeval assemblages from other regions of the globe. The abundance of sauropodomorph dinosaurs at a number of localities in the Arnstadt Formation and its lateral equivalents, especially the Trossingen Formation, may reflect taphonomic biases (Sander 1992). However, these dinosaurs also represent a major element in the Norian- and Rhaetian-age tetrapod assemblages from Argentina and South Africa (chapter 4) and thus provide paleobiogeographic links between Laurasia and Gondwana.

An interesting occurrence of continental vertebrates in shales and limestones of the Upper Triassic Huai Hin Lat Formation at the Chulabhorn Dam in northeastern Thailand shares a number of taxa with the Keuper Group (Buffetaut 1983): the capitosauroid temnospondyl *Cyclotosaurus* (Ingavat and Janvier 1981), the turtle *Proganochelys* (Broin 1984), and phytosaurs similar to *Mystriosuchus* and *Nicrosaurus* (Buffetaut and Ingavat 1982). The vertebrate remains occur in association with conchostracans and plants. This record supports the hypotheses that the Indo-China Block (present-day southeastern Asia), originally part of Gondwana, was accreted to mainland Asia during the Triassic and that the Keuper communities formed part of a much more extensive paleobiogeographic zone.

CHAPTER 6

Late Triassic of Great Britain

Two clusters of unusual Late Triassic localities in the United Kingdom offer important glimpses of terrestrial vertebrate communities from this time interval and thus require separate consideration. One of these clusters comprises several localities in northeastern Scotland and the other fissure fillings in Carboniferous limestones in southwestern England and southeastern Wales. Together these assemblages provide a picture of terrestrial vertebrate communities ranging in time from the Carnian to the Triassic-Jurassic boundary and illustrate some of the faunal changes that were occurring worldwide. Certain "archaic" taxa such as procolophonids and rhynchosaurs were still present, but there were also notable "modern" groups, including sphenodontians, mammalian precursors, and crocodylomorph and dinosauromorph reptiles.

The precise stratigraphic ages of the Lossiemouth Sandstone Formation of Morayshire, northeastern Scotland, and the fissure deposits in southwestern Britain remain contentious. The former has been variously dated as Carnian and Norian based exclusively on comparisons of its tetrapod assemblage with those from other regions of the world (Benton and Walker 1985). Although apparently nobody has ever argued that all fissure fillings are comparable in age, the suggested range of dates for these deposits is quite varied. Traditionally, the fissure fillings were thought to range in time from the Late Triassic (Norian) to the Early Jurassic (Hettangian or Sinemurian). More recently, however, some authors suggested that certain fissure fillings could be as old as Carnian (e.g., Simms, Ruffell, and Johnson 1994), whereas others argued that even the oldest deposits are Rhaetian in age (e.g., Marshall and Whiteside 1980; Whiteside and Marshall 2008). Although there is yet no consensus concerning the age of the deposits, there exists a general consensus that fissure fillings predominantly containing skeletal remains of mammaliaforms and nonmammalian cynodonts are stratigraphically younger than those that are dominated by remains of a diversity of diapsid reptiles (Robinson 1957a, 1971; Fraser 1986). In the absence of reliable means for chronostratigraphic dating, the tetrapod remains from the fissure fillings cannot offer any detailed insights into the tempo and mode of Late Triassic faunal changes. Yet the exquisite preservation of the disarticulated bones, combined with the remarkable abundance of skeletal remains, provides valuable information on many of the smaller terrestrial tetrapods that lived around the Triassic-Jurassic boundary.

LOSSIEMOUTH SANDSTONE FORMATION OF SCOTLAND

The tetrapod fossils of the Lossiemouth Sandstone Formation surely rank among the visually least appealing. Occasionally traces of permineralized bone remain, but the fossils commonly form rather indistinct natural molds in a yellowish, buff-colored and coarse-grained sandstone. Consequently, even the overall shape of the skeletal remains for smaller animals is not immediately apparent. In fact, many of the smaller fossils are easily overlooked and mistaken for nothing more than a badly eroded pebble. However, it is possible to use flexible casting compounds to prepare high-quality casts from the natural molds of bones (Walker 1961, 1964; Benton and Walker 1985).

Research into the tetrapod fossils from the Lossiemouth Sandstone Formation has a long history.

During the nineteenth century, the sandstone was extensively quarried for use as building material. Beginning in 1836, fish remains were found in the various quarries, and this ultimately led to the view that all the sandstones in the region of Lossiemouth and Elgin were part of the Old Red Sandstone series, which is Devonian in age. Then, in 1844, a sandstone block containing rows of large "scales" was discovered at the Scaat Craig quarry. The town clerk of Elgin, Patrick Duff, who was also an avid naturalist, pronounced this find to be the scales of a large fish. The famous Swiss naturalist Louis Agassiz concurred and duly described them as a new form, *Stagonolepis robertsoni* (Agassiz 1844; figure 6.1). As quarrying continued, additional material of *Stagonolepis* was recovered, including limb bones. Thomas Henry Huxley interpreted these remains as belonging to a precursor of crocodylians (Huxley 1875, 1877).

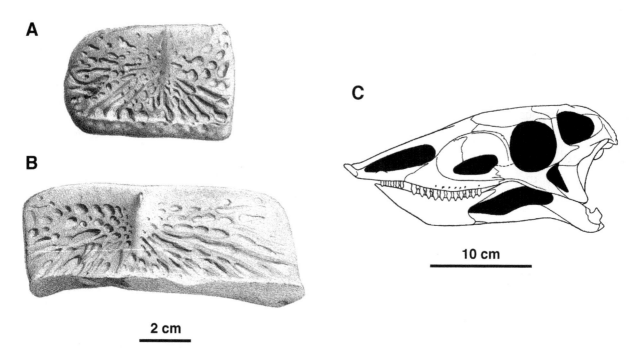

Figure 6.1.   Dorsal osteoderms of the aetosaur *Stagonolepis robertsoni*. (*A*) Right lateral dorsal osteoderm; (*B*) left median dorsal osteoderm. (From Huxley 1877) (*C*) Reconstructed skull of *Stagonolepis robertsoni* in lateral view. (Modified from Walker 1961, by permission of The Royal Society)

Stagonolepis is now placed in the crurotarsan clade Aetosauria (Walker 1961; Heckert and Lucas 1999). Over the next few decades, skeletal remains of a variety of additional reptiles, ranging from the small sphenodontian *Brachyrhinodon* to the large carnivorous archosaur *Ornithosuchus*, were discovered and described.

The sandstone sequences of the Lossiemouth Sandstone Formation have been interpreted as having formed in dune fields (Benton and Walker 1985), where many reptiles became trapped and buried by windblown sand. Such a paleoenvironment contrasts sharply with those represented by the Late Triassic fluviolacustrine deposits of the American Southwest and central Europe, and thus it is not surprising to find a rather different tetrapod assemblage in the Lossiemouth Sandstone Formation.

In addition to *Stagonolepis*, the tetrapod assemblage from the Lossiemouth Sandstone Formation includes the medium-sized rhynchosaur *Hyperodapedon* and the large carnivorous archosaur *Ornithosuchus*. Smaller reptiles include the poorly known possible dinosauriform *Saltopus*, a procolophonid (*Leptopleuron*), a sphenodontian (*Brachyrhinodon*), and two distinctive small archosaurs, *Erpetosuchus* and *Scleromochlus* (Benton and Walker 1985).

There are no known remains of plants or terrestrial invertebrates.

Like other aetosaurs (chapters 5 and 8), *Stagonolepis* is characterized by a cuirass of sculptured osteoderms that covered most of the body, ensheathing even the tail (Huxley 1877; Walker 1961). The dorsal armor comprised four longitudinal rows of osteoderms (figure 6.1*A*). *Stagonolepis robertsoni* reached a length of up to 2.7 meters, with a proportionately small skull (figure 6.1*B*). The leaf-shaped teeth could have served to nip off plant matter. The blunt, edentulous tip of the snout is somewhat expanded and may have been employed for rooting around in the soil for food. The specimens of *Stagonolepis* from the Lossiemouth Sandstone Formation fall into two discrete size classes (Walker 1961), with approximately equal numbers of individuals in each class. As the morphological differences between the two groups are slight, these size classes probably reflect sexual dimorphism.

The rhynchosaur *Hyperodapedon* (chapter 4) reached a length of up to 1.5 meters (Benton 1983b) and has a barrel-shaped trunk (figure 6.2).

The procolophonid *Leptopleuron* attained a length of about 25 centimeters. The greatly enlarged orbital opening extends posteriorly beyond the level of the

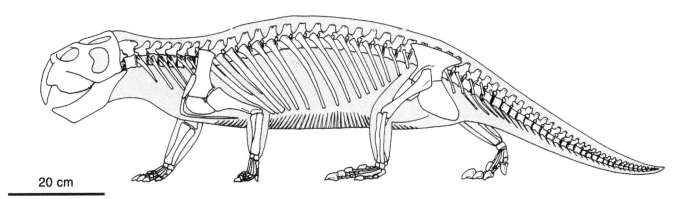

20 cm

Figure 6.2. Lateral view of the reconstructed skeleton of the rhynchosaur *Hyperodapedon gordoni*. (Modified from Benton 1983b, by permission of The Royal Society)

pineal foramen and close to the posterior margin of the skull roof (Huene 1912; Benton and Walker 1985). The premaxillary and anterior dentary teeth of *Leptopleuron* are incisor-like, whereas the maxillary and posterior dentary teeth have transversely broad crowns.

The sphenodontian *Brachyrhinodon* is known from a number of specimens (Huene 1910b; Fraser and Benton 1989). Sphenodontians are considered the sister-taxon of Squamata (lizards and snakes) and are common in Late Triassic and Jurassic strata worldwide. Today, only two species of the tuatara, *Sphenodon*, survive on a few small islands off the coast of New Zealand. Based on early studies, *Sphenodon* gained the status of a "living fossil," but this is a misinterpretation since Mesozoic sphenodontians are a highly diverse group that includes specialized herbivores (e.g., Throckmorton, Hopson, and Parks 1981) and even marine forms (Carroll and Wild 1994). One sphenodontian, recently described from the Lower Jurassic of Mexico, has been interpreted as capable of venom delivery through grooved "fanged" caniniform teeth (Reynoso 2005). The extant *Sphenodon* itself is a highly specialized animal that appears adapted to life in relatively cool temperatures (Gans

1983). All sphenodontians share certain characteristic features such as a prominent posterior process on the dentary and a row of enlarged teeth on the palatine extending closely parallel to the maxillary tooth row.

Studying the skeletal structure of the smaller tetrapods from the Lossiemouth Sandstone presents considerable challenges because features such as sculpturing on bones, sutures, and small teeth are often smaller than the size of a grain of the rather coarse-grained sandstone. Consequently, such details can be virtually impossible to discern even on the casts prepared from the natural molds. The interpretation of two distinctive small archosaurs from the Lossiemouth Sandstone Formation has proved particularly challenging.

The first of these is *Erpetosuchus* (Newton 1894; Benton and Walker 2002). It is a lightly built, small crurotarsan with dorsal armor composed of pairs of small osteoderms. Its skull closely resembles those of crocodylomorph reptiles, especially in the forward inclination of the quadrate and the deeply emarginated otic region (figure 6.3). Olsen, Sues, and Norell (2001) and Benton and Walker (2002) placed *Erpetosuchus* as a sister-taxon of Crocodylomorpha. It differs

Figure 6.3. Lateral view of the skull of the holotype of the crurotarsan archosaur *Erpetosuchus granti* (Natural History Museum, London, BMNH R3139). (Modified from Newton 1894)

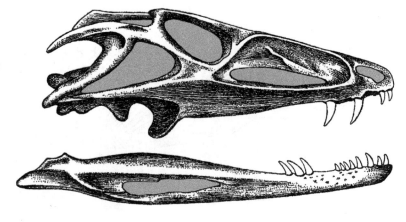

2 cm

from the latter in that the quadrate apparently does not extend back under the squamosal. Furthermore, the radiale and ulnare of *Erpetosuchus* are not elongated, and its coracoid lacks an attenuated posterior process.

One of the most distinctive features of *Erpetosuchus* is its marginal dentition. In the upper jaw, teeth are restricted to the premaxilla and the anterior region of the maxilla. In the mandible, in addition to the anterior teeth, there are at least two or three acutely conical teeth situated further posteriorly on each dentary. A partial skull of *Erpetosuchus* from the Newark Supergroup in Connecticut was first identified on the basis of this distinctive dentition (Olsen, Sues, and Norell 2001; chapter 7).

Scleromochlus is also a small, lightly built archosaur. It attained a length of only about 20 centimeters (Benton 1999). Huene (1914) interpreted it as an arboreal animal capable of leaping from branch to branch. He went so far as to speculate that *Scleromochlus* could have parachuted using skin folds on its forelimbs and perhaps elsewhere along the body. Walker (1970) noted that the short body and unusually long hind limbs of *Scleromochlus* are consistent with jumping (saltatorial) locomotion but interpreted this reptile as a ground-dwelling rather than arboreal form. Benton (1999) concurred but considered *Scleromochlus* cursorial.

Woodward (1907) initially considered *Scleromochlus* a dinosaur, but Broom (1913) and Huene (1914) first interpreted it as a pseudosuchian archosaur, and all later authors (Walker 1970; Benton and Walker 1985) have accepted this placement. Padian (1984) and Sereno (1991) both considered *Scleromochlus* an ornithodiran, although they disagreed on its phylogenetic position among Ornithodira. Padian hypothesized it as the sister taxon to Pterosauria, as first suggested by Huene (1914), but Sereno regarded the evidence as equivocal and argued that *Scleromochlus* could be more closely related to Dinosauria. Based on a revision of all known specimens, Benton (1999) supported ornithodiran affinities for *Scleromochlus* and considered it the sister-taxon to a clade comprising Dinosauria and Pterosauria.

Saltopus is known only from a single, poorly preserved partial skeleton that Huene (1910a) and Benton and Walker (1985) interpreted as that of a small theropod dinosaur. The proportions of its long, slender hind limbs are consistent with cursorial habits, and the tightly bundled metatarsals indicate a digitigrade posture. The specimen provides too little information to permit confident referral to any particular ornithodiran group.

Ornithosuchus was the top predator of the Lossiemouth Sandstone Formation assemblage. It attained a length of up to 3.5 meters and is closely related to *Riojasuchus* from the Los Colorados Formation (Bonaparte 1972) and *Venaticosuchus* from the Ischigualasto Formation of northwestern Argentina (Bonaparte 1970). Walker (1964) reinterpreted *Ornithosuchus* as an early theropod dinosaur, but other authors have rejected his views. Sereno (1991) considered the Ornithosuchidae most closely related to Suchia, but Nesbitt (2007) placed this group closer to "rauisuchians." The proportionately large skull of *Ornithosuchus* has distinctly bulbous premaxillae and holds relatively few teeth, with tall, recurved crowns bearing serrated cutting edges. The dorsal armor comprises pairs of small, keeled osteoderms that are sutured to each other along the midline. The sharply bent cervical osteoderms bear dorsolaterally projecting spines posteriorly. The forelimb of *Ornithosuchus* is shorter than the hind limb; the combined length of the humerus and radius is only about two-thirds that of the femur and tibia, suggesting facultative bipedality (Walker 1964).

The tetrapod assemblage from the Lossiemouth Sandstone Formation shares some taxa with other known Late Triassic communities. *Hyperodapedon* is well documented from the lower part of the Maleri Formation in India (chapter 4) but is the only known faunal element shared by this assemblage with that from Elgin. *Stagonolepis* is present in the Chinle and Tecovas formations of the American Southwest (Long and Murry 1995; chapter 8), and sphenodontian and procolophonid remains are also known from these units (e.g., Heckert 2004). *Leptopleuron* is most closely related to *Hypsognathus* from Norian-age units of the Newark Supergroup in eastern North America (Sues et al. 2000; chapter 7). As discussed in chapter 8, the Chinle and Dockum assemblages comprise a broad range of geological ages and depositional environments, and thus direct comparison to the Elgin assemblage is difficult. The Ischigualasto Formation of northwestern Argentina (chapter 4) has yielded the rhynchosaur *Hyperodapedon*, the ornithosuchid *Venaticosuchus* (Bonaparte 1970), and the aetosaur *Aetosauroides* (Casamiquela 1962). *Aetosauroides* is closely related to, if not congeneric with, *Stagonolepis*, and Heckert and Lucas (2002) even went so far as to suggest that the Ischigualasto and Lossiemouth Sandstone aetosaurs were conspecific. Dzik and Sulej (2007) recently reported *Stagonolepis* from the Drawno Beds of Krasiejów, Poland (chapter 5).

FISSURE FILLINGS OF SOUTHWESTERN ENGLAND AND WALES

The fissure-filling deposits in the Bristol Channel area of southwestern England and southeastern Wales are at once among the most frustrating and most rewarding Triassic vertebrate-bearing occur-rences (figure 6.4). The mixture of completely disassociated bones and frequent lack of anything but the broadest stratigraphic age constraints for the sedimentary fills can prove challenging, but the great abundance of often exquisitely preserved skeletal elements and considerable diversity of taxa represented by these remains offer detailed insights into Late Triassic terrestrial life.

In southwestern Britain, extensive limestone uplands were left exposed by the Hercynian Orogeny at the end of the Carboniferous Period. During the early Mesozoic, these terrains underwent episodes of tectonic activity and karstification, which led to the formation of fissures and caves that subsequently were filled in with bone-bearing surface sediments (Robinson 1957a; Halstead and Nicoll 1971; Simms 1990; Fraser 1994; Wall and Jenkyns 2004). The occurrences of the various fissure localities follow the distribution of the limestone quarries in Avon and southeastern Wales (figure 6.4). The discovery of some of the localities dates back to the early part of the twentieth century when quarrying practices were still rather laborious. Typically, quarrying at that time involved setting small explosive charges in shallow holes drilled into the quarry face. Each blast loosened relatively small amounts of rock, which were loaded by hand or small mechanical shovels. On encountering a fissure system, the quarrymen either worked around the fissures or dumped the infill in an abandoned part of the quarry. Consequently, researchers could return year after year to collect samples of bone-bearing sediment. Fissure systems in quarries such as Cromhall (formerly Slickstones; figure 6.5A), Emborough, Windsor Hill, Ruthin, and Pant-y-ffynon were explored in this manner over many years. This situation changed dramatically with the expansion of the highway system in Britain after World War II, the associated increased demand for limestone as road metal, and increased mechani-

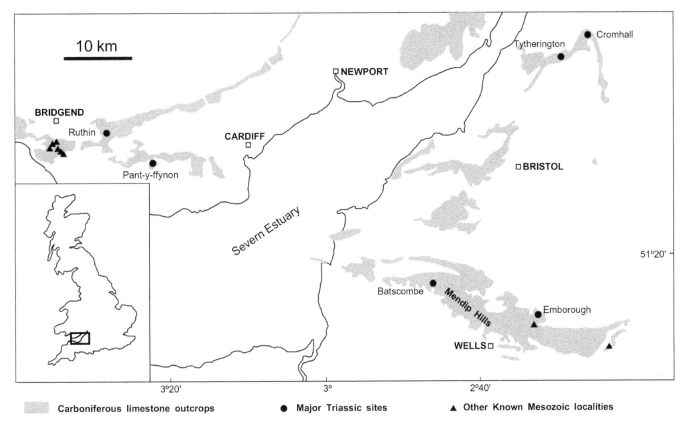

Figure 6.4. Bristol Channel region with locations of the principal vertebrate-bearing Triassic karst systems. (Map by NCF)

zation of quarrying operations. Quarries expanded at a rapid pace, and entire fissure systems could now be removed with a single explosive charge. As a consequence, many new fissures with Triassic sediments were exposed during the 1970s, but they tended to disappear as quickly as they appeared. The first signs of a new fissure system might be high up on a crumbling quarry face that was never stabilized before the next explosive charge obliterated it. Nevertheless, some of these large new quarries yielded significant new vertebrate remains. One of these was Tytherington, where researchers from the University of Bristol repeatedly sampled more than 20 bone-bearing fissures during the 1980s. Even smaller quarries such as Cromhall contain as many as 10 individual bone-bearing fissures.

Most of these early Mesozoic fissure localities have been broadly assigned to two categories (Halstead and Nicoll 1971; Robinson 1957a, 1971; Fraser 1986, 1988a, 1994; Evans and Kermack 1994). First, there are those that formed as karst conduits. Such features have well-defined, solution-etched margins and sedimentary fills containing only reptilian remains. Second, there are fissures that formed as the result of tectonic events (Wall and Jenkyns 2004). They are narrower, more slotlike features, and their sedimentary fills contain mammaliaform remains in addition to those of reptiles; these fissures have also yielded a characteristic assemblage of pollen and spores. Typically, the former occurrences have been considered Norian in age (Robinson 1971), whereas the latter were dated as earliest Jurassic. Although a

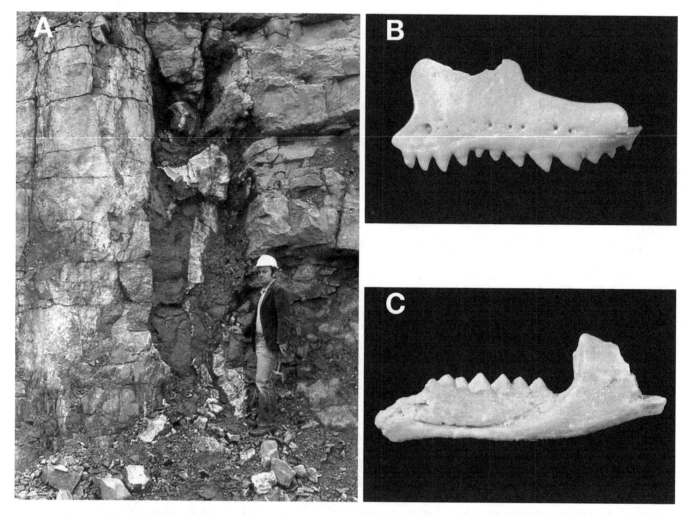

Figure 6.5. (A) Fissure filling in Carboniferous limestones at Tytherington, Gloucestershire. David Whiteside posing as scale. (B) Left maxilla of *Planocephalosaurus robinsonae* (Aberdeen University, AUP 11061, holotype). Length 8.2 millimeters. (C) Left dentary of *Sigmala sigmala* (AUP 11082, paratype). Length 1.6 centimeters. (Photographs by NCF)

diagnostic suite of palynomorphs has since confirmed the ages of the Jurassic sediments, the virtual lack of palynomorphs in the remaining fissure fillings has served only to fuel the controversy over their precise age. However, one particular fissure system containing only reptilian bones yielded a palynomorph assemblage indicative of a Rhaetian age (Marshall and Whiteside 1980). An argument was then made that all the "sauropsid-only" sedimentary fills were no older than Rhaetian (Marshall and White-

side 1980; Whiteside and Robinson 1983; Whiteside and Marshall 2008). Since then the debate has continued. Crush (1984) indicated that some of the infills might even be latest Carnian in age, a view not ruled out by Simms, Ruffell, and Johnson (1994) and supported by Edwards and Evans (2006). Simms, Ruffell, and Johnson (1994) argued that conduit caves would develop to a maximum extent under humid climatic conditions and that there is considerable evidence that the Carnian was a time interval with significantly increased humidity worldwide (see also Prochnow et al. 2006). It is thus conceivable that conduits at least began to form during the Carnian, if not fill in with surface sediment. Nevertheless it should be emphasized that the features did not necessarily all form and fill at the same time, and some of the caves and fissures may be millions of years older than others. Most recently, again citing the same specific locality at Tytherington Quarry, Whiteside and Marshall (2008) published additional data to support their view that the sedimentary infills and associated reptilian assemblages are Rhaetian in age. More significantly, these authors presented evidence that the solution features also largely formed during the Rhaetian and are consistent with fluctuating saline to freshwater environment. In other words, this was a marginal marine environment comprising an archipelago of islands. Based on the presence of semi-fusain and inertinite in some of the sediments, Whiteside and Marshall (2008) argued that frequent forest fires probably caused the death of many of the animals (see also Harris 1958). It is probable that the fissure fillings and their faunal assemblages do not represent only two time intervals but rather a range of time, extending from at least the Norian to the final marine transgression during the Early Jurassic.

At the end of the Triassic, the advancing sea gradually inundated southwestern Britain. As a consequence, what was initially a low-lying coastal plain dotted with slightly elevated limestone hills gradually became an island archipelago, which ultimately disappeared beneath the waves. The Early Jurassic fissure fillings thus contain remains of animals that were living on a chain of small islands. At some localities, bones and teeth of fishes and even marine reptiles attest to the inundation by the Rhaetic Sea.

Although skeletal remains of tetrapods are the primary fossil component of the fissure fillings, there also are invertebrate fossils including conchostracans, myriapods, and a beetle. Robinson (1957a) recorded the conchostracan *Euestheria minuta* from Cromhall Quarry. Fraser (1985) illustrated a diplopod recovered from the cover sediments overlying the fissure deposits at Cromhall.

Although tetrapod bones comprise the bulk of the vertebrate fossils from the fissures, fish remains are also occasionally encountered. At Cromhall Quarry, scales and teeth of *Gyrolepis* and scales of *Pholidophorus* are relatively common in the cover sequences and in the slot fissures in the eastern part of the quarry. Walkden and Fraser (1993) interpreted these sequences as the youngest Mesozoic strata at this locality. The occurrence of remains of *Gyrolepis* and *Pholidophorus* in many of the fissures at Tytherington lends support to the claim that most of the sediments in these systems are Rhaetian in age (Marshall and Whiteside 1980; Whiteside and Robinson 1983; Whiteside and Marshall 2008). The slightly younger fissure fillings at Holwell, which have been definitively assigned to the Rhaetian, are well-known for their abundant fish remains, particularly teeth of sharks such as *Lissodus*, *Palaeospinax*, and *Polyacrodus*. However, rare remains of terrestrial elements such as teeth of haramiyids and bones of the sphenodontian *Clevosaurus* also occur at Holwell.

Most tetrapod remains from the fissure fillings represent relatively small animals, such as sphenodontians, procolophonids, small archosaurs, and mammaliaforms. Larger Triassic tetrapods, such as phytosaurs, "rauisuchians," and metoposaurs, are conspicuously absent. Whether this reflects taphonomic biases or the original composition of the tetrapod communities is not clear. Certainly narrow surface openings of some of the fissures would have been a controlling factor in bone accumulation, but it is equally plausible that the paleoenvironment of this region was not favorable to larger animals. Elsewhere, metoposaurs and phytosaurs typically occur in floodplain settings, and, by the end of the Triassic, such depositional environments apparently were no longer to be found in southwestern Britain. Instead, a maritime climate prevailed over a series of increasingly shrinking islands.

For the most part, the vertebrate fossils comprise completely disassociated bones, but these remains typically occur in dense concentrations. Rare finds of articulated skeletal remains are known from only two localities, Cromhall Quarry in southwestern England (figure 6.5A) and Pant-y-ffynon Quarry in South Wales. In some instances, bone accumulations are dominated by a single tetrapod taxon, but, more commonly, as many as 15 distinct species may be recovered from a single fissure filling. The processes leading to the formation of such large accumulations of bones are still not fully understood. It has been suggested that the activities of predators were partly responsible, and that rainwater rushing over the ancient land surface and into the underground passages washed through middens left by these predators (e.g., Evans and Kermack 1994). The occurrence of large numbers of shed carnivore teeth at many sites lends some support to this scenario. However, it does not provide a comprehensive explanation because the significant size range of animals, together with the presence of many bones of the putative predators, is inconsistent with this interpretation.

Sphenodontian lepidosaurs (figures 6.5B and C, 6.6, and 6.7) are usually the most common faunal element in the fissure fillings, often accounting for more than two-thirds of the recovered vertebrate remains. *Clevosaurus hudsoni* was the first sphenodontian reported from the fissures (Swinton 1939) and is now the best known, based on countless isolated bones as well as a number of articulated specimens (Robinson 1973; Fraser 1988b; figure 6.7B). It

could reach a length of 25 centimeters. To date, two additional species of *Clevosaurus* have been described from Triassic-age fissures in England: *C. minor* (Fraser 1988b) and *C. latidens* (Fraser 1993). These forms are part of a more inclusive group of sphenodontians known as Clevosauridae, which is characterized by a rather blunt snout with an ascending posterior process of the premaxilla excluding the maxilla from the margin of the external naris (Fraser 1988b; Bonaparte and Sues 2006). Jones (2006) added the presence of a long dorsal process on the jugal extending back to contact the squamosal and a suborbital fenestra bounded only by the ectopterygoid and palatine as additional diagnostic features for *Clevosaurus*. A noteworthy feature in *Clevosaurus* is the retention of a distinct, splint-like supratemporal.

Clevosaurids are probably the most widespread group of early Mesozoic sphenodontians. *Clevosaurus* is now known from the Upper Triassic of Brazil (Bonaparte and Sues 2006) and the Lower Jurassic of Wales (Säilä 2005), Yunnan, China (Wu 1994), eastern Canada (Sues, Shubin, and Olsen 1994) and South Africa (Sues and Reisz 1995). The additional teeth on the maxilla and dentary of *Clevosaurus* have pronounced posterior flanges on the maxillary and anterior flanges on the dentary teeth.

Two other sphenodontian taxa, *Planocephalosaurus* (Fraser 1982) and *Diphydontosaurus* (Whiteside 1986), are very common in the older fissure sediments, whereas a third taxon, *Gephyrosaurus*, is abundant in fissure fillings of unquestionably Early Jurassic age (Evans 1980, 1981; Edwards and Evans 2006). *Planocephalosaurus* has an acrodont dentition (figures 6.5B, 6.6, and 6.7A), but the anterior teeth in *Diphydontosaurus* are pleurodont, and *Gephyrosaurus* has a fully pleurodont dentition. The latter two taxa represent more basal rhynchocephalians but already share key features of Sphenodontia, including the prominent

Figure 6.6. Articulated skeleton of *Planocepha-losaurus robinsonae* in matrix from fissure filling. (Photograph by NCF)

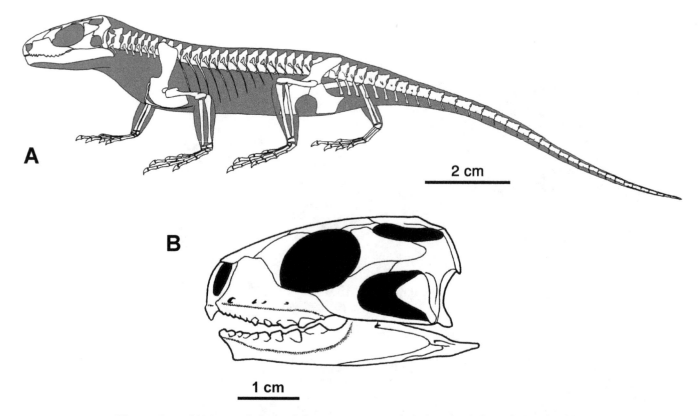

Figure 6.7. (*A*) Lateral view of the reconstructed skeleton of the sphenodontian *Planocephalosaurus robinsonae*. (Modified from Fraser and Walkden 1984) (*B*) Lateral view of the reconstructed skull of the sphenodontian *Clevosaurus hudsoni*. (Modified from Fraser 1988b)

posterior process on the dentary and a row of enlarged teeth on the lateral side of the palatine.

Other sphenodontian taxa are much less common. *Pelecymala* and *Sigmala* are known only from jaw fragments (Fraser 1986; figure 6.5*C*). The latter has a relatively deep mandible and broad, conical teeth similar to those of *Opisthias* from the Upper Jurassic Morrison Formation of the western United States.

Archosaurian reptiles comprise most of the remaining vertebrate remains in the putative Triassic-age fissure fillings, with the basal crocodylomorph *Terrestrisuchus* being the most widespread (figure 6.8). This lightly built reptile attained a length of less than 1 meter and is typically reconstructed in a qua-

drupedal pose (Crush 1984; Sereno and Wild 1992), but the relative proportions of the forelimbs to the hind limbs and the length of the trunk are also consistent with facultative bipedality. Its coracoid has a distinct, posteroventral process. In the carpus, the radiale and ulnare are elongated as in other crocodylomorphs. Known from immature specimens, *Terrestrisuchus* is closely related to, and possibly congeneric with, *Saltoposuchus* from the Löwenstein Formation of Baden-Württemberg, Germany.

There are possibly two species of *Terrestrisuchus* within the suite of fissure assemblages. Crush (1984) described *Terrestrisuchus gracilis* on the basis of material from Pant-y-ffynon Quarry, including associated and partially articulated skeletal remains. Similar,

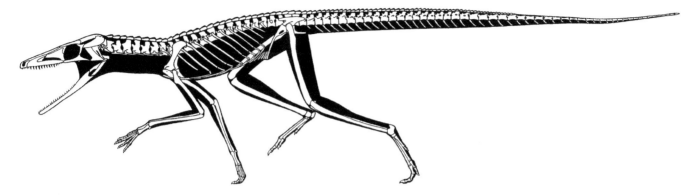

Figure 6.8. Lateral view of the reconstructed skeleton of the basal crocodylomorph *Terrestrisuchus gracilis*. Length about 50 centimeters. (Drawing courtesy and copyright of G. S. Paul)

albeit completely disarticulated elements from Cromhall Quarry show differences in the structure of certain cranial bones, especially the squamosal and basioccipital (Allen and Fraser, in preparation). Furthermore, one articulated specimen from Cromhall exhibits different limb proportions, with relatively shorter hind limbs. It is even conceivable that this find represents a third taxon. Based on the mean lengths of a number of complete humeri and femora, the ratio of forelimb length to hind-limb length among the disassociated remains is consistent with that of *Terrestrisuchus gracilis*.

The fissure openings were not always too small to admit introduction of dinosaurian remains. A small sauropodomorph dinosaur (Yates 2003a; recently placed in a new genus, *Pantydraco*, by Galton, Yates, and Kermack 2007) and an indeterminate coelophysoid theropod dinosaur occur at Pant-y-ffyon. Fraser et al. (2002) named *Agnosphitys* based on an ilium and referred bones and considered it a dinosauromorph. The presence in this taxon of both a distinct ascending process on the astragalus and a well-developed brevis fossa on the ilium support referral to the Dinosauria.

Two other still undescribed taxa of small crurotarsan archosaurs are known from Cromhall, and a third has been recorded from Pant-y-ffynon. One of these has been recovered from only a single fissure filling, but multiple specimens for many of its bones are known. It is a gracile form with at least superficial resemblances to basal crocodylomorphs. The second Cromhall crurotarsan is similar in size to the first but has a very lightly built skeleton, and its skull is characterized by extraordinarily large orbits and antorbital fenestrae. Its affinities remain obscure; Fraser (1994) hinted at possible relationships with aetosaurs, but the Cromhall taxon is clearly different from taxa such as *Stagonolepis* from the Lossiemouth Sandstone Formation.

A single osteoderm from Cromhall has been referred to *Aetosaurus* (Lucas et al. 1999) and definitely establishes the presence of aetosaurs in the region. Additional probably aetosaurian remains, including a partial maxilla, have been identified (Walker, personal communication, 1983) but are uncommon. They all represent relatively small individuals; it is unclear whether they represent juveniles or a small-bodied form.

Skeletal remains of procolophonid parareptiles are known from at least two localities, Cromhall and Ruthin quarries (Fraser 1986, 1988a; Edwards and Evans 2006). The remains from Cromhall and Ruthin possibly represent more than one taxon. Associated remains indicate the presence of a leptopleurine with three prominent bony spikes on each quadratojugal. The Ruthin procolophonid has transversely broad "cheek" teeth with a sharp occlusal ridge, and there is also evidence for the presence of dermal armor. Edwards and Evans (2006) consider it similar to *Scoloparia* from the Wolfville Formation of Nova Scotia (Sues and Baird 1998; chapter 7).

Another distinctive faunal element comprises archosauromorphs with transversely broad maxillary and posterior dentary teeth, which are similar to those of *Trilophosaurus* from the Chinle Formation and Dockum Group (chapter 8): *Tricuspisaurus* from Ruthin Quarry and *Variodens* from Emborough Quarry (Robinson 1957b; Edwards and Evans 2006).

The localities of Emborough and Batscombe are notable for the occurrence of numerous remains of a group of gliding reptiles, the Kuehneosauridae (figure 6.9). Robinson (1962) originally placed material from both localities in the same genus, *Kuehneosaurus*, distinguishing two species, *K. latus* from Emborough and *K. latissimus* from Batscombe. In addition to the two species of *Kuehneosaurus*, *Icarosaurus* has been described on the basis of a single partial but articulated skeleton from the Lockatong Formation of New Jersey (Colbert 1966, 1970a; chapter 7). These three taxa are clearly closely related to each other and share relatively short necks and greatly elongate thoracic ribs that articulate on elongated transverse processes of the dorsal vertebrae and presumably supported a gliding membrane (figure 6.9).

The Kuehneosauridae were first interpreted as basal squamates (Robinson 1962; Colbert 1970a), but recent phylogenetic analyses have demonstrated that they lack the derived features diagnostic for that group. Most recently, Evans (2003) placed kuehneosaurs as a sister-taxon to Lepidosauria among the Lepidosauromorpha.

Following Colbert's (1966) original description of *Icarosaurus*, Robinson reassigned *K. latissimus* from Batscombe to a new genus, *Kuehneosuchus* (Robinson, 1967). The differences between the two British taxa appear to be trivial and are difficult to assess. The Emborough taxon has been reported as having shorter anterior ribs, and some of these have expanded distal ends. Furthermore, none of the ribs displays ventral curvature of the shaft. The ribs of the Batscombe taxon possess rod-shaped shafts (Robinson 1967; Stein et al. 2008), and in all but the first pair the ribs have a moderate ventral curvature of the distal portion of the shaft. Stein et al. (2008) conducted an aerodynamic analysis of the aerofoil shape in the two forms and argued that *Kuehneosuchus*, with its more elongate "wings," was a glider, whereas *Kuehneosaurus* was a parachutist. It is important to reiterate that no articulated remains are known from either locality, and much of the material is at best loosely associated. Thus, determining precise details of the aerofoil shape is fraught with uncertainties.

Stein et al. (2008) considered the differences in the shape of the aerofoil sufficient to distinguish two genera. Yet they conceded that the material from the two British localities might merely represent a single sexually dimorphic species. Stein et al. (2008) even suggested that the gliding *Kuehneosuchus* could have used its larger, perhaps brightly colored membranes to display to the parachuting female (*Kuehneosaurus*). We consider the differences between the kuehneosaurs from the British fissure fillings and between these forms and *Icarosaurus* rather trivial and see no reason to maintain distinctions at the generic level. Although the generic nomen *Kuehneosaurus* has clear priority over *Icarosaurus*, it is important to note that when Robinson (1962) first named *Kuehneosaurus*, she did not provide a diagnosis or designate a holotype, effec-

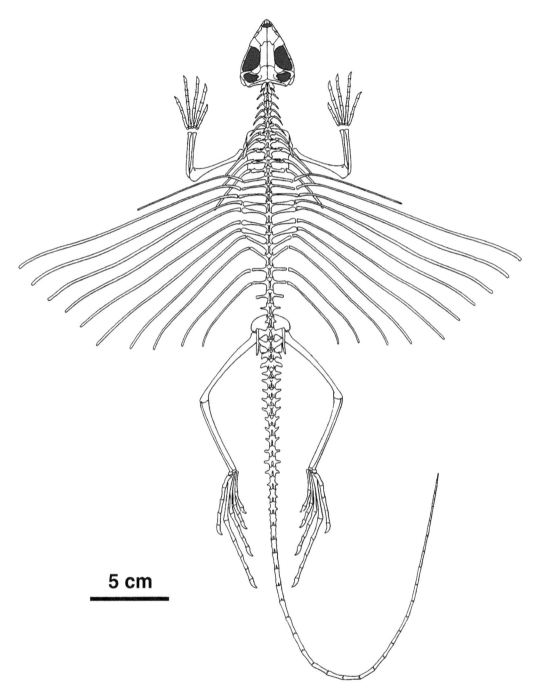

5 cm

Figure 6.9. Reconstruction of the skeleton of the gliding lepidosauromorph reptile *"Kuehneosaurus"* (=*Icarosaurus*) in dorsal view. (Drawing courtesy of the late P. L. Robinson)

tively rendering *Kuehneosaurus* a nomen nudum. Thus, the more evocative name *Icarosaurus* should be applied to both *I. siefkeri* from the Newark basin and the kuehneosaurids from the British fissure fillings. A few isolated jaws and dorsal vertebrae with greatly elongate transverse processes of a kueh-

neosaurid have been recovered from Cromhall Quarry. Why the Emborough and Batscombe assemblages are dominated by kuehneosaurid remains is unclear, but this faunal difference probably reflects differences in depositional environment rather than age.

Two other taxa of Triassic gliding reptiles, *Sharov-ipteryx* from Kyrgyzstan and *Mecistotrachelos* from the eastern United States, are not closely related to kuehneosaurids and are further discussed in chapter 9.

Although kuehneosaurid remains are predominant in the Emborough fissure deposits, there are also bones and teeth of other tetrapod taxa, including two teeth of the mammaliaform *Kuehneotherium*. The still poorly known *Kuehneotherium* is the oldest known taxon possessing molars exhibiting the reversed triangle configuration between upper and lower teeth that is so characteristic of therian mammals (Kielan-Jaworowska, Cifelli, and Luo 2004). Fraser, Walkden, and Stewart (1985) dated the fissure fillings at Emborough as Norian in age, and thus the occurrence of *Kuehneotherium* at this locality would represent the oldest known record for this mammaliaform. Subsequently, however, other authors argued that this find indicated a Rhaetian or even an earliest Jurassic date for the deposits.

Much still unprocessed fissure-filling material is housed in a number of museum collections. As work on these samples continues, the known faunal diversity for many of the fissure assemblages will increase despite the fact that many of the quarry localities are no longer accessible for collecting. For example, the first bones of drepanosaurs (chapter 5) and "rhamphorhynchoid" pterosaurs have been identified only quite recently (Fraser and Unwin 1990; Renesto and Fraser 2003).

The Late Triassic strata located at opposite ends of the British Isles represent two very different and uncommon types of depositional environment. The eolian sandstones of the Lossiemouth Sandstone Formation indicate a rather arid environment, whereas the solution-etched cave and fissure fillings of southwest Britain reflect more humid conditions, probably on an island archipelago. Not surprisingly, the respective tetrapod communities differ in their composition, possibly further enhanced by some difference in age.

The reptilian assemblage from the Lossiemouth Sandstone Formation shares some taxa with other Late Triassic tetrapod communities elsewhere. *Hyperodapedon* is well known from the lower part of the Maleri Formation in India as well as from southern Brazil and northwestern Argentina (chapter 4). *Stagonolepis* occurs in the Drawno Beds of Silesia, Poland (chapter 5), as well as the Chinle Formation and Dockum Group of the American Southwest (chapter 8). The closely related *Aetosauroides* is known from the Ischigualasto Formation of northwestern Argentina (chapter 4). Sphenodontian and procolophonid remains have been recorded from both the Chinle Formation and Dockum Group. By comparison with these assemblages, the absence of metoposaurs and phytosaurs from the Lossiemouth Sandstone Formation presumably reflects the peculiar depositional environment for this unit.

It is always important to exercise particular caution when reconstructing the biota of a region based solely on one or two particular fossil deposits. The fissure fillings of southwestern Britain have long been interpreted as representing a rather distinct depositional environment. The restricted openings of the solution features probably prevented introduction of significant skeletal remains of any large-bodied tetrapods such as phytosaurs. Nevertheless, some bones of small aetosaurs occur in the fissure sediments, and procolophonid and sphenodontian remains are common.

CHAPTER 7

Triassic of the Central Atlantic Margin System

The initial breakup of Pangaea occurred along the junction between present-day North America and Africa, from Greenland to Mexico, during the Middle Triassic. Strong extensional forces caused continental crust to stretch and thin along the rift axis, resulting in basin formation. These basins were filled by continental sedimentary strata, which were capped by or interbedded with extensive basalts and often intruded by diabase dikes or sills. Today a long chain of rift basins, closely following the trend of the Appalachian Mountains, extends along the East Coast of North America, from the Canadian Maritimes in the north to the Carolinas in the south. The remnants of some thirty basins are exposed on land, while others lie buried under the younger sediments of the Atlantic Coastal Plain or offshore (figure 7.1). The city of Newark, New Jersey, lies in the largest completely exposed basin, which today extends across three states. The famous Civil War battlefields at Manassas, Virginia, and Gettysburg, Pennsylvania, are located in the smaller Culpeper and Gettysburg basins, respectively. The apparently largest rift basin, the Fundy basin, now lies for the most part under the Bay of Fundy in Nova Scotia and New Brunswick, Canada, but some of its thick fill of sedimentary and igneous rocks is exposed along the bay's

scenic shorelines (figure 11.3). Additional rift basins occur across the Atlantic Ocean, on the Iberian Peninsula, and especially in Morocco. The entire rift province, known as the central Atlantic margin system, spans a broad range of paleolatitudes and more than some 30 million years (Olsen 1997).

Redfield (1856) coined the term Newark Group in reference to the eroded and faulted sedimentary rock fills of the rift basins in eastern North America. Olsen (1978) first introduced the currently used designation Newark Supergroup. For many years, the entire Newark succession was considered Triassic in age. Using radiometric dates for the basalts from these sections and palynological data for the sedimentary formations overlying the basalts, however, Cornet, Traverse, and McDonald (1973) and Olsen and Galton (1977) demonstrated that the upper portions of the Newark Supergroup are, in fact, Early Jurassic in age.

Most rift basins of the Newark Supergroup are half-grabens that are asymmetrically bounded by major border faults. The sedimentary and igneous fills of these basins record a considerable span of geological time—in the case of the most extensive known succession, from the Fundy basin, ranging from the ?Middle Triassic well into the Early Jurassic. The

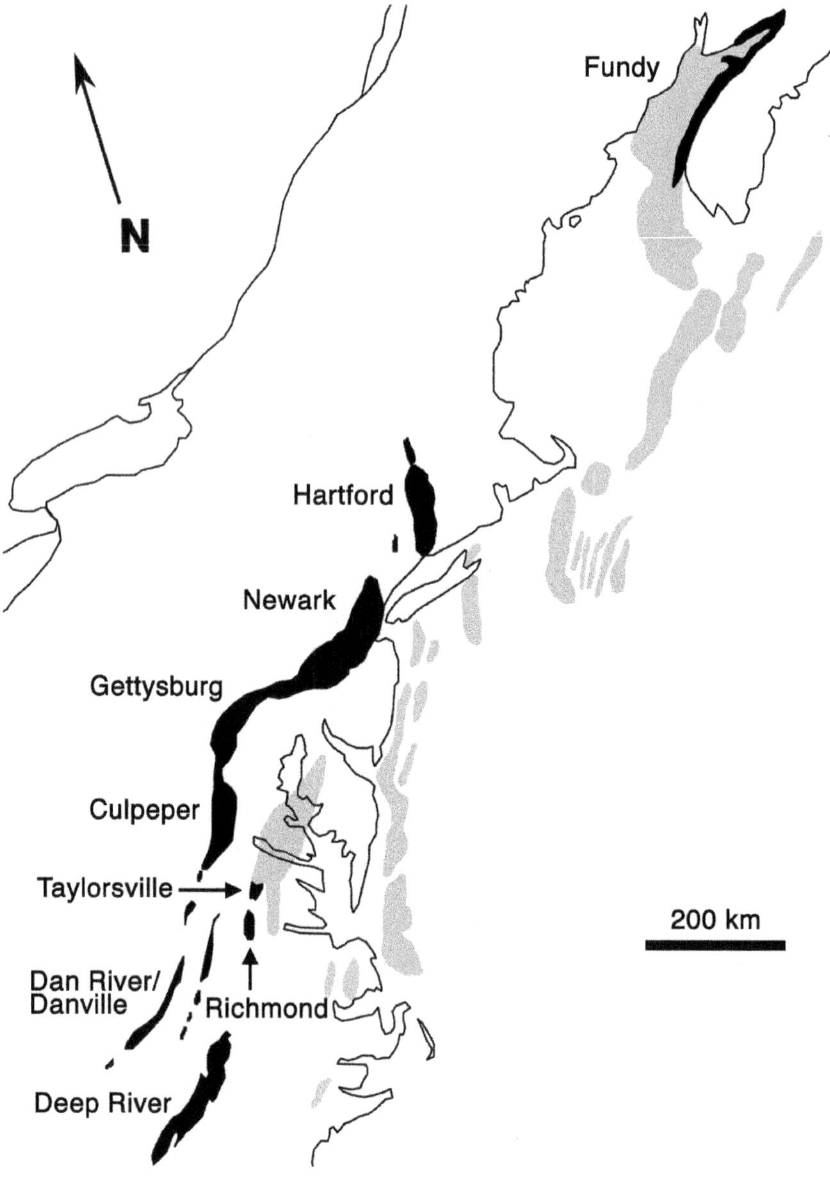

Figure 7.1. Distribution of Newark Supergroup rift basins in eastern North America. Black indicates exposed strata and gray inferred extent of strata. (Modified from Olsen, Schlische, and Gore 1989)

strata preserved in the more northern basins span the Triassic-Jurassic boundary and much of the Early Jurassic, but sedimentation in the southern basins apparently already ceased during the Late Triassic.

All Newark Supergroup rift basins show the same tripartite succession of sedimentary fill. Early in each basin's development, rivers and streams introduced large amounts of sediment into the basin from surrounding highlands. This fluvial phase of

sedimentation coincided with a period of slow subsidence of the basin floor. Then the rate of subsidence increased. As a result, the outflow of water from the basin became restricted or entirely interrupted, and a large, often deep lake came to occupy the rift basin. For a period of time, sedimentation would no longer track basin subsidence. Eventually, as the subsidence rate decreased again, the lake would fill up with sediment, and fluvial sedimentation would resume. Schlische and Olsen (1988) established that the transition from fluvial to lacustrine sedimentation occurred in different rift basins at different times.

Over the decades, researchers noted that, in addition to this pattern of sedimentation, another set of factors appears to have influenced the formation of the basin fills. Repetitive cycles of sedimentation indicate that lake levels continuously waxed and waned over time. This process resulted in cyclical sequences of lacustrine strata, which reach a total thickness of some 5,000 meters in the Newark basin (Olsen 1997). These sedimentary cycles, known as Van Houten cycles in honor of their discoverer, have been linked to periodic variations in rainfall and temperatures (Olsen 1986, 1988). Such cyclical changes in climate are now thought to be under the control of certain regular variations in Earth's orbit around the Sun, the so-called Milankovitch cycles. As Earth spins and moves in its path around the Sun, its axis wobbles slowly. This wobble occurs on an approximately 21,000-year cycle, also known as the precession of equinoxes, and apparently forces climates through intervals of relatively high precipitation alternating with drier periods. Olsen and his collaborators also identified additional types of Milankovitch cycles in the deposition of Newark Supergroup sedimentary rocks. They observed compound cycles of approximately 40,000 years, but also 100,000-year, 400,000-year, and even 2-million-year and 6-million-year cycles. Several of these cycles can be related to different components in Earth's orbit around the Sun. The 40,000-year cycle (obliquity cycle) is associated with regular variations in the tilt of Earth's axis. The 100,000- and 400,000-year cycles are linked to changes in the elliptical path described by Earth around the Sun. The eccentricity of Earth's orbit continuously changes over a period of roughly 100,000 years. Eccentricity is a measure of how much the shape of an ellipse deviates from a perfect circle. Thus, the eccentricity of a circle is zero; as the eccentricity of an ellipse approaches one, the ellipse becomes increasingly more elongated. As the eccentricity of Earth's orbit changes, the distances from the closest and most distant points along the planet's orbit to the Sun change, affecting the amount of solar radiation received on Earth at various times during the year.

At the end of the Triassic a giant mantle plume under the rift zone melted its way through the crust, emplacing vast quantities of basaltic lavas (chapter 11). During a rather short period of time, with radiometric dates for the tholeiitic basalts tightly clustering with a peak around 199–200 Ma (Nomade et al. 2007), the lavas and associated rocks came to cover an area of some 7 million square kilometers, extending from Brazil and West Africa to the eastern seaboard of the United States, the Canadian Maritimes, and western France. The Palisades along the western bank of the Hudson River in New Jersey, just across from Manhattan, and the Holyoke Range in western Massachusetts represent well-known examples of these flood basalts in the eastern United States. Together, these lava flows form part of what Marzoli et al. (1999) termed the Central Atlantic Magmatic Province (CAMP) and which may represent the largest igneous province in Earth's history (see chapter 11). It was not until later during the Early Jurassic that marine sediments first accumulated in the primary rift, which was to form the basin of the Atlantic Ocean. Europe, Africa, and North America have continued to drift apart ever since.

Given the temporal range of the strata of the Newark Supergroup, one would suppose them to be an ideal setting for studying the succession of early Mesozoic continental ecosystems. Following the first reported discovery of dinosaurian footprints in the early nineteenth century, however, fossil vertebrate remains other than tetrapod trackways and fish were found only infrequently, and the rocks of the Newark Supergroup soon came to be seen as mostly devoid of vertebrate fossils. Starting in the 1970s, fieldwork by Paul E. Olsen (Columbia University) and others (including the authors) has demonstrated that this reputation was unfounded and initiated a new, highly productive phase of paleontological exploration that continues to the present day.

A challenge confronting any student of the Newark Supergroup is the relatively poor outcrop situation. The eastern United States has long been densely settled and, furthermore, has a fairly humid climate. Even where bedrock is not buried under the ever-expanding urban sprawl, it is usually concealed under dense vegetation. Thus, fieldwork is largely confined to quarries, roadside exposures, stream banks, and temporary outcrops created by housing and road construction.

In this chapter and in chapter 9, we review the principal occurrences of continental vertebrates and associated organisms now known from Triassic fluviolacustrine strata of the Newark Supergroup. The discussion proceeds from the stratigraphically oldest to the youngest assemblages.

?MIDDLE TRIASSIC

The possibly oldest known occurrence of vertebrate fossils in the Newark Supergroup is in fluvial sandstones and clay-pebble conglomerates exposed along the shoreline of the Bay of Fundy at Carrs Brook near Lower Economy, Nova Scotia (Olsen, Schlische, and Gore 1989). Huber, Lucas, and Hunt (1993a) used this tetrapod assemblage as the basis for the Economian LVF. Dissociated bones occur as clasts in the conglomerates and typically are broken and

waterworn, which renders their identification difficult. Tetrapod fossils recovered from Lower Economy to date include jaws of small procolophonids, an elongated cervical vertebra of a protorosaur, and fragmentary cranial bones of capitosauroid and trematosaurid temnospondyls. Some of these fossils were tentatively assigned to taxa known elsewhere from Anisian-age formations such as the Moenkopi Formation of the American Southwest and the Upper Buntsandstein of central Europe (chapter 3) and employed in support of an Anisian date for the Lower Economy strata (Olsen, Schlische, and Gore 1989; Lucas and Huber 2003). However, because the more specific taxonomic identifications of the tetrapod remains are problematical, the age of the Lower Economy beds has yet to be constrained.

To date, no strata of indisputably Ladinian age are known from eastern North America (Lucas and Huber 2003).

CARNIAN

The probably oldest set of Late Triassic vertebrate assemblages, of possibly early Carnian age, comes from swamp and shallow-water lacustrine deposits in the Richmond and Taylorsville basins of Virginia. Based on conchostracans, Kozur and Weems (2007) dated much of the succession in the Richmond basin and correlative units in the Taylorsville basin as early Carnian (Cordevolian). The Richmond basin contains productive coal measures, which were exploited from the eighteenth century until the 1930s. During one of his visits to North America, the eminent Victorian geologist Sir Charles Lyell already obtained specimens of the redfieldiid actinopterygian *Dictyopyge* from the Black Heath Mine (Lyell 1847). *Dictyopyge* is very common throughout the succession in the Richmond basin (Schaeffer and McDonald 1978). The occurrence of abundant coals in the Richmond basin indicates a warm, humid climate, which is consistent with the reconstructed position of the region at the paleo-equator during the early Late Triassic (Cornet and Olsen 1990). The coals and associated sedimentary rocks have yielded a wealth of often beautifully preserved plant fossils, which document a rich, diverse wetland flora comprising ferns, equisetale-ans, bennettitaleans, and ginkgophytes (Fontaine 1883; Ward 1900; Ash 1980; Cornet and Olsen 1990; figures 7.2–7.5). Many of these plants have very large leaves (Ash 1980). Ferns are particularly

common and comprise Dipteridaceae, Marattia-ceae, and Osmundaceae.

The Tomahawk locality near Midlothian, just southwest of Richmond, is a former roadside expo-sure in the Tomahawk Member of the Vinita For-mation (LeTourneau 1999; formerly Turkey Branch Formation of Cornet and Olsen 1990). It has yielded a rich faunal assemblage including a variety of fishes (including the hybodont shark *Lissodus* and the redfieldiid *Dictyopyge*), reptiles, and cynodont the-rapsids (Sues and Olsen 1990; Sues, Olsen, and Kroehler 1994). The vertebrate remains are associ-ated with abundant root traces and conchostracans. The most common tetrapod is the traversodont

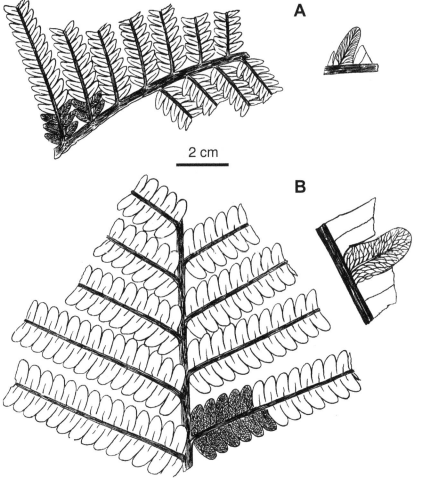

Figure 7.2. Foliage of (*A*) the osmun-dacean fern *Cladophlebis* and (*B*) *Lonchopteris*. (From Fontaine 1883)

2 cm

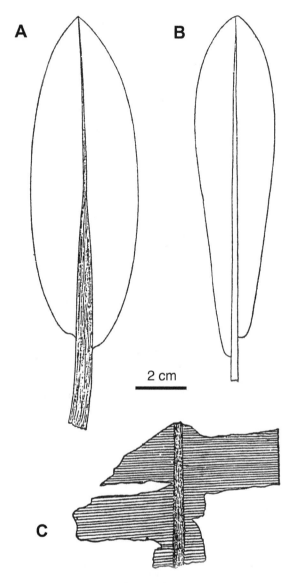

Figure 7.3. Outlines of complete leaves (*A* and *B*) and leaf fragment with details of venation (*C*) of the cycadophyte *Macrotaeniopteris*. (From Fontaine 1883)

cynodont *Boreogomphodon*, which is known from several partial skulls as well as many jaws and teeth of mostly juvenile individuals. Rather unexpectedly, *Boreogomphodon* is most closely related to basal traversodonts from the early Middle Triassic rather than to the derived, Ladinian- and Carnian-age taxa from the Southern Hemisphere (Sues and Hopson, in review; chapter 4).

The Tomahawk assemblage also includes the small mammal-like cynodont *Microconodon* (Sues 2001), which was first described from the Cumnock Formation of the Dan River basin in North Carolina (Emmons 1857; Simpson 1926). Its postcanine teeth have three or four mesiodistally aligned cusps and incipiently divided roots but lack cingula. *Microconodon* is known only from dentaries and isolated teeth, and its affinities may lie with small, mammal-like cynodonts recently discovered in the Caturitta Formation of southern Brazil (chapter 4).

Archosaurian reptiles are represented by skeletal remains of the enigmatic *Euscolosuchus* (Sues 1992) and teeth of unidentified taxa. *Euscolosuchus* has dorsal armor composed of paired rows of sculptured osteoderms that bear large lateral spines.

The Tomahawk locality has also yielded abundant if fragmentary remains of small tetrapods, including a sphenodontian and possibly other lepidosauromorphs, which are among the geologically oldest known records of their groups. Particularly noteworthy are the teeth of *Uatchitodon* (Sues 1991; figure. 7.6). The bladelike, recurved crowns of these teeth have finely serrated edges and bear deep grooves on the labial and lingual surfaces, which closely resemble the venom-conducting grooves on the fangs of certain extant snakes and on the teeth of the gila monster (*Heloderma*). *Uatchitodon* represents the geologically oldest known example of venom-conducting teeth in a reptile.

Siltstones of the correlative Poor Farm Member of the Falling Creek Formation in the Taylorsville basin, which was connected to the Richmond basin during Late Triassic times, have yielded well-preserved plant remains, unionid clams, conchostracans, and some vertebrate remains (LeTourneau 2003). Tetrapods are represented by the heavily armored archosauriform *Doswellia* (Weems 1980), which appears to be most closely related to the South American

A

B

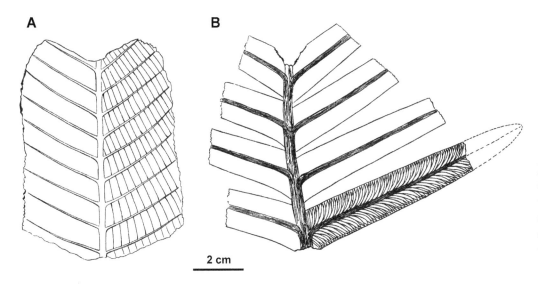

2 cm

Figure 7.4. Foliage of (*A*) the marattiacean fern *Pseudodanaeopsis* and (*B*) the dipteridacean fern *Clathropteris*. (From Fontaine 1883)

Figure 7.5. Frond of the bennettitalean *Pterophyllum* with detail showing venation. (From Fontaine 1883)

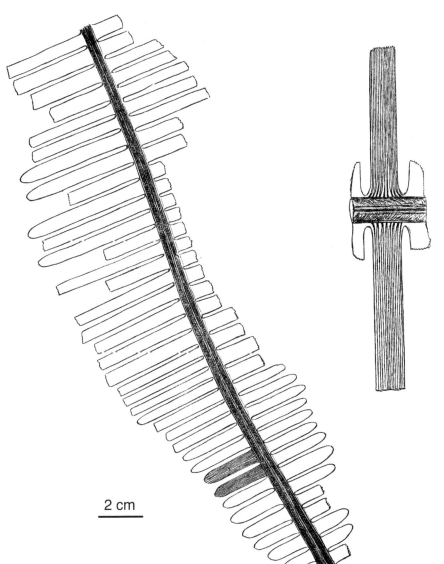

2 cm

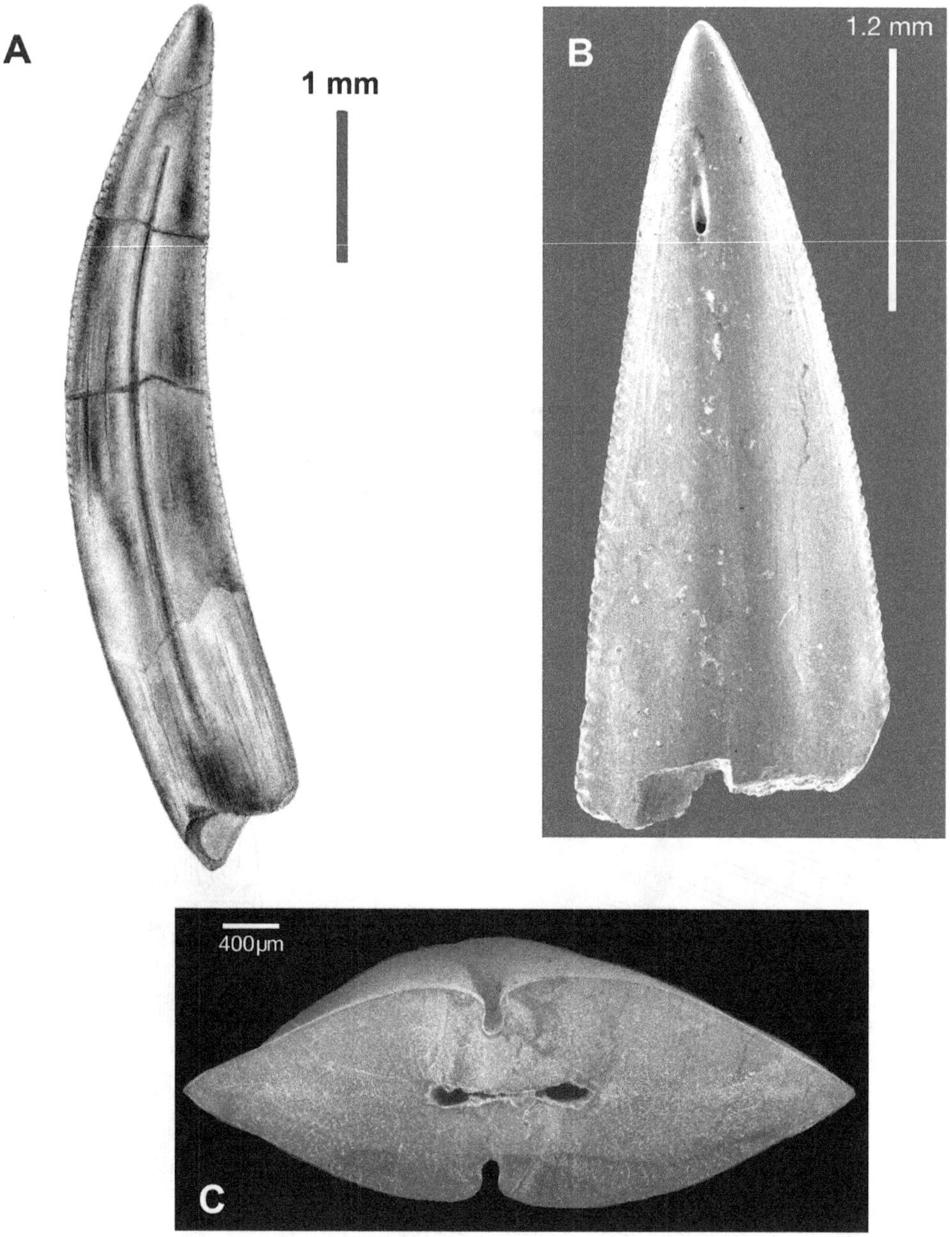

Figure 7.6. (*A*) Tooth of *Uatchitodon kroehleri* (National Museum of Natural History, USNM 448624) in side view. Note the open groove, which presumably served for venom conduction. (From Sues 1996) (*B*) Oblique side view of a tooth crown of cf. *Uatchitodon* (Museum of Northern Arizona, MNA V3680) from the Chinle Formation. Note complete closure of the venom-conducting groove and presence of a small opening near the tip of the tooth crown. (Modified from Sues 1996) (*C*) Transverse section of a tooth crown of *Uatchitodon kroehleri* showing the invaginations of the venom-conducting grooves on the labial and lingual surfaces of the crown.

Proterochampsidae (Dilkes and Sues 2009), and a few isolated archosaurian teeth.

The sandstones and mudstones of the Pekin Formation in the Sanford subbasin of the Deep River basin in North Carolina, primarily exposed in various commercial clay pits, have yielded skeletal remains representing aetosaurs, phytosaurs, and "rauisuchians" as well as dicynodont therapsids similar to *Placerias* from the Chinle Formation of the American Southwest (chapter 8). Recent discoveries by Vince Schneider (North Carolina Museum of Natural Sciences) have added *Boreogomphodon* and a basal crocodylomorph to this list. Fine-grained sandstones and siltstones of the Pekin Formation, especially in the Boren Clay Pit near Gulf, North Carolina, frequently yield excellently preserved plant fossils, including bennettitaleans (figure 7.5), conifers (including an early representative of the Pinaceae), cycads, equisetaleans (*Neocalamites*), and a variety of ferns including *Cladophlebis* (figure 7.2*A*) and *Todites* (Delevoryas and Hope 1973, 1975; Ash 1980). The strata of this unit formed under humid climatic conditions. Ash (1980) placed the floral assemblages from the Pekin Formation and Richmond basin in the *Eoginkgoites* zone, named for a superficially ginkgo-like bennettitalean (Ash 1976).

The next oldest set of tetrapod assemblages from the Newark Supergroup has been dated as late Carnian (Tuvalian) (Lucas and Huber 2003). Huber, Lucas, and Hunt (1993a) based the Sanfordian LVF on the tetrapod assemblage from the Pekin Formation, which Lucas and Huber (2003) subsequently expanded to accommodate the older assemblages discussed above (see also Kozur and Weems 2007). The most diverse of the Sanfordian tetrapod assemblages is from the Wolfville Formation of the Fundy basin in Nova Scotia (Baird in Carroll et al. 1972; Hopson 1984; Sues, Hopson, and Shubin 1992; Sues and Baird 1998; Sues 2003). This formation grades upward from conglomerates to pebbly sandstones to sandstones and mudstones, which were deposited by a braided river system in a semiarid climate (Hubert and Forlenza 1988). However, abundant, often large unionid clams and metoposaurid temnospondyls attest to at least seasonally humid conditions. Today, strata of the Wolfville Formation crop out along the shores of the Bay of Fundy, which is famous for the world's highest tides. The daily pounding and scouring of the shorelines by the waves continuously generate fresh rock exposures, which can be prospected for fossils at low tide. Although root casts are common in some horizons, no identifiable plant macrofossils have been recovered from the Wolfville Formation to date. Vertebrate remains typically comprise fragmentary, often waterworn bones and teeth,

which are sparsely distributed through the conglomeratic layers (Baird in Carroll et al. 1972). From time to time, well-preserved partial skulls and skeletons are found in more fine-grained sandstones (e.g., Sues 2003). The occurrence of *Koskinonodon bakeri* in the Wolfville Formation provides the only available means for correlation at present (figure 7.7); this taxon is elsewhere known from strata of the Dockum Group in Texas considered late Carnian in age (Hunt 1993). The most common faunal elements are procolophonid parareptiles, especially the basal leptopleuronine *Scoloparia* (Sues and Baird 1998; figure 7.8*A*). *Scoloparia* is distinguished by a cluster of bony spines on the "cheek" region of the skull and (in the holotype) a cervical "collar" composed of small osteoderms. Archosaurian reptiles are represented by as yet unidentified aetosaurs, "rauisuchians," and a crurotarsan similar to *Revueltosaurus* (chapter 8). Additional faunal elements include an undescribed rhynchosaur and the unusual archosauromorph reptile *Teraterpeton*. The latter is characterized by the possession of a long edentulous beak followed by rows of complex, transversely broad teeth at the back of the jaws (Sues 2003; figure 7.8*B*); it is most closely related to *Trilophosaurus* from the Chinle Formation and Dockum Group. Contrary to published claims (Baird in Carroll et al. 1972), no indisputably dinosaurian remains have been recovered from the Wolfville Formation to date. Cynodont therapsids are represented by the large traversodont *Arctotraversodon* (Hopson 1984; Sues, Hopson, and Shubin 1992). The Wolfville assemblage is noteworthy for the absence of phytosaurs and scarcity of metoposaurs, both of which are very common in presumably coeval faunal assemblages elsewhere in North America (chapter 8) and in Europe (chapter 5). This faunal difference may reflect the higher paleolatitude of the Fundy basin and related drier climatic conditions.

Figure 7.7. Impression of the skull roof of a juvenile specimen of *Koskinonodon bakeri* (Peabody Museum of Natural History, Yale University, YPM-PU 21742) from the Wolfville Formation of Nova Scotia. Scale bar = 2.5 cm. (Photograph courtesy of D. Baird)

NORIAN

A slightly younger set of vertebrate-bearing formations that Huber, Lucas, and Hunt (1993a) grouped together as the Conewagian LVF includes the mid-dle New Oxford Formation of the Gettysburg basin in Pennsylvania, the Lockatong Formation of the Newark basin in New Jersey, the Cumnock Formation in the Sanford subbasin of the Deep River basin in North Carolina, and the Cow Branch For-

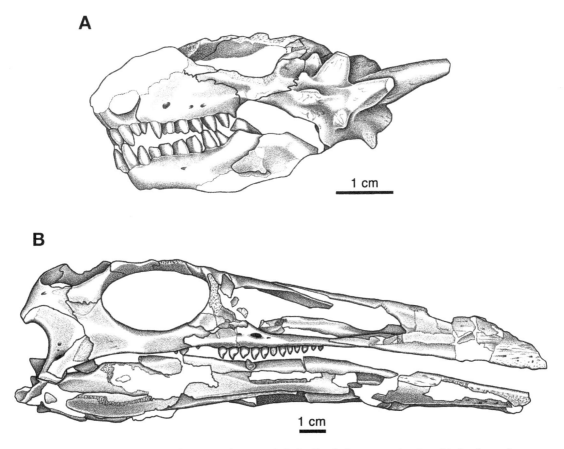

Figure 7.8. (*A*) Lateral view of a partial skull of the procolophonid *Scoloparia glyphanodon* (Royal Ontario Museum, ROM 47484). (*B*) Lateral view of the skull of the holotype of the archosauromorph *Teraterpeton hrynewichorum* (Nova Scotia Museum of Natural History, NSM 999GF041). (Modified from Sues 2003)

mation of the Dan River basin in Virginia and North Carolina (which is discussed in chapter 9). Previously these formations were considered late Carnian in age (Olsen et al. 1989; Huber, Lucas, and Hunt 1993a). The revised chronostratigraphy by Muttoni et al. (2004) would redate all of them as early Norian, although Kozur and Weems (2007) continued to support a late Carnian (Tuvalian) age for the Conewagian LVF based on conchostracans.

The coal measures of the Cumnock Formation of the Sanford subbasin of the Deep River basin in North Carolina have yielded vertebrate bones, coprolites, conchostracans, ostracodes, and conifer wood. At the now abandoned New Egypt coal mine, vertebrates are represented by coelacanth and redfieldiid fishes, the slender-snouted phytosaur *Rutiodon*, unidentified metoposaurs, and the small cynodonts *Dromatherium* and *Microconodon* (Emmons 1856, 1857; Simpson 1926). Emmons (see Ward 1900) reported plant remains from the Cumnock Formation, but their affinities to other Late Triassic plant assemblages are uncertain (Ash 1980). Fraser (2006) noted that the palynomorphs from the Cumnock show a distinct reduction in the percentage of fern spores and an increase in the abundance of gymnosperms with cypresslike growth habits, which he interpreted as indicating raised water levels.

An exposure of siltstones of the middle New Oxford Formation along Little Conewago Creek at Zions View in York County, Pennsylvania, has yielded skulls and postcranial bones of the slender-snouted phytosaur *Rutiodon* and the metoposaurid temnospondyl *Koskinonodon perfectum* (figure 7.9), along with remains of unidentified large phytosaurs and the redfieldiid *Synorichthys* as well as conchostracans (Sinclair 1918; Schaeffer and McDonald 1978; Doyle and Sues 1995). Ash (1980) assigned a plant assemblage reported by Ward (1900) from the New Oxford Formation southwest of York Haven, Pennsylvania, to the *Dinophyton* floral zone, which is characterized by the enigmatic gymnosperm *Dinophyton* and is younger than the *Eoginkgoites* zone. Conifers are particularly abundant in this assemblage.

The Lockatong Formation of the Newark basin comprises gray and black fine-grained clastic deposits of vast lakes, which were apparently deep at times but repeatedly dried up (Olsen 1986). Excavation of a locality at Weehawken, New Jersey, by Paul Olsen and teams from the Peabody Museum of Natural History at Yale University led to the recovery of thousands of fish fossils. Two sedimentary cycles, numbered 5 and 6, yielded very different fish assemblages (Olsen, Schlische, and Gore 1989). In the older Cycle 6, the "holostean" actinopterygian *Semionotus* is

Figure 7.9. Skull of the metoposaurid *Koskinonodon perfectum* (State Museum of Pennsylvania, SMP VP-44) from the New Oxford Formation of Pennsylvania in dorsal view. Skull length about 46 centimeters (measured along midline). (Photograph by H-DS)

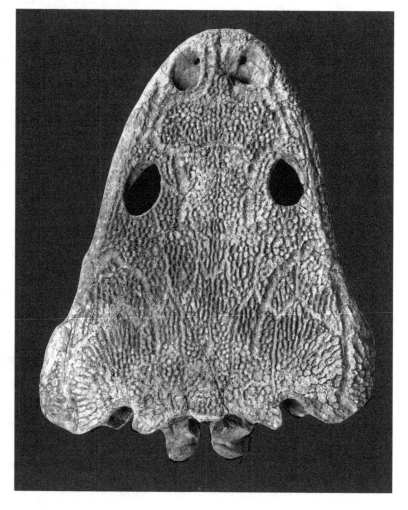

the dominant fish taxon, followed in abundance by *Synorichthys*. Semionotids are absent in Cycle 5, where the palaeoniscoid *Turseodus*, which was not present in Cycle 6, is the most common form. The coelacanth *Diplurus* occurs in both cycles but is more abundant in Cycle 5.

The Lockatong Formation has occasionally yielded skeletal remains of reptiles. The most important locality for such specimens was the now reclaimed Granton Quarry in North Bergen, New Jersey, which yielded specimens of the slender-snouted phytosaur *Rutiodon* (Colbert 1965), the small tanystropheid *Tanytrachelos* (chapter 9), and the drepanosaurid *Hypuronector* (Colbert and Olsen 2001). *Hypuronector* is distinguished by the possession of a deep, laterally flattened tail with tall neural spines and long haemal arches, similar to those of certain present-day swimming vertebrates such as newts, but it also has long, slender limbs suitable for locomotion on land. As discussed in chapter 5, Renesto and Binelli (2006) considered all drepanosaurs arboreal in habits, but the overall structure of the skeleton of *Hypuronector* is clearly more suggestive of aquatic habits.

The Granton Quarry has also yielded the only known skeleton of the lepidosauromorph reptile *Icarosaurus* (Colbert 1966, 1970a). *Icarosaurus* is probably congeneric with *Kuehneosaurus* (including *Kuehneosuchus*) from Late Triassic fissure fillings of southwestern England (chapter 6). Like the latter, it has greatly elongated, rather straight ribs that articulate with prominent transverse processes on the dorsal vertebrae and probably supported a gliding membrane.

The record of fossil vertebrates from Norian- to Rhaetian-age strata of the Newark Supergroup is much less extensive than that for the Carnian stage, but new discoveries are beginning to change this picture. Lucas and Huber (2003) distinguished two

sets of Norian-Rhaetian tetrapod assemblages. The older group, grouped together as the Neshanician LVF (Huber, Lucas, and Hunt 1993a) and considered early to middle Norian in age, includes assemblages from Lithofacies Association II of the Durham subbasin of the Deep River basin, the lower Passaic Formation of the Newark basin, and the lower to middle New Haven Formation of the Hartford basin in Connecticut. The younger set, grouped together as the Cliftonian LVF (Huber, Lucas, and Hunt 1993a) and dated as middle Norian to Rhaetian, comprises assemblages from the middle to upper Passaic Formation of the Newark basin, the upper New Haven Formation of the Hartford basin, and the base of the Blomidon Formation of the Fundy basin. The dating of both sets of vertebrate assemblages is well constrained by pollen and spores and by magnetostratigraphy.

The best-known example of an early Norian tetrapod assemblage from the Newark Supergroup is a remarkable death assemblage discovered in strata of Lithofacies Association II in the Durham subbasin of the Deep River basin in Durham County, North Carolina. Here, a well-preserved partial skeleton of the "rauisuchian" *Postosuchus* (Peyer et al. 2008; chapter 8) was found together with much of an articulated skeleton of the basal crocodylomorph *Dromicosuchus* curled up beneath its pelvic region (Sues et al. 2003). Closer examination of the specimen of *Postosuchus* revealed a rich assortment of vertebrate remains in gut content preserved inside the gastralia: the snout and some bones of a small traversodont cynodont, a partial skeleton of the small aetosaur *Stegomus*, phalanges of a dicynodont, and what appears to be a bone fragment of a temnospondyl. These bones differ in color from those of *Postosuchus* and *Dromicosuchus*. Their surfaces frequently have a distinctly corroded appearance, and some bear bite marks. In addition, the head and neck of the specimen of *Dromicosuchus* show two areas of conspicuous damage, which can be matched to teeth of *Postosuchus*.

The interval from the middle Norian to the Rhaetian is characterized by the presence of the leptopleuronine procolophonid *Hypsognathus* (Gilmore 1928; Colbert 1946; Sues et al. 2000). *Hypsognathus* is most closely related to *Leptopleuron* from the Upper Triassic Lossiemouth Sandstone Formation of northeastern Scotland (chapter 6) and shares with it the presence of a cluster of bony spines on the quadratojugal (figure 7.10).

To date, few other tetrapod remains are known from this interval of the Newark Supergroup. An important record is a partial skull of the crocodile-like crurotarsan *Erpetosuchus* from the lower New Haven Formation in Cheshire, Connecticut (Olsen, Sues, and Norell 2001). *Erpetosuchus* was previously known only from the Lossiemouth Sandstone Formation of Scotland, which is usually considered Carnian in age based solely on its tetrapod assemblage (see chapter 6), but U-Pb dating of calcrete samples from the fossiliferous horizon at the Connecticut locality yielded an age of 211.9±2.1 Ma (Wang

Figure 7.10. (*A*) Palatal view of the snout of a juvenile specimen of the procolophonid *Hypsognathus fenneri* (SMP VP-2160) from the Passaic Formation of Pennsylvania. Note differentiation of the dentition into incisiform premaxillary and molariform maxillary teeth. (*B*) Lateral view of the skull of a large adult specimen of *Hypsognathus fenneri* (Peabody Museum of Natural History, Yale University, YPM 55831) from the upper portion of the New Haven Formation of Connecticut. The apices of the quadratojugal spines not preserved. (Modified from Sues et al. 2000)

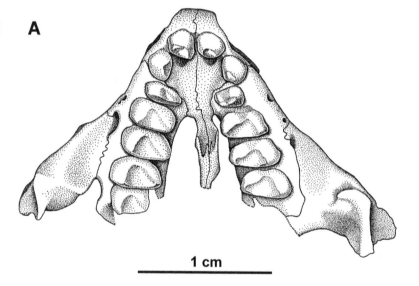

A

1 cm

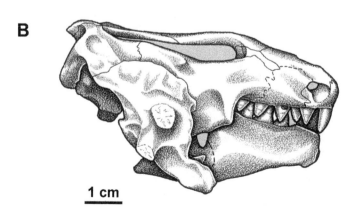

B

1 cm

et al. 1998), placing this occurrence well within the Norian. The upper New Haven Formation of the Hartford basin near Meriden, Connecticut, has yielded an incomplete skull of an indeterminate sphenodontian (Sues and Baird 1993). Unidentified phytosaur remains have been reported from the upper Passaic Formation of Pennsylvania (Lea 1851) and the middle New Haven Formation of Connecticut (Marsh 1893). The latter unit has also produced a natural mold of the dorsal armor of a small aetosaur, which Marsh (1896) designated as the holotype of *Stegomus arcuatus*. Lucas et al. (1999) synonymized *Stegomus* with *Aetosaurus* from Norian-age strata in Europe and Greenland (chapter 5), but Sues et al. (2003) and Schoch (2007) have not accepted this synonymy.

Although the record of tetrapod skeletal remains from this unit is meager, various horizons in the Passaic Formation of the Newark basin in New Jersey and Pennsylvania have produced abundant tetrapod trackways (Olsen, Schlische, and Gore 1989; Silvestri and Szajna 1993). Typically Late Triassic ichnogenera such as *Brachychirotherium* (which was produced by crurotarsan archosaurs) are found up to within a few meters of the palynologically recognized Triassic-Jurassic boundary in the Jacksonwald syncline of the Newark basin in Pennsylvania (Silvestri and Szajna 1993).

Thus, the Newark Supergroup, long considered virtually devoid of tetrapod fossils, is emerging as one of the most important sequences of Triassic-Jurassic fossiliferous continental strata in the world. It will be critical for any future discussions concerning early Mesozoic biotic changes.

MOROCCO

Exploration of the related early Mesozoic rift basins across the present-day Atlantic, on the Iberian Peninsula and in Morocco, is still at an early stage. During the Triassic, northwestern Africa faced present-day Nova Scotia, and both were bounded to the north by a major tectonic feature, the Cobequid-Chedabucto fracture zone (Medina 2000). From 1962 to 1975, Jean-Michel Dutuit (formerly Muséum National d'Histoire Naturelle, Paris) conducted paleontological fieldwork in the Argana Valley, which is located southwest of Marrakech in the western part of the High Atlas Mountains of Morocco. The Argana Valley contains part of the large Essaouira-Agadir basin. An extensive succession of Permian and early Mesozoic strata was deposited in a complex half-graben, which is bounded by an E-dipping fault. The sedimentary fill has been divided into eight units, which were numbered T1–T8 and of which T4 and T5 have yielded tetrapod fossils (Dutuit 1976; Jalil 1996; Tourani et al. 2000). Dutuit discovered a diverse tetrapod assemblage in the lower portion of the Irohalene Member (T5) of the Timezgadiouine Formation. Furthermore, a second assemblage with a somewhat different faunal composition appears to be present in the upper part of the Irohalene Member (Hunt 1993). The Irohalene Member reaches a thickness of 200 meters and comprises cyclical, interbedded red to pink mudstones and sandstones, which Tourani et al. (2000) interpreted as a succession of point bar sandstones and floodplain mudstones related to meandering fluvial channels.

Triassic tetrapods from the Argana Valley include metoposaurid and other temnospondyls, dicynodont therapsids, and archosaurian reptiles including phytosaurs, aetosaurs, and the "rauisuchian" *Arganasuchus* (Dutuit 1976, 1989; Jalil 1996; Jalil and Peyer 2007). The large dicynodonts "*Moghreberia*" and "*Azarifeneria*" are very similar to, and possibly synonymous with, *Placerias* from the Chinle Formation of the American Southwest (Cox 1991; Lucas 1998). Lucas (1998) has used the presence of these dicynodonts and other faunal elements to argue for a late Carnian age for the Irohalene Member. Although the Irohalene Member has only a poor palynological record to date, the unconformably overlying Tadart Ouadou Sandstone Member (T6) of the Bigoudine Formation has yielded pollen and spores that indicate a late Carnian (Tuvalian) to Norian age (Tourani et al. 2000). The Argana Valley was located at about 22° north paleolatitude (Kent and Tauxe 2005) and closely resembles the neighboring Fundy and Newark basins of the Newark Supergroup in many geological features (Olsen 1997; Medina 2000).

The strata of the Newark Supergroup in eastern North America are among the most thoroughly studied Triassic continental deposits anywhere in the world. Employing magnetostratigraphy and cyclostratigraphy, Paul Olsen and Dennis Kent are developing a high-resolution temporal framework for much of the Newark Supergroup. This, together with a growing number of fossil discoveries, has facilitated investigation of regional changes in terrestrial vertebrate assemblages during the Late Triassic and across the Triassic-Jurassic boundary. As further discussed in chapter 11, the Newark Supergroup is becoming critical for discussions concerning the tempo and mode of terrestrial biotic changes during the early Mesozoic.

Not only do the strata of the Newark Supergroup record an extensive period of geological time, but they also cover a considerable range of paleolatitudes. Thus, they should record any provincialism that might have been present in the faunal assemblages. There is a hint of such provincialism in the occurrence of traversodont cynodonts in the Richmond and Deep River basins, which were located at the paleoequator during the early Late Triassic. At the same time, there exists a remarkable similarity between the tetrapod taxa from Laurasia and Gondwana at higher paleolatitudes. Perhaps these forms were not adapted to the climatic conditions in the interior of equatorial Pangaea and preferred the coastal areas in the tropics. This scenario assumes that the equatorial climate in the western interior of Pangaea was rather hot with only seasonal precipitation as predicted by the megamonsoon model (chapter 1).

CHAPTER 8

Late Triassic of the Western United States

Late Triassic continental strata are widely exposed across the western United States, especially in the Four Corners region of Arizona, Colorado, New Mexico, and Utah and in the Panhandle of Texas (figure 8.1). Over the years geologists have applied a plethora of regional lithostratigraphic names to these deposits, resulting in a very complex stratigraphic nomenclature that frequently obscures continuity of deposition across this vast region. Lucas (1993) first argued that the Late Triassic continental strata in the western United States represent but a single large depositional basin. This idea received support from the discovery of detrital zircon crystals of identical radiometric age in strata from northwestern Texas to western Nevada, indicating the existence of a major river system extending across the region during the Late Triassic (Riggs et al. 1996). An additional argument in support of continuity of deposition across a single large basin is the similarity in faunal composition between tetrapod assemblages from widely separate localities (Lucas 1993).

The Chinle-Dockum basin extended at least from northern Wyoming in the north to southwestern Texas in the south and from southeastern Nevada in the west to northwestern Oklahoma in the east, covering an area of some 2.3 million square kilometers (Lucas and Huber 2003). It was situated inland of a magmatic arc that was associated with the subduction zone off the West Coast of North America (Dickinson et al. 1983). The Late Triassic continental strata of the American Southwest have traditionally been divided into the Chinle Formation (or Group) on the Colorado Plateau and the Dockum Group (or Formation) of eastern New Mexico and northwestern Texas, the depositional settings of which were thought to be separated along the trend of the Rio Grande rift. Even where strata from the Chinle and Dockum are obviously correlative, most geologists have retained this nomenclatural separation.

The Chinle Formation has a maximum thickness of about 550 meters and comprises mostly red siliciclastic strata, although some horizons are white, blue, lavender, yellow, greenish, and gray. (Lucas [1993] elevated the Chinle Formation to the status of Group, but this usage has not yet been generally adopted.) The Chinle Formation unconformably overlies the Moenkopi Formation and is divided into up to seven members in northeastern Arizona: (from oldest to youngest) Shinarump, Cameron, Blue Mesa, Sonsela, Petrified Forest, Owl Rock, and Rock Point (figure 8.2). (In many locations, the unit between the Shinarump and

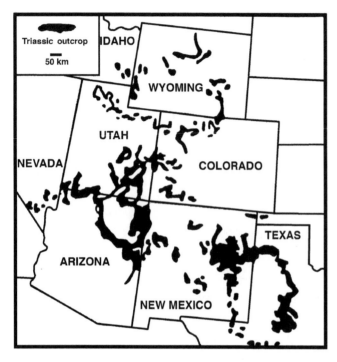

Figure 8.1. Distribution of exposed strata of the Chinle Formation and Dockum Group in the western United States. (From Lucas and Huber 2003)

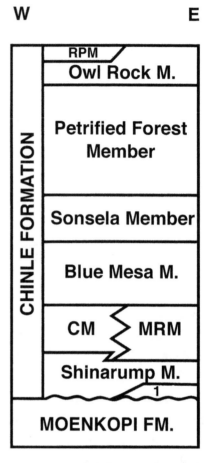

Figure 8.2. Succession of the members of the Chinle Formation across Arizona (from west to east). *1*, unit known as "Mottled Strata"; *CM*, Cameron Member; *MRM*, Mesa Redondo Member; *RPM*, Rock Point Member. (Modified from Parker 2005)

Blue Mesa members is referred to as the Bluewater Creek Member or Mesa Redondo Member.) Except in the youngest units, the mudstones of the Chinle are typically rich in bentonite, which formed by diagenetic alteration of volcanic ashes introduced from adjoining highlands. Sandstones typically represent fluvial channel deposits and vary considerably in maturity. Clasts in the conglomerates were either transported into the depositional basin from adjoining regions (Paleozoic limestones), had their origin within the basin (ripped-up calcretes and mudstones), or represent a mixture of material from both sources (Lucas and Huber 2003). Lacustrine deposits frequently include carbonates. The strata of the Chinle mainly comprise fluvial sediments but also include lacustrine and eolian deposits. These rocks provide the palette of dazzling colors in the Painted Desert of Arizona (figure 8.3) and other landscapes in the American Southwest.

Lehman and Chatterjee (2005) distinguished two major upward-fining fluviolacustrine depositional sequences within the Dockum Group of northwestern Texas. The lower sequence comprises the Santa Rosa Sandstone and the Tecovas Formation. Separated by a major unconformity (Tr-4), the much thicker and more widespread upper sequence consists of the Trujillo Sandstone and the Cooper Canyon Formation. The Santa Rosa Sandstone and Trujillo Sandstone comprise coarse-grained clastics representing channel-related facies, whereas the Tecovas and Cooper Canyon formations are character-

Figure 8.3. Exposures of the Petrified Forest Member of the Chinle Formation in the Painted Desert, viewed from the outlook at Chinde Point, Petrified Forest National Park, Arizona. (Photograph by H-DS)

ized by fine-grained clastics representing overbank facies, with some lacustrine deposits nested within that facies.

Lucas (1993, 1997; Lucas and Huber 2003) has presented comprehensive syntheses of the lithostratigraphy of the Chinle and Dockum. He distinguished a series of regionally extensive stratigraphic intervals on the Colorado Plateau, which he designated as A, B, C′, C, and D in ascending order. (Gregory [1917] first proposed a four-part division of the Chinle For-

mation but labeled the intervals A–D in descending order, excluding the Shinarump conglomerate.) When combined with the stratigraphic scheme for the Chinle Formation on the Colorado Plateau, Lucas's interval A corresponds to the Shinarump Member and its correlatives; B to the Blue Mesa Member and its correlatives; C′ to the Sonsela Member and its correlatives; C to the Petrified Forest and Owl Rock members and their correlatives; and D to the Rock Point Member and its correlatives.

Lucas used two major unconformities (known as Tr-4 and Tr-5) to distinguish three principal depositional sequences within the Chinle Formation. The oldest sequence is the Shinarump–Blue Mesa sequence, which Lucas considered late Carnian in age. However, new U-Pb dating of single zircon crystals from a tuffaceous sandstone at the base of the Blue Mesa Member in western New Mexico have yielded a weighted mean $^{206}Pb/^{238}U$ age of 219.2±0.7 Ma, which is considered a maximum age because the horizon does not represent an original ashfall and places the unit well within the Norian stage (Mundil and Irmis 2008). With the possible exception of the Shinarump Member, it would appear now that the Chinle Formation comprises only strata of Norian and Rhaetian age (see also Zeigler and Geissman 2008). The depositional succession of the Shinarump–Blue Mesa sequence begins with quartzose sandstones and conglomerates (stratigraphic interval A) that unconformably rest on older strata including the Moenkopi Formation. The overlying lower part of stratigraphic interval B comprises variegated mudstones, sandstones, and some limestones. It, in turn, is followed by mudstone-dominated deposits (upper portion of stratigraphic interval B) with extensive modification due to soil formation.

The Tr-4 unconformity separates the second sequence of the Chinle Formation, the Sonsela–Owl Rock sequence, from the underlying Blue Mesa Member. The beginning of the deposition of the Sonsela–Owl Rock sequence is marked by widespread intrabasinal conglomeratic sandstones (stratigraphic interval C′) that unconformably overlie older Chinle deposits. Interval C′ is followed by redbeds of fluvial and floodplain origin, which make up most of stratigraphic interval C. These redbeds, in turn, are overlain by carbonate-siltstone strata forming the Owl Rock Member, which is restricted to the region of the Colorado Plateau.

The third or upper sequence of the Chinle Formation is the Rock Point sequence, which Lucas considered late Norian to Rhaetian in age. The Tr-5 unconformity marks its base. The Rock Point Member comprises repetitive strata of siltstone, sandstone, and some carbonates. It has long been assumed that a major unconformity separates the Rock Point sequence from the overlying Lower Jurassic Glen Canyon Group (Pipiringos and O'Sullivan 1978). Lucas and Tanner (2007b) suggested that the Triassic-Jurassic transition is present in the sedimentary succession on the Colorado Plateau. Furthermore, based on new geological mapping and paleomagnetic data, Zeigler and Geissman (2008) argued that the uppermost Chinle Formation in northern New Mexico is Rhaetian to possibly even earliest Jurassic (earliest Hettangian) in age.

The Chinle Formation and Dockum Group preserve a remarkably rich record of Late Triassic continental life, including plants, invertebrates, fishes, and tetrapods. Although paleontologists have amassed vast quantities of fossils since the late nineteenth century, fieldwork continues to yield new, often unexpected finds (e.g., Irmis et al. 2007). Indeed, the prospect for discoveries of Late Triassic animals and plants in Chinle and Dockum strata appears virtually inexhaustible, and space confines us here to reviewing a selection of particularly well-known occurrences.

In general terms, the deposition of the Chinle and Dockum strata occurred at paleolatitudes of about 5° to 15° north of the paleoequator (Cleveland, Atchley, and Nordt 2007). Using isotopic data from pedogenic carbonates, Prochnow et al. (2006) inferred more humid climatic conditions in the lower portion of the Chinle, followed by a drying trend in the upper part. Cleveland et al. (2008) also provided evidence in support of a significant increase in mean annual temperature and aridity in the upper part of the Chinle Formation in New Mexico. A typical

Chinle depositional setting comprises floodplains and braided streams, which is reflected by the abundance of aquatic and semiaquatic tetrapods, especially metoposaurs and phytosaurs (Long and Murry 1995). Rich and varied vegetation thrived along bodies of water. Countless petrified trunks of conifers (mostly belonging to the invalid form taxon *Araucarioxylon*) attest to the occurrence of stands or even forests of tall trees. The wood of *"Araucarioxylon"* displays characteristic features of extant Araucariaceae, and mature trees could attain heights of almost 60 meters and a trunk diameter of up to 3 meters (Ash and Creber 2000). The bark on *"Araucarioxylon"* trunks was thin, and the tree had a massive root system. Ash (1986) noted that the abundance of ferns, equisetaleans, and lycopsids at many sites attests to at least locally humid conditions. Demko, Dubiel, and Parrish (1998) interpreted such occurrences as strata deposited within incised valley systems during wetter intervals. Frequently, cycads, bennettitaleans, and ginkgoes are found along with these plants, but they are thought to have thrived in adjacent, somewhat drier habitats. More dry-adapted (xeric) plants like certain conifers likely flourished in other areas, and the fragmentary preservation and less common occurrence of their fossil remains are consistent with their introduction into the depositional setting from more remote locations (Ash 1986).

The rich fossil record of the Chinle and Dockum strata, especially pollen and spores as well as tetrapods, has been extensively used for biostratigraphic correlation with other Late Triassic continental sequences (Litwin, Ash, and Traverse 1991; Lucas 1993, 1998, 1999; Lucas and Huber 2003). Although no direct lateral contacts can be established between the Chinle Formation and the Dockum Group to the east and extensive Triassic marine strata to the west, Lucas and Marzolf (1993) and Lucas and Huber (2003) have developed a sequence-stratigraphic model to facilitate correlation between units of the Chinle Formation and the marine sequences of the upper portion of the Star Peak Group and the Auld Lang Syne Group in northwestern Nevada.

As previously discussed, the Chinle Formation can be divided into three depositional sequences that are separated by major unconformities. At the base of the Shinarump–Blue Mesa and the Sonsela–Owl Rock sequences, respectively, are conglomeratic sandstones, which were deposited in a vast basin characterized by extensive channel formation and subaerial exposure during intervals when no deposition occurred (Lucas and Huber 2003). These sandstones are overlain by deposits of fluviolacustrine origin. Each sequence is topped by carbonates and siltstones that were deposited in marshy settings and show signs of subaerial exposure prior to the deposition of the overlying sequence. Lucas and Marzolf (1993) and Lucas and Huber (2003) interpreted the basal sandstones as low-stand systems tracts where deposition took place at the beginning of a marine transgression. They considered the overlying fluviolacustrine clastic deposits transgressive systems tracts, which were followed by carbonates and siltstones that represent high-stand systems tracts. Lucas and his colleagues then identified comparable systems tracts in the marine strata of northwestern Nevada and attempted to correlate them with the continental Chinle deposits. This effort provides direct ties into the standard zonation of the Triassic by means of marine invertebrates, primarily ammonoid cephalopods (Tozer 1984). Only Chinle stratigraphic interval D appears to lack a marine equivalent, and thus it cannot be reliably dated.

Lucas and Hunt (1993) proposed a biostratigraphic zonation of the Chinle Formation with four LVFs based on the respective first and last occurrences of specific tetrapod taxa, especially metoposaurs, phytosaurs, and aetosaurs (Lucas 1998; Lucas and Huber 2003). Subsequent studies have challenged the utility of various proposed index taxa and demonstrated some overlap between the stratigraphic ranges of others (e.g., Irmis 2005; Parker 2005, 2006; Rayfield et al. 2005; Rayfield, Barrett, and Milner 2009).

By contrast, Irmis (2005) recognized two principal tetrapod assemblages in the Chinle Formation of Arizona—a lower assemblage in the Mesa Redondo and Blue Mesa members and an upper one in the Petrified Forest and Owl Rock members, with faunal change occurring within the Sonsela Member. He characterized the lower assemblage by the presence of the metoposaur *Koskinonodon*, the dicynodont *Placerias*, the phytosaur *Leptosuchus*, the aetosaurs *Acaenasuchus*, *Desmatosuchus haplocerus*, and *Stagonolepis*, and the "rauisuchian" *Poposaurus*. The upper assemblage includes the metoposaur *Apachesaurus*, the phytosaur *Pseudopalatus*, the crurotarsan *Revueltosaurus*, and the aetosaurs *Desmatosuchus smalli*, *Rioarribasuchus*, and *Typothorax*.

In the rest of this chapter we examine the four basic stratigraphic intervals of the Chinle (and their correlates in the Dockum) in turn, with special reference to major localities. Where appropriate we discuss the possible utility of tetrapod taxa as biostratigraphic markers.

CHINLE STRATIGRAPHIC INTERVAL A

Lucas and Huber (2003) regarded Chinle stratigraphic interval A as corresponding to the Otischalkian LVF, which Lucas (1998) characterized by the presence of *Metoposaurus*, the phytosaurs *Parasuchus* (formerly known under its subjective junior synonym *Paleorhinus*) and *Angistorhinus* (=?*Rutiodon*), and the aetosaur *Longosuchus*. As an additional index fossil, Lucas proposed the archosauriform *Doswellia*. This unusual taxon was first described from the Richmond basin

of the Newark Supergroup in Virginia (Weems 1980; Dilkes and Sues 2009) and was later reported from Texas (Long and Murry 1995). The Otischalkian LVF is based on a tetrapod assemblage from the Dockum Group at sites just north of the abandoned community of Otis Chalk, near Big Springs in Howard County, Texas. Lucas (1998) considered the Otischalkian late Carnian in age, which is consistent with previously reported palynological information (Litwin, Ash, and Traverse 1991).

Hunt (1993) considered *Metoposaurus bakeri* distinct from *Koskinonodon* (under the preoccupied name *Buettneria*), but Sulej (2002) referred it to the latter genus.

Phytosaurs referred to *Parasuchus* have a much greater stratigraphic range than previously assumed (Parker 2005). It is important to emphasize that *Parasuchus* represents a grade of basal phytosaurs that share the plesiomorphic position of the external nares in front of the antorbital fenestrae, and the phylogenetic relationships of the various taxa assigned to "*Parasuchus*" remain yet to be determined. Basal phytosaurs referred to "*Parasuchus*" also share dorsoventrally flattened skulls with largely dorsally facing orbits and slender snouts (chapter 5). By contrast, *Angistorhinus* has a deeper skull with the external nares situated level with the antorbital fenestrae, as is the case in all other phytosaurs more derived than those assigned to "*Parasuchus*."

The aetosaur *Longosuchus* reached a length of about 3 meters. Its dorsal dermal armor is distinguished by lateral osteoderms that form distinct, spikelike structures throughout the neck, trunk, and anterior tail region (Sawin 1947; Hunt and Lucas 1990).

According to Lucas (1998), the Otischalkian LVF comprises the oldest Late Triassic tetrapod assemblages in the American West. The Popo Agie Formation of Wyoming has yielded a very similar assemblage, with the metoposaurid *Koskinonodon*, the phytosaurs

"*Parasuchus*" and *Angistorhinus*, the aetosaur *Desmatosuchus*, the "rauisuchian" *Poposaurus*, a hyperodapedontine rhynchosaur, and the dicynodont *Placerias* (Lucas 1994).

CHINLE STRATIGRAPHIC INTERVAL B

Lucas and Huber (2003) regarded Chinle stratigraphic interval B as corresponding to the Adamanian LVF, which Lucas (1998) characterized by the occurrence of the phytosaurs *Leptosuchus* and *Smilosuchus* (figures 8.4 and 8.5) and the aetosaurs *Desmatosuchus* and *Stagonolepis*. Considered latest Carnian in age by Lucas, this interval has since been radiometrically dated as Norian (Mundil and Irmis 2008). The type tetrapod assemblage of the Adamanian is from the Blue Mesa Member of the Chinle Formation in Petrified Forest National Park, near a defunct railroad siding at Adamana, Arizona. Tetrapod-bearing strata and localities assigned to

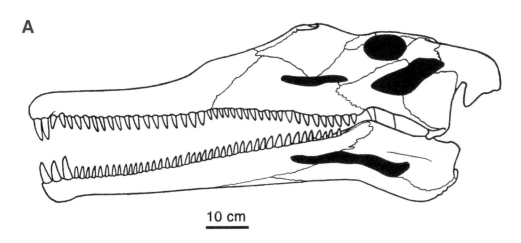

10 cm

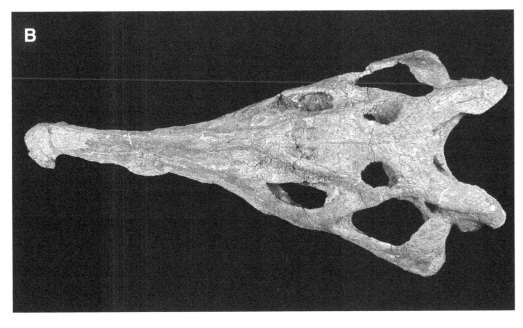

Figure 8.4. (*A*) Lateral view of the reconstructed skull of the phytosaur *Leptosuchus adamanensis*. (Modified from Camp 1930) (*B*) Dorsal view of a skull of the phytosaur *Pseudopalatus* sp. (Petrified Forest National Park, PEFO 31219). Length 1.09 meters. (Courtesy of National Park Service)

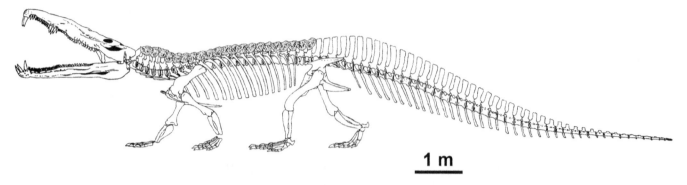

1 m

Figure 8.5.   Composite reconstruction of the skeleton of a large phytosaur in lateral view, based on the skull of *Smilosuchus* and postcranial bones of *Leptosuchus* and related taxa. (From Long and Murry 1995)

this faunachron by Lucas (1998) occur widely across the American Southwest and include the famous *Placerias* Quarry in Arizona. In addition to the index taxa, Lucas's Adamanian tetrapod assemblages include the metoposaurs *Apachesaurus* and *Koskinonodon*, the aetosaur *Paratypothorax*, the "rauisuchian" *Postosuchus*, the crocodylomorph *Hesperosuchus*, and the dicynodont *Placerias*. Langer (2005) and Rayfield and colleagues (Rayfield et al. 2005; Rayfield, Barrett, and Milner 2009) have pointed out that *Angistorhinus*, *Koskinonodon*, "*Parasuchus*," *Placerias*, and *Stagonolepis* all occur in both the Otischalkian and Adamanian LVFs and that the two LVFs cannot be clearly distinguished from each other.

The *Placerias* Quarry is located in the Mesa Redondo Member, near Romero Springs, about 10.4 kilometers northwest of St. Johns, Arizona. It has yielded one of the richest, most diverse assemblages of Late Triassic continental vertebrates anywhere. To date, some 80 taxa have been recorded (Kaye and Padian 1994; Long and Murry 1995; Irmis 2005; Parker 2005). The name of the locality refers an accumulation of more than 1,600 mostly disarticulated and commingled bones representing at least 39 individuals of the large kannemeyeriid dicynodont *Place-*

rias (Camp and Welles 1956; Cox 1965). The skull of *Placerias* is proportionately enormous. Its maxillae bear prominent, bladelike processes, which contain small, presumably no longer functional canines. *Placerias* is rare elsewhere in the Chinle Formation.

The *Placerias* Quarry probably represents an accumulation of tetrapod bones on a floodplain with a high water table (Fiorillo, Padian, and Musikasinthorn 2000) rather than a marsh setting as originally assumed by Camp and Welles (1956). The bones occur mostly in mudstones and a layer with pedogenic carbonate nodules. They show few signs of postmortem damage. In addition to *Placerias* itself, the quarry has yielded skeletal remains of three taxa of "rauisuchian" crurotarsans and up to four taxa of aetosaurs (Parker 2005). Other noteworthy finds include limb bones of a coelophysoid theropod dinosaur and indeterminate dinosauriforms (Nesbitt, Irmis, and Parker 2007). More terrestrial animals dominate the tetrapod assemblage, and phytosaurs are rare. Screen washing of matrix from the *Placerias* Quarry has yielded a wealth of skeletal remains of small temnospondyls, reptiles, and possibly cynodonts (Kaye and Padian 1994).

In Petrified Forest National Park in eastern Arizona, the Blue Mesa Member is richly fossiliferous above the Newspaper Rock bed, yielding abundant petrified wood ("*Araucarioxylon*"), occurring as both logs and stumps preserved in place, and other plant, vertebrate, and invertebrate fossils (Parker 2006). In addition to conifers, the plant assemblage comprises a great diversity of ferns (*Cladophlebis, Clathropteris, Phlebopteris, Todites, Wingatea*), the enigmatic gymnosperm *Dinophyton*, equisetaleans, the ginkgoalean *Baiera*, and the bennettitalean *Zamites*. Its composition corresponds to Ash's (1980) *Dinophyton* floral zone. Both the metoposaurid *Koskinonodon* and the phytosaur *Leptosuchus* ("*Machaeroprosopus*") are represented by abundant skeletal remains (Camp 1930; Parker 2006). Additional tetrapods include a dinosauriform and the aetosaurs *Desmatosuchus* and *Typothorax*.

The Tecovas Formation of the Dockum Group in northwestern Texas has yielded a similar tetrapod assemblage, which includes two additional noteworthy faunal elements. *Adelobasileus* is known from an isolated braincase and has been interpreted as the oldest known mammaliaform (Lucas and Luo 1993). This interpretation is plausible, but additional remains of this intriguing taxon are needed to resolve its phylogenetic position with more confidence. *Adelobasileus* may prove to be closely related to the mammal-like probainognathian cynodonts *Brasilodon* and *Brasilitherium* from the Caturrita Formation of southern Brazil (chapter 4).

The distinctive archosauromorph reptile *Trilophosaurus* is best known from a large sample of skeletal remains from two localities near Otis Chalk (figure 8.6). Gregory (1945) has documented its skeletal structure in detail. *Trilophosaurus* is characterized by the possession of transversely broad maxillary and dentary teeth with three transversely aligned cusps, indicating a herbivorous diet. Its premaxillae and anterior portions of the dentaries are devoid of teeth and may have been covered by a keratinous beak in life. The skull of *Trilophosaurus* has only large upper temporal fenes-

trae, which are separated by a tall parietal crest; a deep, solid bony bar forms the cheek region (figure 8.6B). The limbs are long and slender, and the unguals are compressed mediolaterally. Additional species of this taxon have been described from the *Placerias* Quarry (Murry 1987) and the Sonsela Member in Petrified Forest National Park (Mueller and Parker 2006).

CHINLE STRATIGRAPHIC INTERVALS C' AND C

Lucas and Huber (2003) regarded Chinle stratigraphic intervals C' and C as corresponding to the Revueltian LVF, which Lucas (1998) and Hunt (2001) characterized by the presence of the phytosaur *Pseudopalatus* and the aetosaur *Typothorax*. However, recent discoveries have established that *Typothorax* had a much longer stratigraphic range than previously assumed (Parker 2007). The type vertebrate assemblage of the Revueltian LVF comes from the Bull Canyon Formation in Guadalupe and Quay counties, east-central New Mexico.

The Sonsela Member unconformably overlies the Blue Mesa Member and includes the famous Rainbow and Crystal forests in Petrified Forest National Park, both of which preserve vast accumulations of logs of petrified wood. Its strata have yielded phytosaurs (*Leptosuchus* and *Pseudopalatus*; figure 8.4) and aetosaurs (*Paratypothorax, Stagonolepis,* and *Typothorax*; Parker 2006; Parker and Irmis 2006).

In addition to *Pseudopalatus* and *Typothorax*, the tetrapod assemblage from the Bull Canyon Formation includes the aetosaurs *Desmatosuchus* and *Paratypothorax*, the "rauisuchians" *Postosuchus* and *Shuvosaurus*, the enigmatic archosaur *Revueltosaurus*, and the thin-shelled basal turtle *Chinlechelys* (Long and Murry 1995; Lucas 1998; Joyce et al. 2009).

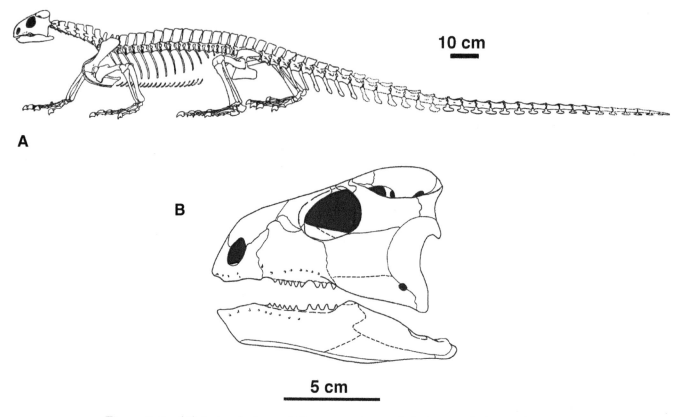

10 cm

A

B

5 cm

Figure 8.6. (*A*) Lateral view of the reconstructed skeleton of the archosauro-morph *Trilophosaurus buettneri*. (From Gregory 1945) (*B*) Lateral view of the re-constructed skull of *Trilophosaurus buettneri*. (Combined from Gregory 1945 and Parks 1969)

Hunt (1989) first named *Revueltosaurus* on the basis of isolated teeth with denticulated cutting edges, which resemble those of basal ornithischian dinosaurs such as *Lesothosaurus*. Thus, he plausibly interpreted these teeth as representing a basal ornithischian. Hunt and Lucas (1994) further argued in support of ornithischian affinities for *Revueltosaurus* and also named additional taxa based on similar isolated teeth. Parker et al. (2005) reported the discovery of hundreds of cranial and postcranial bones representing at least six individuals of *Revueltosaurus* from the middle portion of the Petrified Forest Member in Petrified Forest National Park, Arizona. These remains establish that *Revueltosaurus* does not represent an ornithischian but a previously unrec-

ognized lineage of crurotarsan archosaurs. It has a crocodile-like ankle joint and dorsal armor composed of subrectangular osteoderms that have anterior bars and bear sculpturing composed of more or less circular pits, which are arranged randomly or in a weakly radial pattern. Although assessment of the phylogenetic position of *Revueltosaurus* must await detailed description of the new material, various features indicate a close relationship to the Stagonolepididae.

Sankar Chatterjee (Texas Tech University) and his crews have recovered a remarkable vertebrate assemblage from a massive red mudstone in the correlative Cooper Canyon Formation of the Dockum Group at a locality 14.4 kilometers southeast of Post in Garza County, Texas. Aquatic and semiaquatic animals dominate the assemblage at this locality, known as the Post Quarry, and include a variety of fishes, the metoposaurid *Apachesaurus*, the small temnospondyl *Rileymillerus* (Bolt and Chatterjee 2000), and the phytosaur *Pseudopalatus* (Long and Murry 1995). The Post Quarry also yielded skeletal remains of a considerable diversity of terrestrial crurotarsans, including the aetosaurs *Desmatosuchus, Paratypothorax,* and *Typothorax* (Small 1989, 2002) and the "rauisuchians" *Poposaurus* (Weinbaum and Hungerbühler 2007), *Postosuchus* (Chatterjee 1985), and *Shuvosaurus* (Chatterjee 1993; Long and Murry 1995; Nesbitt 2007). Dinosauriform archosaurs, known only from fragmentary remains, comprise a coelophysoid theropod, a basal saurischian similar to *Staurikosaurus*, and a more basal dinosauriform (Nesbitt and Chatterjee 2008).

Chatterjee (1991, 1999) described what he considered the oldest known bird, *Protoavis*, based on two incomplete skeletons from the Post Quarry as well as referred isolated bones from a site known as the Kirkpatrick Quarry in the underlying Tecovas Formation. This discovery caused a worldwide sensation because it predated the oldest known undisputed bird, *Archaeopteryx*, from the Upper Jurassic (Tithonian) Solnhofen Formation of Bavaria (Germany), by some 70 million years. Most authors have not accepted Chatterjee's claims and argued that *Protoavis* is based on a mixture of bones from a variety of tetrapod taxa. The braincase, femur, and astragalocalcaneum referred to *Protoavis* probably belong to a theropod (Nesbitt, Irmis, and Parker 2007). Furthermore, Renesto (2000) noted a close similarity between the cervical vertebrae assigned to *Protoavis* and those of drepanosaurid archosauromorphs (chapter 5). However, a comprehensive revision of all the material assigned by Chatterjee to *Protoavis* has yet to be undertaken.

Desmatosuchus differs from other aetosaurs in the presence of long, recurved bony spikes projecting

from the cervical lateral osteoderms (Case 1922; Long and Murry 1995; Parker 2008; figure 8.7). The sculpturing on its paramedian osteoderms comprises an irregular pattern of grooves, pits, and ridges. *Desmatosuchus* reached a length of at least 4 meters. Also known from the Löwenstein Formation of southern Germany (chapter 5), *Paratypothorax* attained a length of 3 meters and has an unusually wide trunk region covered by straplike median dorsal osteoderms, which bear sculpturing composed of ridges radiating from a posteromedial boss (Long and Ballew 1985; Long and Murry 1995). Finally, *Typothorax* reached a length of up to 4 meters. Its trunk is wide, but its paramedian dorsal osteoderms differ from those of *Paratypothorax* in their distinctly pitted ornamentation and other features (Long and Murry 1995).

Postosuchus was the apex predator in the Post Quarry and many other Late Triassic tetrapod assemblages from North America (Chatterjee 1985; Long and Murry 1995; Peyer et al. 2008; figure 8.8). It reached a length of at least 5 meters. The skull of *Postosuchus* is large, narrow, and deep (figure 8.8*B*). Its neck and trunk are relatively short. The tail is long and mediolaterally flattened. A noteworthy feature of the pelvic girdle is the long pubis, which terminates in a bootlike process distally. The overall structure of the pelvis suggests that *Postosuchus* had a more upright posture than most other crurotarsan archosaurs (Long and Murry 1995). Although Chatterjee (1985) reconstructed *Postosuchus* with a bipedal posture, it has a robust pectoral girdle and its limb proportions are comparable to those in other "rauisuchian" crurotarsans.

Nesbitt and Norell (2006) demonstrated that *Shuvosaurus*, initially interpreted as the oldest known ornithomimosaurian dinosaur (Chatterjee 1993), and the unusual "rauisuchian" *Chatterjeea* (Long and Murry 1995) are, in fact, based on the skull and postcranial

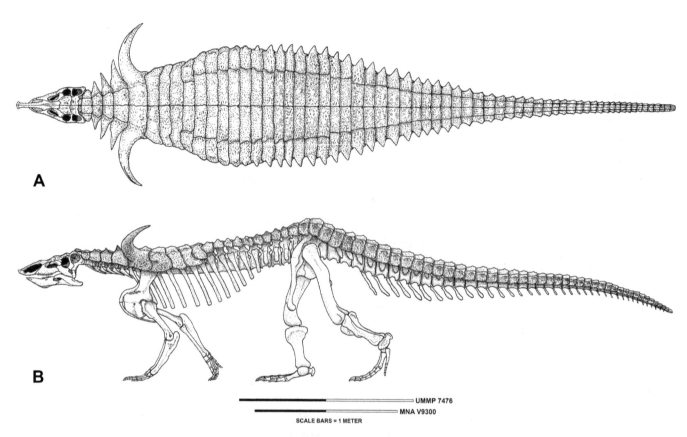

Figure 8.7. Reconstructed skeleton and dorsal dermal armor of the aetosaur *Desmatosuchus spurensis* in (*A*) dorsal and (*B*) lateral views. Scale bars are based on two different-size specimens (Museum of Northern Arizona, MNAV9300, and University of Michigan Museum of Paleontology, UMMP 7476). (Reconstruction by J. W. Martz from Parker 2008; courtesy of W. G. Parker)

skeleton, respectively, of the same animal—a bipedal, superficially rather dinosaur-like "rauisuchian." *Shuvosaurus* attained a length of about 2 meters (Nesbitt 2007). Its snout lacks teeth and may have been covered by a keratinous beak in life. The postcranial skeleton of *Shuvosaurus* is lightly built. Its neck is long and slender. The shoulder girdle is unusually robust, with a large, platelike coracoid, and the forelimbs of *Shuvosaurus* are relatively long. Its long, slender pubis terminates in a prominent distal "foot." The pes bears large, mediolaterally compressed unguals. The habits of *Shuvosaurus* must have been very different from those of its large-headed, predatory relatives such as *Postosuchus*.

Chatterjee (1984) described a new ornithischian dinosaur, *Technosaurus*, based on jaw fragments and postcranial bones from the Post Quarry. Nesbitt, Irmis, and Parker (2007) argued that this material represents a mixture comprising a jaw fragment referable to *Shuvosaurus*, indeterminate postcranial bones, and jaw elements of an indeterminate archosauriform.

Finally, Chatterjee (1983) referred a tooth-bearing jaw fragment from the Post Quarry to the tritheledontid cynodont *Pachygenelus*, which is otherwise known only from the Lower Jurassic of South Africa and Nova Scotia (Shubin et al. 1991). Shubin et al. (1991) argued that the specimen does not represent a cynodont, but its affinities remain to be resolved.

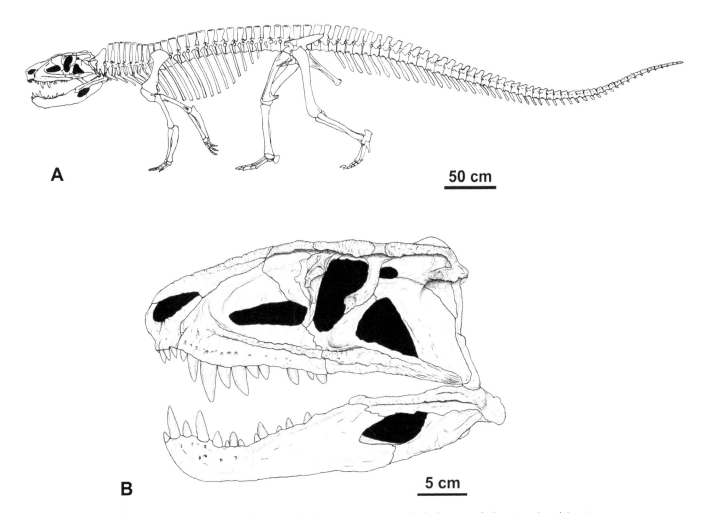

A   **50 cm**

B   **5 cm**

Figure 8.8.   (*A*) Lateral view of the reconstructed skeleton of the "rauisuchian" *Postosuchus kirkpatricki*. (From Long and Murry 1995) (*B*) Lateral view of the reconstructed skull of *Postosuchus kirkpatricki*. (Drawing courtesy and copyright of J. C. Weinbaum)

Another noteworthy tetrapod assemblage comes from the Snyder Quarry in the Petrified Forest Member of the Chinle Formation near Ghost Ranch in Rio Arriba County, north-central Mexico. It includes the phytosaur *Pseudopalatus* (Zeigler, Heckert, and Lucas 2003), the aetosaurs *Rioarribasuchus* (Lucas, Hunt, and Spielmann 2006 = *Heliocanthus* Parker 2007) and *Typothorax*, and a coelophysoid theropod (Nesbitt, Irmis, and Parker 2007). *Pseudopalatus* is represented by skeletal remains of mostly subadult individuals, including 11 skulls, which

range in length from 30 centimeters to about 1 meter. Zeigler, Heckert, and Lucas (2003) presented compelling evidence that this fossil occurrence formed as the result of a catastrophic event. Only some of the remains are still preserved in articulation, suggesting postmortem transport by water. What is unusual is the occurrence of a significant amount of charcoal within the sediments, pointing to a wildfire during which the animals may have perished. Subsequently, heavy rains caused flash floods that swept through the burned areas and eventually deposited a mixture of animal remains, charcoal, and sediment.

Irmis et al. (2007) reported on a new locality with tetrapod remains, now known as the Hayden Quarry, in the Petrified Forest Member at Ghost Ranch. The assemblage includes *Pseudopalatus*, *Rioarribasuchus*, and *Typothorax*, but also a dinosauriform, the basal saurischian *Chindesaurus*, and a coelophysoid theropod. Most noteworthy is the occurrence of the small dinosauromorph *Dromomeron*, which is most closely related to *Lagerpeton* from the Ladinian-age Chañares Formation of northwestern Argentina (chapter 4) and demonstrates the coexistence of basal dinosauromorphs and dinosaurs into the Late Triassic.

CHINLE STRATIGRAPHIC INTERVAL D

Lucas and Huber (2003) regarded Chinle stratigraphic interval D as corresponding to the Apachean LVF, which Lucas (1998) considered late Norian to Rhaetian in age (Lucas and Tanner 2007b) and characterized by the occurrence of the phytosaur *Redondasaurus* and the aetosaur *Redondasuchus*. The type tetrapod assemblage for this faunachron comes from the Redonda Formation in Guadalupe and Quay counties, east-central New Mexico. *Redondasaurus* is closely related to *Pseudopalatus*, and Long and Murry (1995) even synonymized it with the latter (but see Hungerbühler 2002). *Redondasuchus* is probably synonymous with *Typothorax* (Long and Murry 1995; Parker 2007; Parker, Stocker, and Irmis 2008).

The best-known example of a tetrapod assemblage from Chinle stratigraphic interval D comes from the *Coelophysis* Quarry, which is located in the "Siltstone Member" of the Chinle Formation at Ghost Ranch. To date, this locality has yielded skeletal remains of at least 1,000 individuals, comprising all age classes, of the theropod dinosaur *Coelophysis* (Colbert 1989; Schwartz and Gillette 1994). Cope (1887a, 1887b, 1889) first described *Coelophysis* based on fragmentary postcranial bones apparently collected in the vicinity of Ghost Ranch. In 1947, a field crew from the American Museum of Natural History led by Edwin H. Colbert discovered the vast accumulation of well-preserved skeletal remains of *Coelophysis* at Ghost Ranch. Colbert and his crews continued to work the site in 1948, 1949, 1951, and 1953. In 1981 and 1982, a joint team from the Carnegie Museum of Natural History (Pittsburgh), the Peabody Museum of Natural History at Yale University, and the Museum of Northern Arizona (Flagstaff) reopened the *Coelophysis* Quarry and recovered much additional material. So densely packed together are the skeletons of *Coelophysis* that the remains can be recovered only as large blocks of bone-bearing siltstone for more detailed preparation in the laboratory. Many of the specimens represent articulated or partially articulated skeletons. The remains show few signs of scavenging or other postmortem damage, indicating rapid burial of the animals. It appears likely that large flocks of these dinosaurs perished in sudden flood events. Various authors have interpreted the occurrence of a gracile and a robust morph as evidence for sexual dimorphism in *Coelophysis*. Two of the larger skeletons of *Coelophysis* collected by Colbert's teams preserve reptilian bones in their body cavities, which has led to speculation that these were the remains of cannibalized juveniles (Colbert 1989) or even that *Coelophysis* gave birth to live young (Fraser 2006). However, Nesbitt et al. (2006) demonstrated that the identifiable bones among these remains belong, in fact, to a crocodylomorph and probably represent gut contents.

A lightly built form, *Coelophysis* attained a length of up to 3 meters (Colbert 1989; figure 8.9). Its skull is long, narrow, and low (figure 8.9*B*). The length of the antorbital fenestra exceeds 25 percent of total skull length. A distinct gap separates the premaxillary and maxillary teeth below the external naris. The slender neck of *Coelophysis* is longer than the trunk. The tail is very long. The gracile forelimb of *Coelophysis* has a manus with three functional digits and a reduced fourth digit comprising only the metacarpal and a single phalanx. The tibia and fibula are slightly longer than the femur. The pes is functionally tridactyl; its third digit is the longest, the first

digit is reduced in length, and the metatarsal of the fifth digit is a mere splint of bone.

Most tetrapod remains recovered from the *Coelophysis* Quarry are referable to *Coelophysis*, but a few specimens document the presence of other reptilian taxa. An additional taxon of saurischian dinosaurs is currently under study. Crurotarsan archosaurs include the sphenosuchian crocodylomorph *Hesperosuchus* (Clark et al. 2001), the "rauisuchian" *Effigia* (Nesbitt 2007), and the phytosaur *Redondasaurus* (Hunt and Lucas 1993; Hungerbühler 2002). *Effigia* is closely related to *Shuvosaurus* (Nesbitt 2007; figure 8.10). Archosauromorph reptiles are represented

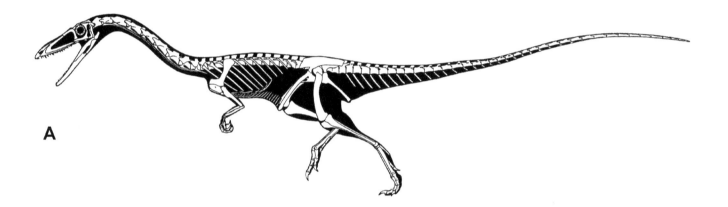

Figure 8.9. (*A*) Lateral view of the reconstructed skeleton of the coelophysoid theropod *Coelophysis bauri*. (Drawing courtesy and copyright of G. S. Paul) (*B*) Lateral view of the skull of an adult specimen of *Coelophysis bauri* (Carnegie Museum of Natural History, CM 31374). (Photograph courtesy of R. R. Reisz)

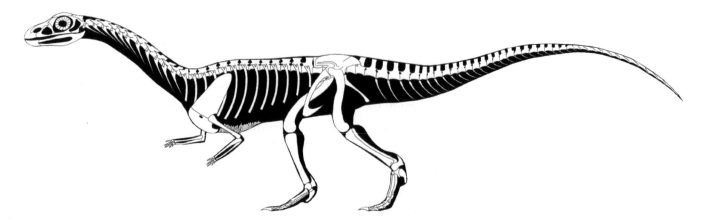

Figure 8.10. Lateral view of the reconstructed skeleton of the "rauisuchian" crurotarsan *Effigia okeeffeae*. (Drawing courtesy of S. J. Nesbitt)

by drepanosaurids (Harris and Downs 2002; Nesbitt, personal communication, 2008) and the enigmatic, armored archosauriform *Vancleavea* (Hunt et al. 2002). Heckert et al. (2008) reported a sphenodontian, *Whitakersaurus*, from the *Coelophysis* Quarry. In addition to tetrapods, the *Coelophysis* Quarry has also yielded remains of various fish including an unidentified actinopterygian similar to *Synorichthys* and a coelacanth (Schaeffer 1967).

Zeigler, Kelley, and Geissman (2008) have argued that the "Siltstone Member" of the Chinle Formation shares a paleo pole position with the lower portion of the Moenave Formation of the Glen Canyon Group, which is generally considered mostly Jurassic in age. Thus, Zeigler and Geissman (2008) considered the *Coelophysis* Quarry late Rhaetian to earliest Hettangian rather than Norian in age.

Like the Newark Supergroup in eastern North America, the Chinle Formation and Dockum Group of the American Southwest preserve a succession of diverse Late Triassic terrestrial communities. Not unexpectedly, some faunal resemblances exist between the tetrapod assemblages from the two regions, but there are also numerous differences. The latter probably reflect not only differences in depositional settings but also different locations within Pangaea. As a broad generalization, the Newark Supergroup is dominated by lacustrine sequences, whereas the Chinle Formation and the Dockum Group comprise mainly fluvial deposits. However, there are also significant fluvial strata in the Newark Supergroup and well-developed lacustrine deposits in the Chinle and Dockum. The tetrapod assemblages share some taxa such as some aetosaurs, the metoposaur *Koskinonodon*, and the "rauisuchian" *Postosuchus*. Furthermore, various taxa of freshwater fishes occur in the Chinle Formation and Dockum Group as well as in the Newark Supergroup (Schaeffer 1967; Schaeffer and McDonald 1978; Huber, Lucas, and Hunt 1993b).

The American Southwest is renowned for occurrences of Late Triassic theropod dinosaurs. The most famous example is the *Coelophysis* Quarry at Ghost Ranch with its vast accumulation of skeletons of the coelophysoid theropod *Coelophysis*. The virtual absence of skeletal remains of Triassic dinosaurs from the Newark Supergroup is partially compensated for by an abundance of trackways that can be attributed to theropod dinosaurs (Baird 1957; Weems 1987). Whether *Coelophysis* or related forms produced any of these tracks cannot be determined at present, but the differences in the fossil record highlight differences in the depositional settings between the two regions. Surprisingly, no skeletal remains referable to sauropodomorph dinosaurs are known to date from the Late Triassic of either the American Southwest or the Newark Supergroup, although sauropodomorphs are common in coeval strata in the Germanic basin and Greenland (chapter 5) as well as Argentina and South Africa (chapter 4).

The Chinle Formation and Dockum Group are arguably the best-known and most heavily sampled of all the Late Triassic fossiliferous continental sequences. They contain a series of exceptionally rich and diverse animal and plant assemblages. Thus, it is even more remarkable that the prospect for new paleontological discoveries in these strata appears to be undiminished. Recent work on microvertebrate remains (Heckert 2004) has shown great promise. The continuing stream of new discoveries shows that our knowledge of terrestrial biodiversity during this time interval is still rather incomplete.

CHAPTER 9

Two Extraordinary Windows into Triassic Life

From time to time, discoveries of occurrences of exceptionally preserved fossils, known as conservation *Lagerstätten* (*Konservat-Lagerstätten*), shed new light on the past diversity of life. (The German word *Lagerstätte*—lode place or, more freely translated, mother lode—denotes a deposit of a commercially valuable substance such as ore or coal.) Due to unusual depositional conditions, these rare occurrences preserve animals and plants that are otherwise poorly documented or unknown in the fossil record, as well as, often, nonmineralized parts of organisms. Famous examples of conservation *Lagerstätten* include the Middle Cambrian Burgess Shale of British Columbia, Canada, with its diverse communities of soft-bodied metazoans, the Upper Jurassic Solnhofen Formation of Bavaria, Germany, renowned for exquisitely preserved specimens of the oldest known bird, *Archaeopteryx*, pterosaurs, and insects, and the Lower Cretaceous Yixian Formation of Liaoning, China, which has yielded feathered theropod dinosaurs, birds, mammals, and the earliest undisputed angiosperms. Two conservation *Lagerstätten* are noteworthy for their remarkable records of Late Triassic life on land: the Solite Quarry on the state line between Virginia and North Carolina and the Madygen Formation of southwestern Kyrgyzstan.

SOLITE QUARRY

The Solite Quarry at Cascade, on the state line between Virginia and North Carolina, has had only a brief history of paleontological exploration. Opened in the mid-1950s, its exposures initially attracted the attention of sedimentologists and structural geologists, and its strata were even reported to be virtually devoid of fossils (Meyertons 1963). This situation changed when Paul Olsen first visited the quarry in 1974 and discovered skeletons of a new small aquatic protorosaur, *Tanytrachelos*, as well as a number of fish remains. Thus encouraged, Olsen embarked on more extensive study of the black shales exposed in the quarry and ultimately undertook extensive excavations. This work led to the recovery of numerous additional specimens of *Tanytrachelos* (Olsen 1979), a variety of fishes including a coelacanth, tetrapod tracks, and a diversity of plant remains. However, the most significant discoveries were some 300 insect specimens (Olsen et al. 1978), which represented the first extensive record of complete insect fossils for the entire Triassic Period. Until that time, only a few Triassic insect-bearing formations were known, and they had yielded mostly isolated wings. Thus, it is surprising that it took another two decades before exploration of the Solite Quarry recommenced. This time, teams from the Virginia Museum of Natural History, Lamont-Doherty Earth Observatory of Columbia University, and the American Museum of Natural History recovered more than 3,000 insect and many other fossils (Fraser et al. 1996; Fraser and Grimaldi 2003). This collecting effort continues to the present day (figure 9.1).

Although fossils are common at Solite, insect specimens are restricted to two or three particular horizons. Because of their frequently minute size, they are easily mistaken for flecks of mica or organic detritus when first spotted with the naked eye. Additionally, the skeletal remains of fishes and reptiles are usually obscured by thin, tightly adhering layers of sediment, which are virtually impossible to remove without damage to the underlying bones. Finally, the mostly fragmentary plant material does not preserve cuticles and comprises silvery carbonaceous films that are barely visible under natural light.

Figure 9.1. Excavation in exposures of the Cow Branch Formation in the Solite Quarry near Cascade. Note the dip of bedding planes of the black shales. (Photograph by NCF)

However, the hidden fossil riches of the Solite Quarry can be revealed using a combination of imaging and preparation techniques. When immersed in alcohol and viewed through a binocular microscope with illumination from a fiber-optic ring light, the extraordinary preservation of the insect fossils quickly becomes apparent (Fraser and Grimaldi 2003; Grimaldi et al. 2005). Examining the vertebrate remains requires the use of more advanced technology. Although the impressions of complete skeletons are often visible on a bedding plane, it is virtually impossible to remove the dolomite-rich dark gray sediment without damage to the underlying bones. Occasionally, it is possible to use an airbrasive machine to remove the sedimentary covering in a few places, but the technique is ineffective in the interbedded clay and coarser siltstone horizons that have yielded the majority of fish fossils and two specimens of a remarkable new gliding reptile. The first specimen discovered was initially misidentified as a partial skeleton of a coelacanth because the faint impressions of the elongated thoracic ribs resemble the bony rays of the caudal fin in a coelacanth fish. Even under close scrutiny the hind limbs and skull are almost impossible to see, and there are no apparent traces of the forelimbs (Fraser et al. 2007). Only CT scanning generated clear images of both specimens, showing well-developed forelimbs. The same technique has also been employed to elucidate anatomical details of *Tanytrachelos*.

The Dan River–Danville basin is part of the rift system of the Newark Supergroup (chapter 7). It forms an elongate half-graben bordered by the southeast-dipping Chatham fault zone. Although the basin is almost 170 kilometers long, it is only between 3 and 15 kilometers wide. Its sedimentary fill, termed the Dan River Group, is about 4,000 meters thick and unconformably overlies basement (Kent and Olsen 1997).

The Solite Quarry is located near the middle of the Dan River–Danville basin where it straddles the North Carolina–Virginia state line. It actually comprises three separate quarries that were worked over a period spanning more than four decades for black shales, which were used primarily for the production of lightweight aggregate. The first pit was opened in 1957 and is now completely overgrown. A second quarry was established in the late 1950s and operated until the early 1980s, when a third pit was opened. Operations in the latter ceased in 2002. Fraser and Grimaldi (2003) have designated the three quarries as A, B, and C. Together, these sites expose several hundred meters of strata representing the upper member of the Cow Branch Formation, which Kent and Olsen (1997) interpreted as an approximately 1,950-meter-thick sequence of cyclical gray to black mudstones representing a deepwater lacustrine facies with Milankovitch cycles. Olsen and Johansson (1994) suggested that lake levels may have fluctuated between 200 meters or more in depth to complete exposure during arid phases in one cycle. They also argued that the finely laminated, organic-rich shales probably represent periods when the lake was at or close to its deepest levels. Recent work focusing on the geochemistry (Liutkus et al., in preparation) also indicates cyclical change with an increasingly deep and anoxic depositional environment alternating with periods of subaerial exposure. By contrast with earlier interpretations, however, the new study interprets the microlaminated layer with insect fossils as having formed at the transition from ephemeral lake or lake margin covered by mats of cyanobacteria to standing water marked by periodic carbonate precipitation. Liutkus et al. (in review) surmise that high concentrations of carbon dioxide in the lake sediments may have rendered them inhospitable to scavengers and other benthic organisms, either directly or by promoting the growth of toxic algae. A present-day analog would be the saline, alkaline dolomite lakes of South Australia. The new study also favors the reconstruction of a much more shallow lake, with depths perhaps not exceeding 15 or 20 meters at its deepest point.

Long considered late Carnian in age, the upper member of the Cow Branch Formation, along with correlative strata of the Conewagian LVF (chapter 7), has now been redated as early Norian.

VERTEBRATES

Skeletal remains of a diversity of fishes include at least two taxa of coelacanths and various actinopterygians. The largest form is a coelacanth, probably *Pariostegus*, which, based on isolated bones, may have reached a length of up to 1.8 meters. Unfortunately, no complete specimens have been discovered to date. Smaller coelacanths have been referred to *Diplurus* (Olsen and Johansson 1994). The most common fish, the moderate-sized palaeoniscoid *Turseodus*, is represented by both articulated and disassociated remains and is readily identified on the basis of its characteristically ridged scales. A species of the neopterygian *Semionotus* is also quite common. In addition, the redfieldiid actinopterygian *Synorichthys* is occasionally found. Finally, a single tooth probably belongs to a relatively large freshwater shark.

The most common vertebrate is the small protorosaur *Tanytrachelos* (Olsen 1979; Casey, Fraser, and Kowalewski 2007; figure 9.2*A* and *B*). More than 300 complete and partial skeletons of this reptile have been collected at Solite to date. Although *Tanytrachelos* has also been reported from the *Placerias* Quarry of the Chinle Formation in Arizona (Kaye and Padian 1994; chapter 8), these remains are fragmentary and not diagnostic. With the exception of a few finds from the Lockatong Formation of the

Newark basin in Pennsylvania and New Jersey (Olsen and Johansson 1994; chapter 7), all known specimens of *Tanytrachelos* come from the Solite Quarry. At Solite, they are sometimes found close together on a single bedding plane, indicating incidents of mass mortality.

Tanytrachelos is most closely related to *Tanystropheus* from the Middle to Late Triassic of Europe (Olsen 1979; chapter 3). Although its neck lacks the extraordinary elongation of that in *Tanystropheus*, it is still rather long, equivalent to about two-thirds the length of the trunk (figure 9.2*A*). The long hind limbs of *Tanytrachelos* are frequently preserved in a froglike pose (figure 9.2*B*) and probably provided the main propulsive force during swimming. Some skeletons preserve faint traces of soft tissues, including skin impressions, definitive myotome blocks in the tail, and occasionally remnants of tendons extending along the digits of the foot. Additional fibrous structures on the pes of *Tanytrachelos* may represent the remains of webbing, an interpretation further supported by the presence of tracks in the same formation that preserve what appear to be faint traces of webbing between the impressions of individual digits.

The proportionately small skull of *Tanytrachelos* has a marginal dentition composed of small, needlelike teeth, suggestive of insectivorous habits. Small clusters of insect parts, including macerated wings, may represent remnants of food regurgitated by this reptile.

Tanytrachelos exhibits pronounced sexual dimorphism (Olsen 1979; Casey, Fraser, and Kowalewski 2007). Although adults of the two sexes were similar in size, one sex has two large paired ossifications (heterotopic bones) that are intimately associated, and possibly articulated, with the transverse processes of the fourth and fifth caudal vertebrae. These spikelike, robust bones project forward toward the pelvic

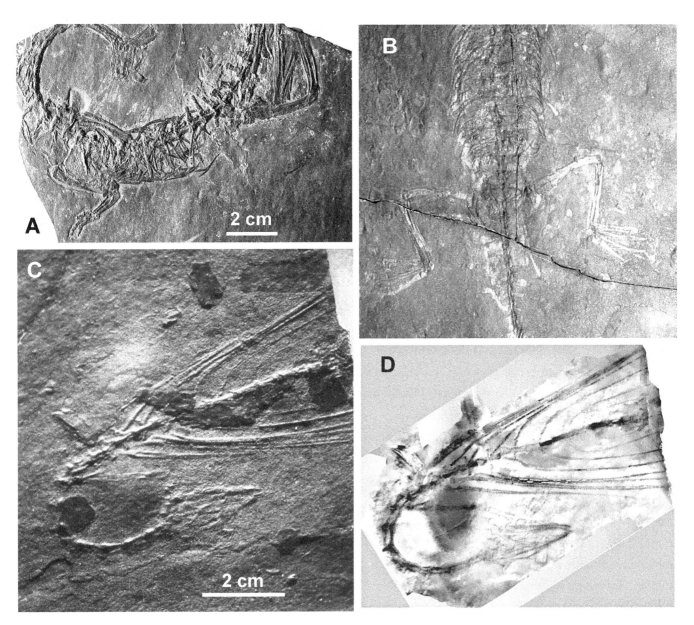

Figure 9.2. Reptiles from the Cow Branch Formation of the Solite Quarry. (*A*) Holotype of the tanystropheid protorosaur *Tanytrachelos ahynis* (Peabody Museum of Natural History, Yale University, YPM 7496). (*B*) Pelvic region and hind limbs of a referred specimen of *Tanytrachelos ahynis* (Virginia Museum of Natural History, VMNH 2830). No scale provided. (*C–D*) Holotype of the gliding reptile *Mecistotrachelos apeoros* (VMNH 3649) as preserved (*C*) and imaged by CT scanning (*D*). (*A–C* photographs by NCF; *D* image courtesy of T. R. Ryan)

girdle. Wild (1973) observed similar ossifications in *Tanystropheus* and argued that they might represent the equivalent of hemipenes in squamates. Thus, he interpreted individuals with these bones as males. However, these elements appear to be far too large to represent hemipenes (Böhme 1988), and their function remains unresolved. Indeed, it is equally conceivable that these elements occurred in females, in which they could have supported some kind of brood pouch.

Another possible insectivore is *Mecistotrachelos*, a still poorly known gliding reptile (Fraser et al. 2007; figure 9.2*C* and *D*). As previously noted, only two specimens have been recovered to date. Although *Mecistotrachelos* has elongated ribs that presumably supported a gliding membrane, it differs significantly from *Icarosaurus* (chapters 6 and 7). It has relatively much more elongate cervical vertebrae, and the elongate ribs articulate with much shorter transverse processes. Most important, a very different aerofoil shape was described for *Icarosaurus*, in which the central elongated ribs had a marked ventral flexure that, in turn, would have resulted in a concave ventral surface of the gliding membrane (Colbert 1970a). The straight distal portions of the thoracolumbar ribs in *Mecistotrachelos* rule out fixed cambering of the aerofoil. However, if differential vertical movements of the ribs were possible, *Mecistotrachelos* could have created a variable-camber aerofoil. Moving an anterior rib down would have increased the camber of the aerofoil, thereby increasing lift and drag. Raising an anterior rib would have flattened the gliding membrane and decreased the camber, thus reducing both lift and drag. According to Fraser et al. (2007), the robust proximal rib heads indicate the possibility of independent movement of the anterior ribs.

The elongate cervical vertebrae of *Mecistotrachelos* hint at archosauromorph, perhaps protorosaurian affinities, but other phylogenetically informative features, such as the structure of the ankle, cannot be clearly distinguished in the known specimens. Furthermore, there is no evidence for the presence of elongate cervical ribs as in protorosaurs. The preserved feet on the referred specimen of *Mecistotrachelos* have relatively short, subequal metatarsals and phalanges and are preserved flexed in a grasping pose. This pose may indicate an ability to grasp slender branches and thus arboreal habits.

The two known specimens of *Mecistotrachelos* exhibit marked differences in limb proportions, which Fraser et al. (2007) interpreted as sexual dimorphism.

INSECTS

As previously noted, few formations worldwide have yielded abundant remains of Triassic insects to date. Even in those units that are relatively rich in insect fossils, such as the Grès à Voltzia of eastern France (chapter 3) and the Molteno Formation of southern Africa (chapter 4), most specimens are isolated wings. As venation patterns of insect wings are diagnostic at least to the family level, much can be inferred concerning insect diversity and paleoenvironmental conditions.

In contrast to other known Triassic insect assemblages, aquatic forms dominate that from the Solite Quarry (figures 9.3 and 9.4). Indeed, it represents the oldest well-known freshwater insect community (Grimaldi and Engel 2005). The most common insects are belostomatid and naucorid heteropterans (water bugs). Whereas the smallest nymphs are only about 2 millimeters long, adult belostomatids can reach a length of 12 millimeters (figure 9.4*A* and *B*). Much like their extant relatives, the Solite belostomatids have strong, grasping forelegs and presumably preyed on other aquatic animals.

Archescytinids are the second most abundant group of insects. These small basal hemipterans have

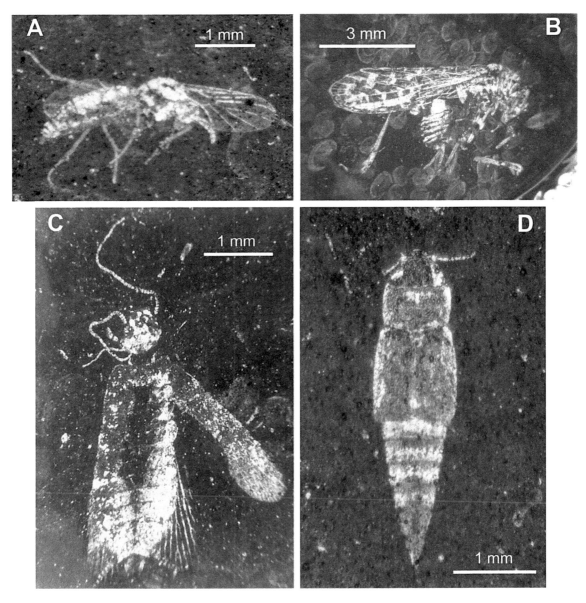

Figure 9.3. Insects from the Cow Branch Formation of the Solite Quarry. (*A*) Stem anisopodid *Crosaphis* sp. (VMNH 731). (*B*) New taxon of elcanid orthopteran (VMNH 785). (*C*) New taxon of small blattoid (VMNH 917). (*D*) New taxon of staphylinid beetle (VMNH 734). (Photographs by D. Grimaldi and NCF)

piercing mouthparts, which may have been used to feed on plants. Archescytinids were previously known only from the Permian of Australia, Russia, and the United States (Rasnitsyn and Quicke 2002).

The most spectacular insects from the Solite Quarry are orthopterans referable to the family Elca-nidae (figure 9.3*B*). They can exceed 20 millimeters in length, and the thorax has the distinctive "hunch-back" shape. The hind legs have a stout femur and long tibia and are clearly suitable for jumping.

The Solite assemblage is noteworthy for the great diversity of dipterans. It establishes that most of the

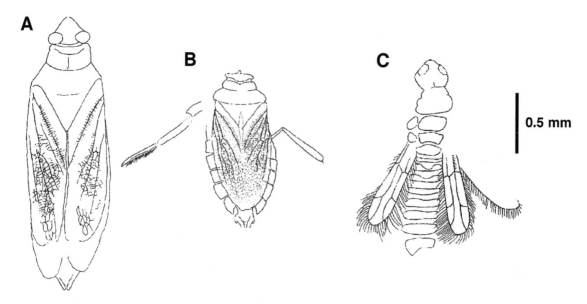

Figure 9.4. Insects from the Cow Branch Formation of the Solite Quarry. (*A–B*) Adults of unidentified belostomid (*A*: YPM 16830; *B*: VMNH 727). Adult length upto 12 millimeters. (*C*) Thysanopteran *Triassothrips virginicus* (holotype, VMNH 747). (Modified from Fraser and Grimaldi 2003)

major clades of Diptera, including Bibionomorpha, Culicomorpha, Psychodomorpha, and Tipulomorpha, had appeared by the Late Triassic (Grimaldi and Engel 2005).

The Tipulomorpha include the Tipulidae (crane flies). Tipulomorph larvae are usually associated with moist habitats and are often found in decomposing plant material; many are aquatic. The Limoniidae, another group of extant tipulomorphs, include *Architipula* and *Metarchilimonia* from the Solite Quarry. The latter taxon is represented by a number of exquisitely preserved specimens, including females with clearly visible ovipositors.

The Psychodomorpha (moth flies) have a surface covering of fine hairs on the wings, which lends them a velvety appearance. The Solite specimens referable to this group already exhibit this characteristic feature of the wings.

The most interesting dipteran fossils recovered from Solite to date represent an early member of the Culicomorpha (biting midges, mosquitoes, and re-

lated forms). There are more than 12,000 described extant species of Culicomorpha, and the larvae of almost all culicomorphs are aquatic or at least prefer moist habitats. Several well-preserved specimens of culicomorphs have been found at Solite, but it is difficult to discern patterns of wing venation, and thus they have not yet been formally described. Significantly, one of the Solite specimens has a long, slender proboscis that is nearly twice as long as the head is tall and strongly resembles the piercing mouthparts of a blood-feeding insect. In various extant species of biting midges, the styletlike mandibles are well developed in the females. The mandibles are used to pierce the skin and thereby allow the females to suck the blood of their hosts. The males have a similar-sized proboscis, but their mandibles are vestigial and they feed on nectar.

The Bibionomorpha are represented by a variety of forms, and in contrast to those of the tipulomorphs, psychodomorphs, and culicomorphs, larvae of bibionomorphs are almost exclusively terrestrial. Some feed on decaying matter or fungi, whereas others subsist on plants and still others are predators.

The stem anisopodid (wood gnat) *Crosaphis* was originally described based on isolated wings from Australia, but complete specimens are now known from Solite (figure 9.3*A*).

The stem brachyceran from Solite is noteworthy in that its wing venation already shows the deflection of the CuA vein toward the CuP vein, meeting it well before reaching the wing margin. However, its antennae were apparently still relatively long, more similar to those of nematocerans. Blagoderov, Grimaldi, and Fraser (2007) assigned the Solite form to a new family Prosechamyiidae.

Today there are some 5,500 species of thrips (Thysanoptera). These mostly minute insects are of commercial interest because many species are pests on fruit blossoms. The minute *Triassothrips* from the Solite Quarry is remarkable for its similarity to extant thysanopterans (Grimaldi, Shmakov, and Fraser 2004; figure 9.4*C*). Its wings are short, with the characteristic fringed margins, and do not extend beyond the penultimate tergite of the abdomen. Both the thorax and abdomen are long and slender.

Originally misidentified as a trichopteran (Fraser et al. 1996), a mecopteroid (scorpion fly) has more recently been referred to the Pseudopolycentropodidae and represents the first record of this group from the Western Hemisphere (Grimaldi et al. 2005).

In other known Triassic insect assemblages, cockroaches and beetles are typically the most common groups. Cockroaches are not very common at Solite, although at least two forms are known (figure 9.3*C*). One represents the geologically youngest known record of the primarily late Paleozoic Phylloblattidae (Vršanský 2003). Beetles are reasonably well represented in the Solite assemblage; judging from the patterns of ornamentation on the elytra, at least eight coleopteran taxa were present. However, only a few specimens are preserved with visible appendages, and it is therefore difficult to undertake more specific taxonomic identification. Particularly noteworthy are specimens of Staphylinidae (rove beetles), which already have short elytra that cover only the proximal region of the abdomen (figure 9.3*D*).

Three specimens of one of the oldest known araneomorph spider have been recovered from the Solite Quarry, one of which preserves considerable detail of the structure of the legs (Selden et al. 1999).

FLORA

Fragments of foliage are abundant throughout the Solite exposures, but finds of more complete fronds are less common and typically restricted to a few horizons.

The most common plant fossils are isolated shoots of cheirolepidaceous conifers, referable to the form genera *Brachyphyllum* and *Pagiophyllum*. Along with various kinds of cone scales, these remains suggest stands of conifers on the slopes overlooking the lake. Remains of bennettitalean foliage are also quite common. These include *Zamites*, *Pterophyllum* (figure 7.5), and *Sphenozamites*. Bennettitaleans appear to be more abundant in the "wetter" parts of each depositional cycle, and conifer shoots become dominant in the "drier," regressive phases.

Ferns are less common, although fragments of foliage can be found in a number of cycles. *Dictyophyllum* is the most common form (figure 5.14*B*). Again, ferns tend to be restricted to intervals with humid conditions. Occasionally remains of ginkgophytes (*Sphenobaiera*; figure 5.14*A*) and equisetaleans have been found.

A biogeographically significant plant fossil is the winged seed *Fraxinopsis* (Axsmith et al. 1997). This form taxon is otherwise known only from the Late

Triassic of Gondwana, where it is always found associated with the gymnosperm foliage taxon *Yabeiella*.

Geochemical analysis of the strata exposed in the Solite Quarry indicates that the lake waters were almost certainly alkaline. Furthermore, the bottom sediments appear to have been anoxic, as evidenced by the absence of bioturbation. The latter factor may well account for the much higher percentage of complete insect specimens than is found at other well-known Triassic insect localities. Unusually high fluorine content of the sediment might indicate the presence of thermal springs feeding into the bottom of the lake.

MADYGEN

Outcrops of the sedimentary strata of the Madygen Formation occur at a number of locations along the northern rim of the Turkestan Range in southwestern Kyrgyzstan and adjoining areas of Tadzhikistan and Uzbekistan in Central Asia (Dobruskina 1994, 1995; Voigt et al. 2006). The approximately 250-kilometer-long Turkestan Range, which is part of the southern Tien Shan Mountains, forms the southwestern margin of the Fergana Valley. The original exposures of the Madygen Formation are in the vicinity of the village of Madygen in the Batken Oblast (district) of southwestern Kyrgyzstan. These deposits were first explored during the 1930s, when the territories of the former Soviet Union were systematically surveyed for natural resources. Although the Madygen Formation contains coal seams, these horizons are too thin to be deemed of economic interest. Starting in 1945, geologists from the University of Tashkent mapped the geology of the region and discovered several localities yielding abundant

plant fossils. Between 1957 and 1967 teams from the Paleontological Institute of the Soviet Academy of Sciences in Moscow undertook a number of expeditions to the Madygen region, amassing extensive collections of plants as well as more than 15,000 specimens of insects and the first specimens of Triassic tetrapods from Central Asia. The Russian paleobotanist Irina Dobruskina revisited the region in 1987 to gather additional plant fossils and geological data (Dobruskina 1995). Since 2005, teams under the direction of Sebastian Voigt from the Bergakademie Freiberg and the Geological Survey of South Kyrgyzstan have undertaken several expeditions to explore the fossil-bearing strata of the Madygen Formation.

Based on the work by Dobruskina (1995) and Voigt et al. (2006), the Madygen Formation comprises up to 600 meters of alternating breccias, conglomerates, and siliciclastic strata with minor coals. Fluviolacustrine deposits of brown, gray, and yellowish color predominate. The Madygen Formation is separated from underlying Paleozoic limestones by an unconformity and is itself unconformably overlain by Jurassic continental strata. Synsedimentary deformation on a meter scale indicates significant tectonic activity in the region during the deposition of the Madygen strata. Voigt et al. (2006) place the depositional setting at a paleolatitude of 35°–40° north and assumed a seasonally humid climate.

The Madygen Formation has yet to be reliably dated. In one region, it was reported as being overlain by strata containing a Rhaeto-Liassic conchostracan assemblage, but the geological context of this particular occurrence is problematical (Voigt, personal communication, 2009). Until the 1970s, there existed a consensus that the Madygen Formation included both Late Permian and Early Triassic horizons, based on age assessments using both insect and plant fossils. Work on the plant material and extensive comparisons with other Triassic floras from Eurasia led Dobruskina (1994, 1995) to propose a Ladinian to Carnian age for the Madygen Formation, which has been widely accepted in the literature. Shcherbakov (2008a) argued for an older, Ladinian age based on the insect assemblage. However, the age of the Madygen Formation has yet to be constrained by other lines of stratigraphic evidence. Attempts by Voigt et al. (2006) and others to extract pollen and spores have proven unsuccessful.

VERTEBRATES

Fishes from the Madygen Formation include the dipnoan *Asiaceratodus* and several actinopterygians, comprising the widely distributed *Saurichthys* and five endemic taxa (Sytchevskaya 1999). In addition, egg capsules document the presence of freshwater sharks (Voigt, personal communication, 2009).

Tetrapod remains are uncommon. Ivakhnenko (1978) described a tiny poorly preserved partial skeleton of what he interpreted as the oldest known urodele, *Triassurus*, but this specimen may well represent a larval temnospondyl. Tatarinov (2005) described an incomplete skeleton of a small cynodont, *Madysaurus*. Unfortunately, the unique specimen was severely damaged during preparation, and comparisons to other known Triassic cynodonts are difficult. Voigt (personal communication, 2009) and his team recently discovered the skull and armor of a new chroniosuchian anthracosaur as well as various remains of reptiles.

The two most famous tetrapods from the Madygen Formation are the enigmatic reptiles *Longisquama* and *Sharovipteryx* (Unwin, Alifanov, and Benton 2000; figure 9.5). Each is known from a single skeleton. Preservation of the bones is poor in both specimens, and the fossils have been liberally coated with a preservative that now obscures much detail. The

fossils were recovered from light grayish yellow shales.

Sharovipteryx, first described by Sharov (1971) under the preoccupied generic name *Podopteryx*, is represented by an almost complete but poorly preserved skeleton preserved on part and counterpart slabs (figure 9.5*A*). It has a total body length of about 24 centimeters. The narrow skull has a long, pointed snout and large orbits located at midlength. The neck of *Sharovipteryx* is equal in length to the short trunk, and its third through seventh cervical vertebrae are elongate. The ilium has an elongate preacetabular process. The hind limbs of *Sharovipteryx* are remarkably long and slender, with the tibia being slightly longer than the femur. Its pes has long, slender digits. The most distinctive feature of *Sharovipteryx* is the presence of prominent membranes that extend between the hind limbs and the proximal portion of the long tail to form a uropatagium. Additional membrane impressions are visible in front of the femur and extend laterally as far as the knee. It is not clear, however, whether they were attached to the forelimb in the manner reconstructed by Sharov (1971). Sharov considered *Sharovipteryx* an arboreal form capable of gliding from tree to tree, using its head and body as a rudder and its tail as a counterweight. Gans, Darevskii, and Tatarinov (1987) proposed a slightly different reconstruction of the gliding apparatus but essentially agreed with Sharov's interpretation. Most recently, Dyke, Nudds, and Rayner (2006) attempted a restoration of *Sharovipteryx* as a delta-wing glider.

Unwin, Alifanov, and Benton (2000) suggested protorosaurian affinities for *Sharovipteryx*, but additional material is needed to clarify the skeletal structure and phylogenetic position of this intriguing reptile.

Longisquama is known from a partial skeleton comprising the skull and the anterior portion of the

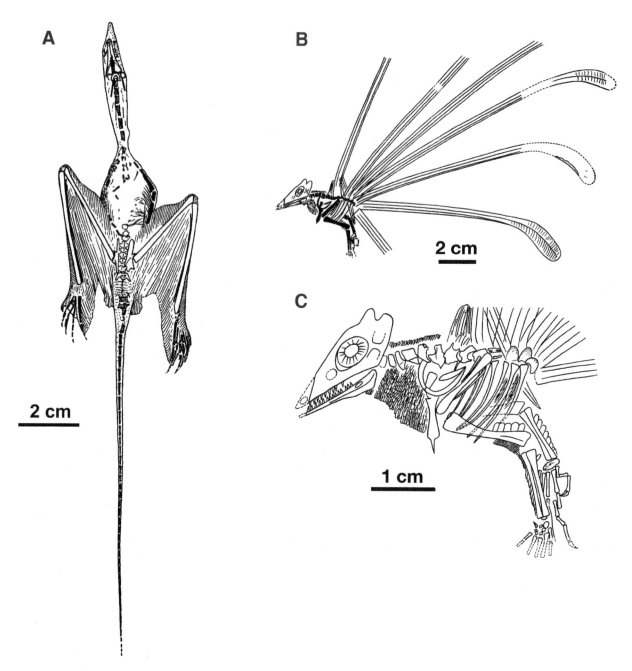

Figure 9.5. Reptiles from the Madygen Formation of Kyrgyzstan. (*A*) Holotype of *Sharovipteryx mirabilis* (Paleontological Institute of Russian Academy of Sciences, PIN 2584/8) in dorsal view. (From Sharov 1971) (*B–C*) Holotype of *Longisquama insignis* (PIN 2584/4). (*B*) Outline drawing and (*C*) detail of skeleton in lateral view. (From Sharov 1970)

body with the forelimbs and associated dorsal appendages (figure 9.5B and C). In addition, a number of isolated dorsal appendages have been collected. *Longisquama* is a small animal, with a skull length of about 2.3 centimeters and forelimb length of approximately 4.4 centimeters. The elongate penultimate phalanges of the manual digits indicate arboreal habits (Unwin, Alifanov, and Benton 2000).

The most striking feature of *Longisquama* is the presence of greatly elongated scale-like appendages along the back of the animal (figure 9.5B). Sharov (1970) reconstructed these structures as forming a single row extending the length of the back. Haubold and Buffetaut (1987) reconstructed paired rows of appendages, which could have served as a gliding or parachuting device when folded down. The dorsal appendages are shaped like hockey sticks and appear to decrease in length toward the pelvic region: the most anterior scale has a length of 150 millimeters, and the most posterior ones have a length of 100 millimeters. It is unclear how these structures were attached to the body of *Longisquama*.

Jones et al. (2000) claimed that the dorsal appendages of *Longisquama* were protofeathers and used this interpretation in their arguments against the prevailing hypothesis that birds were derived from theropod dinosaurs. However, Reisz and Sues (2000b) showed that these "protofeathers" are, in fact, quite unlike avian feathers. Voigt et al. (2009) interpreted the dorsal structures in *Longisquama* as a unique form of integumentary structure. They argued that each appendage consisted of a single-branched internal frame enclosed by a flexible external membrane and supported Sharov's assertion that there was only a single row of these structures along the back of the animal.

The phylogenetic position of *Longisquama* remains unresolved. Its skull has two temporal openings, supporting diapsid affinities, and Sharov (1970) claimed that an antorbital fenestra is present and thus *Longisquama* was a "pseudosuchian." Again, additional, better-preserved specimens are needed to establish the affinities of *Longisquama*.

INSECTS

Insect remains are common in the Madygen Formation, especially at a locality called Dzhailoucho (Rasnitsyn and Quicke 2002; Shcherbakov 2008a). More than 20,000 specimens representing 20 orders have been recovered to date, and portions of these large collections remain unstudied. Although most specimens are only impressions of isolated wings, they are often excellently preserved and many even retain traces of the original color pattern. According to Shcherbakov (2008a), more than 500 taxa of insects have already been described, although some of these may be based on diagenetically or tectonically deformed specimens of other known forms (Béthoux, Papier, and Nel 2005). Ponomarenko and Popov (in Dobruskina 1995) and Shcherbakov (2008a) noted faunal resemblances between the insect assemblage from the Madygen Formation and those from the Middle to Late Triassic of Australia and South Africa.

The most spectacular insects from the Madygen Formation are the Titanoptera, which are now known from some 200 specimens assigned to 21 species (Shcherbakov 2008a). Close relatives of grasshoppers and their kin (Orthoptera), titanopterans such as *Gigatitan* could attain a wingspan of at least 40 centimeters (Sharov 1968; Grimaldi and Engel 2005; figure 9.6A). Their legs were adapted for running rather than jumping. The forelegs bear stout ventral spurs and appear suitable for catching prey. The wings were held flat over the abdomen at rest. The forewings bear large stridulatory structures (figure 9.6B), which resemble those in present-day ensiferan orthopterans,

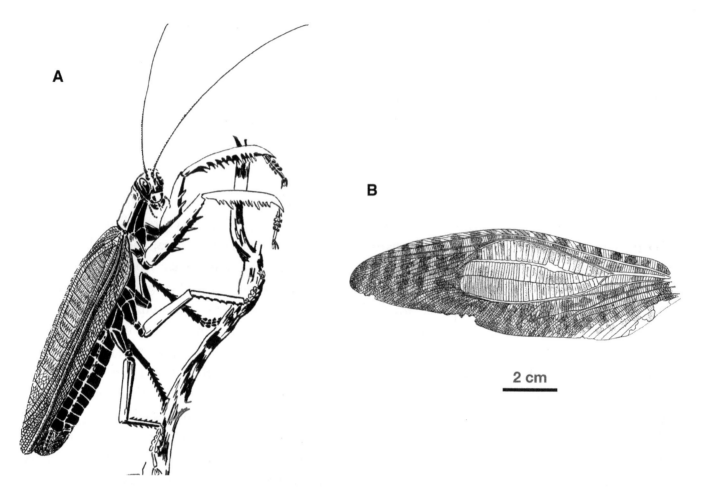

Figure 9.6. (A) Reconstruction of the titanopteran orthopteran *Gigatitan* from the Madygen Formation of Kyrgyzstan. (Modified from Rohdendorf and Rasnitsyn 1980) (B) Forewing of the titanopteran orthopteran *Clatrotitan andersoni* (Australian Museum, AM F.36274) from the Middle Triassic Hawkesbury Sandstone of New South Wales, Australia. Note the large stridulatory area near the center of the wing and the color pattern on wing surface. (From McKeown 1937)

and titanopterans must have been capable of producing deep, resonant sounds (Grimaldi and Engel 2005).

The most common groups are Blattodea, Auchenorrhyncha (cicadas, plant hoppers, and treehoppers), and Coleoptera. Cockroaches are very diverse, with 26 species in 11 genera and five families, including the predominantly Paleozoic Phylloblattidae (Vishnyakova 1998; Papier and Nel 2001). Although less common than cockroaches, the beetles have even greater taxonomic diversity: Arnol'di et

al. (1977) assigned some 3,500 specimens to 63 species in 34 genera and seven families. However, coleopteran remains comprise mostly isolated elytra, and thus the diversity of this group has probably been overestimated. Cupedoid beetles are the most diverse and common (figure 9.7). Although none of the beetles shows adaptations for swimming, some taxa may have preferred habitats close to water or even spent part of their life cycles in water. Auchenorrhyncha are the third most common group, with 17 families documented to date. A noteworthy ele-

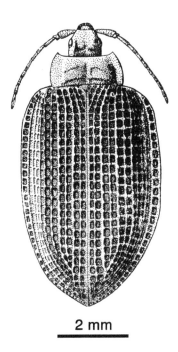

2 mm

Figure 9.7. Reconstruction of the body of the cupedoid beetle *Platycupes dolichocerus* from the Madygen Formation. (From Ponomarenko 1969)

ment of the Madygen insect assemblage are the earliest known Hymenoptera, which are represented by Xyelidae (sawflies). Nematoceran dipterans are documented by 11 species from five families, including the earliest known member of the tipulomorph family Limoniidae.

A handful of insect groups with aquatic nymphs include Ephemeroptera, Odonata, Plecoptera, and Trichoptera. However, remains of nymphs are rare, and even adults of these taxa are uncommon. Voigt et al. (2006) interpreted distinctive subcylindrical burrow structures as having been made by either dipteran or trichopteran larvae, but Shcherbakov (2008a) has questioned that interpretation. Shcherbakov (2008a) excluded a possible record of a naucorid water bug from the Madygen Formation as actually derived from overlying younger strata. He noted that Heteroptera (true bugs) are very rare in the Madygen Formation. Three adults of basal Nepo-

morpha are known, two of which are related to *Heterochterus* from the Carnian-age Mount Crosby Formation of Queensland, Australia.

OTHER INVERTEBRATES

The Madygen Formation has yielded numerous crustacean remains and rare freshwater bivalves. Crustaceans are represented primarily by the Kazacharthra, a group of phyllopods apparently endemic to Mesozoic strata of parts of Asia and characterized by a kidney-shaped carapace enclosing the head and trunk. In addition, there are a variety of ostracodes and an unidentified malacostracan decapod (Voigt et al. 2006).

FLORA

The Madygen Formation has yielded a rich and varied plant assemblage. Some 60 percent of known plant fossils represent pteridosperms, which are represented mainly by foliage and are referred to 15 genera including the peltaspermaceans *Lepidopteris, Peltaspermum,* and *Scytophyllum* (Dobruskina 1994, 1995).

About 10 percent of plant fossils represent equisetaleans. Dobruskina (1995) distinguished three genera, *Equisetites, Neocalamites,* and *Prynaduia.* Locally, rhizomes of *Neocalamites* are still preserved in life position. The remainder of the plant assemblage comprises lycopsids such as *Ferganadendron,* bennettitaleans (including *Pterophyllum*), ferns (*Danaeopsis, Chiropteris, Cladophlebis*), ginkgophytes (*Baiera, Ginkgoites, Sphenobaiera*), conifers (including *Podozamites*), floating hepatics, and thallophytes.

Much remains to be learned about the geology, stratigraphic age, and fossil assemblages of the Madygen

Formation. There is now substantial evidence that most of the strata were deposited in a lake. Voigt et al. (2006) argued that the fragmentary nature of most animal and plant remains indicates their transport from other areas into the lacustrine setting. The excellent preservation of integumentary structures in *Longisquama* and *Sharovipteryx* and foliage with cuticles suggests rapid burial and perhaps microbially mediated preservation of organic material. However, the absence of black mudstones or shales as well as the occurrence of a variety of benthic invertebrates and invertebrate burrows militate against anoxic conditions at the bottom of the lake (Voigt et al. 2006). The widely distributed, evenly laminated, and fine-grained clastic strata suggest calm depositional conditions below the wave zone.

The Madygen Formation shares with the Solite exposures of the Cow Branch Formation often exceptional preservation and a great diversity of animal and plant remains. Yet the two units are very different in the composition of their vertebrate and invertebrate assemblages. The most striking aspects of the Solite biota are the dominance of aquatic animals and the oldest known assemblage of aquatic insects, including numerous nymphs and larvae. By contrast, the Madygen insect assemblage comprises predominantly terrestrial forms. The Solite insect assemblage is also unusual in the dominance of relatively small forms; most insects are less than 5 millimeters in length. Even the cockroaches are relatively small (figure 9.3C). This contrasts sharply with most other known fossil insect assemblages, where larger, more robust forms tend to be most common. At present we cannot offer a satisfactory explanation for this bias toward small size, but is highly improbable that larger species were uncommon in the region at the time of deposition. It is also worth reiterating that much of the known fossil record of insects consists of isolated wings. The relatively tough wings readily survive postmortem changes and transport, and this may well account for the scarcity of complete insects. By contrast, Solite is the only known Triassic insect assemblage that is primarily documented by complete insects—further evidence of a highly unusual depositional environment. The Madygen plant assemblage differs from that at Solite, particularly in the great abundance of pteridosperm foliage and in the diversity of equisetaleans, which are rare at the latter locality. The plant material from Solite comprises mainly conifer and bennettitalean foliage and indicates a rather different environmental setting.

CHAPTER 10

Biotic Changes During the Triassic Period

THE IMPACT OF THE END-PERMIAN EXTINCTION

The end-Permian biotic crisis is generally considered the most severe of the five mass extinctions during the last 600 million years (Benton 2006; Erwin 2006). Indeed, Benton's (2006) book has the eye-catching title *When Life Nearly Died*. It has been claimed that as many as 90 to 95 percent of species of marine invertebrates vanished. A 9 parts-per-thousand negative carbon isotope excursion in the inorganic and organic carbon rock records indicates a substantial disturbance in the global carbon cycle. On land, Benton and Twitchett (2003) claimed that some 60 to 70 percent of vertebrate families became extinct. Rees (2002) reported that global plant diversity at the species level apparently declined more than 50 percent. Finally, Retallack, Veevers, and Morante (1996) and Looy et al. (1999) noted a "coal gap" in much of the Early Triassic sedimentary record, which they attributed to the disappearance of peat-forming (and thus coal-producing) plants such as glossopterid pteridosperms and arborescent lycopods at the end of the Permian. It is not until the Middle Triassic that coals reappear in the geological record, documenting the reestablishment of peat lands in various regions of the world (Greb, DiMichele, and Gastaldo 2006).

What was the impact of the end-Permian extinctions on continental biotas? McElwain and Punyasena (2007) noted that the majority of dominant plant genera and families did not, in fact, vanish at the Permo-Triassic boundary but merely were relegated to less important ecological roles or replaced by related taxa. Knoll (1984) argued that the observed floral changes had taken place at different times in different regions of the world (see also DiMichele et al. 2008). Finally, recent discoveries have established that many plant clades that came to characterize Triassic continental floras, such as the corystospermalean gymnosperm *Dicroidium* (Kerp et al. 2006) and certain cycads (DiMichele et al. 2001), already appeared during the Permian. Indeed Greb, DiMichele, and Gastaldo (2006) viewed the Permo-Triassic floristic transition as an ecological restructuring rather than a mass extinction.

Published figures for tetrapod biodiversity are based on a literal reading of the fossil record and thus overestimate actual extinction rates at the Permo-Triassic boundary. Phylogenetic analyses of procolophonoid parareptiles (Modesto, Sues, and Damiani 2001), archosauromorph reptiles (Modesto and Sues

2004) (figure 10.1), and temnospondyls (Ruta and Benton 2008) show that a significant number of lineages in these clades actually did range across the Permo-Triassic boundary. Nevertheless, several groups of tetrapods did disappear near or at the end of the Permian, notably gorgonopsian therapsids and pareiasaurian parareptiles. Dicynodont therapsids, by far the most abundant and diverse herbivorous tetrapods during the Late Permian, persisted into the Early Triassic with only four known lineages (Fröbisch 2007; Surkov and Benton 2008). However, one of these lineages subsequently underwent a second evolutionary radiation, with the development of numerous large forms, during the Triassic. Cynodont therapsids first appeared during the Late Permian but diversified only during the Triassic (Hopson and Kitching 2001; Kemp 2005).

Béthoux, Papier, and Nel (2005) observed that the known Permian and Triassic insect communities differ significantly in their composition. A number of major lineages of insects disappeared toward or at the end of the Permian, and certain groups with low diversities during the Permian diversified during the Middle and Late Triassic. Béthoux, Papier, and Nel (2005) claimed the existence of a gap of some 15 million years in the fossil record between the latest Permian and earliest Triassic occurrences of insects. Shcherbakov (2008b) refuted this claim and noted that a number of localities, especially from Russia, fill much of this gap. Most of these Induan- or Olenekian-age occurrences have yielded low-diversity insect assemblages comprising beetles, orthopterans and protorthopterans, and auchenorrhynchans.

PLANT EVOLUTION DURING THE TRIASSIC

The record of terrestrial plant macrofossils is still poor for much of the Early and Middle Triassic, the

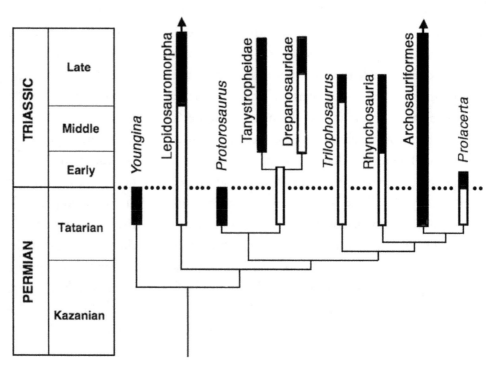

Figure 10.1. Phylogeny of various clades of neodiapsid reptiles and outgroup taxon *Youngina*. Black bars indicate the known fossil record and white bars as yet unrecorded "ghost lineages." Note the inferred diversity around the Permo-Triassic boundary. Arrows at the top of the black bars indicate that the lineages extend to the present day. (Based on Modesto and Sues 2004)

Anisian-age Grès à Voltzia (chapter 3) representing a notable exception. It is much better and more extensive for the Late Triassic, but, even during that interval, occurrences of fossil plants are not as common as, for example, during the Pennsylvanian (Kerp 2000). With the exception of locally very abundant silicified wood, Triassic plant macrofossils usually are preserved as compressions and thus offer only limited anatomical information. Pollen and spores help to fill in some gaps in the record of Triassic plant diversity. As noted earlier, many of the principal groups of Triassic plants had already evolved during the late Paleozoic.

Early Mesozoic floras are characterized by an increasing abundance of conifers, continuing a trend that extends back to at least the Early Permian. Plants also increasingly show structural adaptations to drier climatic conditions, such as smaller fronds (in pteridosperms), larger, more robust leaves (in conifers), and thicker leaf cuticles with sunken stomata (Kerp 2000).

As noted earlier in this chapter, the biotic crisis at the Permo-Triassic boundary led to a significant restructuring of plant communities. During the Triassic, floras composed of archaic ferns, cordaitaleans, arborescent lycopods and sphenophytes, and pteridosperms gave way to those made up of voltzialean and podocarpalean conifers, cycads, new clades of ferns, ginkgophytes, and bennettitaleans (Knoll 1984; Kerp 2000). At the beginning of the Triassic, isoetaleans, especially the ubiquitous *Pleuromeia*, initially dominated lowland floras in much of the world (Looy et al. 1999). The early Anisian Grès à Voltzia of eastern France provides the first extensive record of a more diverse Triassic plant community comprising both wetland forms and elements more adapted to drier conditions (chapter 3).

The next extensive fossil record of plant assemblages comes from the Erfurt Formation (Lower Ke-

uper) of Germany and is late Middle Triassic (Ladinian) in age (chapter 5). Sphenophytes, especially equisetaleans, form the most common element (Kelber 1999). Additional floral elements include a diversity of ferns such as the marattialean *Danaeopsis*, the oldest bennettitaleans (represented by the foliage taxon *Pterophyllum*), the first caytonialeans (*Sagenopteris*), and conifers such as *Podozamites*. Kerp (2000) and Kelber (in Deutsche Stratigraphische Kommission 2005) emphasized the considerable number of plant taxa shared by the Keuper communities with Gondwanan floras, including *Schizoneura*, *Phyllotheca*, *Linguifolium*, *Podozamites* (=?*Heidiphyllum*), and possibly even *Dicroidium*.

The currently known Late Triassic plant fossil record dates mostly from the Carnian and Rhaetian stages. The record of Norian-age plant assemblages is still poor. Carnian-age floras for the most part represent peat-forming wetland vegetation, characterized by an abundance of ferns and equisetaleans (Kerp 2000). Rhaetian plant communities, which are well-known from East Greenland, southern Sweden, and southern Germany, closely resemble those from the Early Jurassic in their taxonomic composition (Dobruskina 1994).

Overall, Middle to Late Triassic terrestrial vegetation is characterized by the successive appearance of new groups such as the bennettitaleans (which are thought to be closely related to angiosperms) as well as extant clades of ferns (Dipteridaceae, Matoniaceae) and conifers (Dobruskina 1994; Kerp 2000). Kerp (2000) noted that several major assemblages of Triassic plants (e.g., Lunz in Austria; chapter 5) discovered during the nineteenth and early twentieth centuries are in urgent need of restudy, and such reassessments will likely lead to important new insights concerning early Mesozoic vegetational history.

ORIGIN OF EXTANT INSECT GROUPS

Examining patterns of distribution and faunal succession among insects during the Triassic is challenging. Although various new occurrences of Triassic insects have been discovered in recent years, such finds are still much less common than those of tetrapod assemblages. Much of the fossil record of insects is confined to a rather small number of conservation *Lagerstätten*, most of which date from the later part of the Triassic (Grimaldi and Engel 2005). Furthermore, these finds represent a variety of depositional environments, come from widely separated locations and different stratigraphic intervals, and thus are rarely closely comparable (chapter 9).

Shcherbakov (2008b) hypothesized three main phases in insect evolution during the Triassic, of which the Early Triassic assemblages discussed earlier in this chapter represent the first. The second phase is made up of insect communities comprising forms restricted to the Triassic and the oldest representatives of major later Mesozoic and Cenozoic groups. The insect assemblage from the Grès à Voltzia of eastern France (chapter 3) represents an early stage of this second phase. Blattodea, Coleoptera, Ephemeroptera, and Auchenorrhyncha are the dominant groups, and the assemblage also includes the oldest known dipterans. Shcherbakov (2008b) regarded the insect assemblages from the Madygen Formation of Kyrgyzstan (chapter 9) and Molteno Formation of southern Africa (chapter 4), together with those from the Carnian-age Mount Crosby and Blackstone formations of Australia, as representing a later stage of the second phase. He noted that the Madygen assemblage includes 28 first known records of insect groups and 16 last known records but also has 12 apparently endemic families. The Carnian insect assemblages from southern Africa and Australia are not as diverse as that from the Madygen Formation

but contain a greater number of groups that survived into post-Triassic times. Shcherbakov (2008b) characterized the third phase of Triassic insect evolution by the disappearance of various characteristic Triassic groups and the first occurrences of many later Mesozoic taxa. The insect assemblage from the Solite Quarry of Virginia (chapter 9) is the best documented example of this phase. Shcherbakov (2008b) related these three phases of insect evolution to floral changes.

The Triassic fossil record includes the earliest representatives of many important groups of extant insects, such as true flies (Diptera), true beetles (Coleoptera), thrips (Thysanoptera), belostomatid and naucorid water bugs, and xyelid hymenopterans (sawflies) (Grimaldi and Engel 2005). The morphological resemblance of some of these Triassic taxa to the extant members of their respective clades is often remarkable. For example, the staphylinid beetles from the Solite Quarry already have a slender and fusiform body, short elytra, exposing the distal portion of the abdomen, and a dense covering of short hairs (Fraser and Grimaldi 2003; Grimaldi and Engel 2005). Due to the scarcity of Early to early Middle Triassic insect assemblages, it is not yet possible to determine whether groups such as dipterans, thysanopterans, and belostomatids appeared in the fossil record rather suddenly or whether they had longer but as yet unrecorded stratigraphic ranges.

The evolutionary history of many insect clades is intimately linked to that of plants (e.g., Mitter, Farrell, and Wiegmann 1988). The vegetational changes during the Triassic likely had a significant impact on the evolutionary diversification of many lineages of insects (Shcherbakov 2008b). Although stem-group coleopteroids date back to the Early Permian, the oldest true beetles are known from the Triassic (Grimaldi and Engel 2005). Early Mesozoic beetles represent basal lineages that included forms presumably adapted for feeding on conifers and cycads. It is thought that the diversification of flowering plants (angiosperms) during the later Mesozoic led to the tremendous diversification of angiosperm-feeding beetles, which comprise the vast majority of the most diverse clade of present-day coleopterans, Polyphaga (Farrell 1998).

Labandeira, Kvaček, and Mostovski (2007) argued that Mesozoic gymnosperms were comparable to angiosperms in the diversity of their growth habits and presented evidence that there already existed a much greater diversity of obligate associations between specific plant and insect groups than had previously been assumed. Such interactions included feeding on pollen and spores, seed predation, and consumption of a variety of other plant tissues (Labandeira 2006). In the rich fossil record of plant-insect interactions from the Molteno Formation (chapter 4), Labandeira (2006) observed the earliest known examples of leaf mining in addition to feeding on foliage, galling, and seed predation. There is still little fossil evidence for consumption of pollen or nectar by Triassic insects. An exception is a report by Klavins et al. (2005) on Middle Triassic pollen-rich coprolites that contain pollen organs very similar to those of extant cycads. Various groups of early Mesozoic insects had mouthparts suitable for boring into ovules or seeds (Labandeira, Kvaček, and Mostovski 2007).

Hasiotis (2000) reviewed a diversity of trace fossils left by Triassic insects and other arthropods, especially from the Chinle Formation. He found evidence for coprophagy and for nest building by soil-dwelling insects. Hasiotis assigned certain nest structures from the Chinle Formation to termites and bees, respectively. According to Grimaldi and Engel (2005), however, both groups of insects definitely date back only to the Cretaceous. Engel (2000) reinterpreted the alleged bee nests from the Chinle as having been constructed by wood-boring beetles. Much remains to be learned about the "hidden" biodiversity of soil arthropods, but there definitely exists evidence for diverse arthropod-soil interactions during the Triassic.

CHANGES IN TERRESTRIAL TETRAPOD COMMUNITIES DURING THE TRIASSIC

The Triassic marked one of the major changes in the history of continental ecosystems, the replacement of therapsids as the dominant terrestrial tetrapods by archosaurian reptiles. Reviewing the fossil record of tetrapods, Romer (1966) proposed an admittedly oversimplified succession comprising three principal Triassic "faunas":

- Early Triassic ("A"): Therapsids still dominant; archosauriform and basal archosaurian reptiles (grouped together as "thecodontians" at that time) and other less common faunal elements
- Middle Triassic ("B"): Gomphodont cynodonts and rhynchosaurs dominant; considerable diversification of archosaurs including dinosaurian precursors
- Late Triassic ("C"): Dinosaurs dominant; accompanied by derived nondinosaurian archosaurs; therapsids only minor faunal element

Even surveying the fossil record known at that time, Romer noted several transitional assemblages as well as regional differences. Furthermore, some of Romer's "C" assemblages are now considered Early Jurassic rather than Late Triassic in age (Olsen and Galton 1977).

Padian (1986b) distinguished three sets of elements in communities of Triassic continental tetrapods. The first set comprised holdovers from the Paleozoic, which still dominated at the beginning of the Triassic. They included therapsids, procolophonid parareptiles, various groups of temnospondyls, and basal diapsids. Diapsid reptiles had already diverged into archosauromorphs and lepidosauromorphs during the Permian (Benton 1985; chapter 1). The second set of faunal elements includes a considerable number of taxa that appear to be restricted to the Triassic, including a variety of crurotarsan archosaurs and therapsids. The third set comprises numerous lineages that first appeared in the fossil record during the Late Triassic and include many of the principal groups of extant tetrapods such as crocodylomorphs, lepidosaurs, mammaliaforms, turtles, and the precursors of birds (which do not appear in the fossil record until the Late Jurassic).

The Early Triassic record of continental tetrapods is more diverse than often suggested. Tetrapod assemblages still dominated by therapsids, especially the ubiquitous dicynodont *Lystrosaurus*, had a wide geographic distribution during the earliest Triassic. However, therapsids form only a minor element in the Early Triassic (Induan) assemblages from Russia and Australia, where temnospondyls predominate.

The late Early (Olenekian) and early Middle Triassic (Anisian) record of tetrapods is still sparse, but the available evidence, especially in the form of trackways, hints at an already considerable diversification among archosaurian reptiles. Although therapsids never regained the abundance and diversity they had attained during the Late Permian, they remained a significant element of continental tetrapod faunas until the early Late Triassic. Two lineages of herbivorous therapsids, gomphodont cynodonts and kannemeyeriid dicynodonts, diversified during the late Early and Middle Triassic and became abundant, especially in Gondwana.

Among archosaurian reptiles, "rauisuchian" crurotarsans were the principal predators in most Middle and Late Triassic continental tetrapod communities, and their fossil record extends to the end of the Triassic. The first dinosaurs date from the late Carnian or early Norian, but closely related ornithodirans such as *Marasuchus* date back to at least the Ladinian. The best-known assemblage of early dinosaurs, from the Ischigualasto Formation of northwestern Argentina, already includes basal saurischians or theropods (*Eoraptor*, *Herrerasaurus*), a basal sauropodomorph (*Panphagia*), an ornithischian (*Pisanosaurus*), and as yet undescribed additional taxa of dinosaurs (Sereno 1997; Langer and Benton 2006; Martinez and Alcober 2009). Clearly, the initial diversification of Dinosauria took place either earlier during the Carnian or even in pre-Carnian times. By the late Norian, dinosaurs had become common faunal elements in both hemispheres and included a variety of large carnivores and herbivores.

Carnian- and Norian-age tetrapod assemblages from Laurasia are characterized by an abundance of semiaquatic to aquatic predators, specifically metoposaurid temnospondyls and phytosaurs. Both of the latter are also known from India and Madagascar but, for reasons not yet understood, were absent from Africa and South America. A single jaw fragment from the Caturrita Formation of southern Brazil has been attributed to phytosaurs (Kischlat and Lucas 2003), but its identification is problematical (Bonaparte and Sues 2006) and it is worth noting that no other phytosaur fossils have been identified to date in the well-documented assemblages of Late Triassic tetrapods in Argentina and Brazil. Toward the end of the Triassic, crurotarsan archosaurs with the exception of crocodylomorphs vanished, and dinosaurs rapidly diversified during the succeeding Jurassic Period. Procolophonid parareptiles and most lineages of temnospondyls also vanished at or close to the Triassic-Jurassic boundary.

The picture of evolutionary diversification among Triassic small tetrapods is still tantalizingly incomplete, probably because systematic reconnaissance of many Triassic continental strata for microvertebrate remains has not yet been undertaken. The presence of the Early Triassic salientians *Triadobatrachus* (from the base of the Middle Sakamena Group in southwestern Madagascar; Rage and Roček 1989) and *Czatkobatrachus* (from an Olenekian-age fissure deposit in the Czatkowice quarry near Cracow, Poland; Evans and Borsuk-Białynicka 1998) suggests that true frogs (Anura) originated during the Triassic. Yet the oldest known anuran, *Prosalirus*, comes from the Lower Jurassic Kayenta Formation of Arizona (Jenkins and Shubin 1998). Similarly, salamanders (Caudata) and caecilians (Gymnophiona) presumably evolved during the Triassic, especially as the former are considered the sister-group of anurans, but again no undisputed Triassic fossils for either lineage have been discovered to date. The oldest known crown-group salamander, *Chunerpeton*, from the Middle Jurassic Jiulongshan Formation of Inner Mongolia (China; Gao and Shubin 2003), is referable to the extant Cryptobranchidae and indicates that the evolutionary history of salamanders extended farther back in time. The earliest known, still limbed caecilian, *Eocaecilia*, is from the Lower Jurassic Kayenta Formation of Arizona (Jenkins, Walsh, and Carroll 2007).

Among lepidosaurian reptiles, sphenodontians have a well-documented fossil record extending back to the Carnian (Fraser and Benton 1989; Sues, Olsen, and Kroehler 1994). Surprisingly, however, there are still no undisputed Triassic representatives of their sister group, lizards (Squamata; Evans 2003). A single jaw of an acrodontan lizard, *Tikiguania*, has recently been reported from the Carnian-age Tiki Formation of India (Datta and Ray 2006), but Evans (personal communication, 2008) has questioned its provenance. During the later Mesozoic, sphenodontians are common in assemblages in which lizards are un-

common and vice versa, suggesting possible competition between these two lepidosaurian clades.

Cynodont therapsids are first known from the Late Permian of South Africa, Germany, and Russia, but most of the evolutionary diversification of this clade took place during the Triassic (Hopson and Kitching 2001; Rubidge and Sidor 2001; Kemp 2005). During the Early Triassic, eucynodonts diverged into two major lineages, Cynognathia and Probainognathia. The former included the carnivorous *Cynognathus* and the predominantly herbivorous Gomphodontia. The oldest known member of the carnivorous Probainognathia is *Lumkuia* from the *Cynognathus* Assemblage Zone of South Africa (Hopson and Kitching 2001). Probainognathians are common in Ladinian- and Carnian-age assemblages from South America. One probainognathian lineage gave rise to mammals and their closest relatives (Mammaliaformes), which first appear in the fossil record during the Norian (Kielan-Jaworowska, Cifelli, and Luo 2004; Kemp 2005). Certain small probainognathians such as *Brasilitherium* and *Brasilodon* from the Caturrita Formation of Brazil (chapter 4) are closely related to basal mammaliaforms such as *Sinoconodon* from the Lower Jurassic Lower Lufeng Formation of Yunnan, China. At the end of the Triassic, synapsids comprise only mammaliaforms (including mammals), the small, presumably insectivorous Tritheledontidae, which ranged into the Early Jurassic, and the herbivorous Tritylodontidae, which persisted into the Early Cretaceous.

Drawing on a diverse set of anatomical and physiological data, Kemp (2005) pictured the first mammaliaforms as small (with a body weight of 5–10 grams), agile, crepuscular or nocturnal insectivores that could have lived on the forest floor and were capable of climbing trees. This particular mode of life would have limited competition with other small, presumably diurnal insectivores, especially lepidosaurs.

Similarly, archosaurian reptiles were presumably mostly day-active animals (like most extant birds), and, due to their usually considerably larger body size, are unlikely to have been potential competitors for ecological resources.

One significant paleoecological aspect of communities of Triassic continental tetrapods is the proliferation of specialized herbivores. A considerable number of reptilian and therapsid taxa developed dental features suitable for oral processing of high-fiber plant fodder, suggesting intensified exploitation of vegetational resources across a wide range of tetrapod lineages (Gow 1978; Reisz and Sues 2000a). Most procolophonid parareptiles and some archosauromorph reptiles such as *Trilophosaurus* have "cheek" teeth with transversely broad crowns, often bearing multiple cusps, in the more posterior portions of the jaw. They employed orthal jaw motion to break down plant fodder, and the upper and lower teeth interdigitated like cogs when the jaws closed. Presumably plant material was trapped between the interdigitating tooth rows and then chopped or shredded along the apical ridges of individual teeth. Derived rhynchosaurs have maxillary tooth plates with multiple tooth rows that flank a deep longitudinal groove into which the tooth-bearing edge of the dentary would have occluded (Benton 1983b). Among gomphodont cynodonts, traversodonts developed molariform postcanine teeth that met in complex, mammal-like occlusion. During mastication, they employed a power stroke during which the shearing surfaces of the lower postcanine teeth were dragged posterodorsally across the matching surfaces on the upper postcanines (Crompton 1972).

Among ornithodirans, herbivory apparently evolved independently in at least three lineages—*Silesaurus*, Ornithischia, and Sauropodomorpha. During the Late Triassic, large sauropodomorphs such as *Plateosaurus* became the first herbivores that could feed at levels well above the ground. Until that time, herbivorous tetrapods (with the exception of some possibly arboreal forms) presumably foraged within 1 meter above the ground.

THE RISE OF THE ARCHOSAURIA

Various scenarios have attempted to explain the transition from still therapsid-dominated continental tetrapod assemblages of the Early Triassic to the archosaur-dominated communities that came to characterize later Mesozoic continental ecosystems. Most invoke some form of competition and interpret certain morphological or inferred physiological innovations, or both, in archosaurs as conferring selective advantages over potential competitors, especially therapsids.

Charig (1984) emphasized the differences in limb posture between therapsids and other "palaeotetrapods," on the one hand, and archosaurian reptiles, on the other. Therapsids were obligatory quadrupeds, most of which retained sprawling postures (Hotton 1980; Kemp 2005). Archosaurs, on the other hand, typically adopted more upright postures, with the limbs increasingly held and moving closer to the body. This postural change would have facilitated more rapid and, even in many early forms, at least facultatively bipedal locomotion.

Robinson (1971) argued that diapsid reptiles including archosaurs replaced therapsids in part because the former were able to excrete nitrogen as urea with little loss of water. Extant reptiles and birds excrete nitrogen in the form of uric acid, either as slurry (in birds) or as a nearly dry pellet (in lizards), whereas extant mammals (and presumably their therapsid precursors) almost exclusively excrete nitrogen as urea, which requires copious amounts of water to be removed from the body. Given increasingly drier climatic conditions during the early Mesozoic, Robinson (1971) and Hotton (1980) argued that the water-saving disposal of nitrogen waste in archosaurs, along with their locomotor specializations, could have conferred a significant competitive advantage on these reptiles.

Charig (1984) argued that the fully upright limb posture in dinosaurs resulted in "generally superior" locomotor abilities and accounted for the evolutionary success of this group. In his scenario, theropod dinosaurs became more efficient predators that, in the fullness of time, outcompeted carnivorous crurotarsans and also eliminated the dominant groups of herbivores, including aetosaurs, kannemeyeriid dicynodonts, gomphodont cynodonts, and rhynchosaurs. The demise of the latter would, in turn, have facilitated and accelerated the evolutionary diversification of the various lineages of plant-eating dinosaurs.

Benton (1983a) challenged the idea that large-scale competition between dinosaurs with upright limb posture and therapsids and other tetrapods with sprawling limb posture over millions of years could have resulted in the former outcompeting the latter. Instead he argued that the extinctions among therapsids and other groups of tetrapods toward the end of the Triassic could have been related to extrinsic factors, specifically changes such as the disappearance of *Dicroidium*-dominated plant communities in Gondwana. However, the fossil record does not bear out Benton's specific correlation of the extinction of the rhynchosaurs to the disappearance of floras dominated by *Dicroidium* (Rogers et al. 1993).

Retallack, Smith, and Ward (2003) and Ward (2006) related the evolutionary success of archosaurian reptiles, particularly dinosaurs, to the inferred composition of Earth's atmosphere during the Triassic. In developing his scenario, Ward drew on a model developed by Berner (2006) that had calculated that

atmospheric levels of oxygen (O_2) were lower than 15 percent (present-day level: 20.95 percent) at certain points during the interval from the Early Triassic to the Middle Jurassic, while levels of carbon dioxide (CO_2) rose to about 5 percent in the Triassic (present-day level: 0.038 percent). However, Belcher and McElwain (2008) have demonstrated that oxygen levels below 15 percent are inconsistent with data from experimental combustion of plant material and numerous records of charcoal from early Mesozoic wildfires. According to Ward (2006), atmospheric conditions low in oxygen and high in carbon dioxide would have exerted important selective pressures, favoring dinosaurs and their relatives (Ornithodira) with their typically bipedal posture and a birdlike air sac system (in saurischian dinosaurs). Most quadrupedal tetrapods other than mammals employing a sprawling gait have a distinct side-to-side sway to their bodies while walking, and this leads to compression of the lungs, inhibiting inhalation (Carrier and Farmer 2000). By contrast, the rib cage and lungs in a bipedal tetrapod are not affected while the animal is walking, and thus respiration is separated from locomotion. Furthermore, birds have a system of air sacs in addition to lungs, and, based on a variety of skeletal features, an avian-style respiratory system was already present in saurischian dinosaurs (O'Connor and Claessens 2005). Pterosaurs independently developed an analogous flow-through breathing system, which enabled them to develop powered flight and first conquer the skies during the Late Triassic—long before the first birds (Claessens, O'Connor, and Unwin 2009). In birds, inhaled air first passes through the air sacs before entering the lungs, from which it is subsequently exhaled. This system of air passing through the respiratory system in one direction facilitates establishment of a countercurrent system in which the blood in the pulmonary vessels can pass in the opposite direction. This would provide a more efficient means for the extraction of oxygen extraction and disposal of carbon dioxide than the respiratory system in other tetrapods where the lungs are involved in both inhaling and exhaling (Ward 2006).

The incompleteness of the fossil record does not permit rigorous testing of the aforementioned scenarios for explaining the evolutionary success of dinosaurs. In their survey of tetrapod diversity in the Ischigualasto Formation of northwestern Argentina, Rogers et al. (1993) found no evidence for the kind of long-term, gradual decline in the diversity of the earlier tetrapod groups predicted by scenarios invoking competitive replacement. Furthermore, it is questionable that the upright posture in dinosaurs would have conferred a competitive advantage. Indeed, certain Late Triassic "rauisuchians" (e.g., *Effigia*, *Postosuchus*; chapter 8) closely resembled dinosaurs in build and presumably habits and coexisted with dinosaurs for millions of years. Independent from dinosaurs and their closest ornithodiran relatives, these forms and basal crocodylomorphs evolved upright posture, but "rauisuchians" disappeared at the end of the Triassic.

Benton (1983a) and Sereno (1997) viewed early dinosaurs as ecological opportunists that merely took advantage of a "lucky break," taking over niches vacated by the demise of crurotarsan archosaurs. Brusatte et al. (2008) recently presented an analysis of morphological disparity and rates of character evolution in both crurotarsans and dinosaurs. They found higher disparity among crurotarsans but indistinguishable rates of evolutionary change in the two clades. Brusatte and his colleagues regarded this as support of their hypothesis that historical contingency, rather than extended competition or inherent superiority, was the principal factor behind the rise of the dinosaurs. However, the situation is less clear-cut. For example, histological work on long bones by Ricqlès, Padian, and Horner (2003) indicated that most ornithodirans grew at much faster rates than crurotarsans, presumably reflecting high basal metabolic rates in the former. Possibly other, as yet unexplored biological differences could account for the evolutionary success of ornithodirans. Testing hypotheses concerning competition often poses considerable challenges even in present-day ecosystems. Thus, it comes as no surprise that it is difficult, perhaps even impossible, to test such hypotheses convincingly in the fossil record.

CHAPTER 11

The End of the Triassic
Out with a Bang or a Whimper?

If, like the month of March, the Triassic came in like a lion, did it go out like a lamb? While some paleontologists argue that it did (e.g., Weems 1992), it has been widely accepted, especially since the work by Raup and Sepkoski (1982), that the Triassic ended with one of the "Big Five" extinctions during the Phanerozoic. In the marine realm, a major extinction occurred among invertebrates, especially most calcareous demosponges and various groups of molluscs (Hallam 2002). A worldwide 2–3 parts-per-thousand negative carbon isotopic excursion indicates significant changes in the global carbon cycle (Hesselbo et al. 2002). In the terrestrial fossil record, both plant macrofossils and pollen and spores from a number of localities spanning the Triassic-Jurassic boundary in eastern North America and Greenland document significant floral changes at the species level (Olsen et al. 2002a, 2002b; McElwain and Punyasena 2007). However, again it is worth noting that apparently no major clade of plants vanished (McElwain and Punyasena 2007). Among terrestrial tetrapods, most groups of temnospondyls and crurotarsan archosaurs as well as the procolophonid parareptiles disappeared near or at the end of the Triassic (Colbert 1958; Olsen and Sues 1986; Benton 1986, 1991, 1994; Olsen et al. 2002a, 2002b). Several hypoth-

eses have been presented to account for the observed faunal and floral changes. In this chapter we discuss the four best-supported ones.

GRADUAL CHANGE?

The traditional view regarding the tempo of end-Triassic faunal change is one of gradual replacement. Charig (1984) argued for a gradual change in Late Triassic tetrapod communities, as "archaic" forms (which he collectively referred to as "palaeotetrapods") such as the phytosaurs, procolophonids, and rhynchosaurs gave way to "modern" groups (Charig's "neotetrapods") such as crocodylomorphs, dinosaurs, and mammaliaforms. As discussed in chapter 10, he regarded the upright posture of dinosaurs and basal crocodylomorphs as "generally superior" to the sprawling posture of basal therapsids and archosauriform reptiles and thought that the latter groups were simply outcompeted over time. By contrast, Benton (1987) considered it more likely that the "palaeotetrapods" disappeared first and that the "neotetrapods" subsequently occupied their vacant niches. If this were indeed the case then one would expect little overlap between the former and the latter. Unfortunately, the difficulties of accurate dating of continental strata inevitably hinder efforts to test the competing hypotheses. Indeed, a radical reassessment of the ages of the sedimentary strata in the Newark Supergroup has already had a major impact on the traditional concept of gradual change. Cornet, Traverse, and McDonald (1973) and Olsen and Galton (1977) redated the stratigraphic age of the upper units of the Newark Supergroup as Early Jurassic. Until then the entire Newark Supergroup had been considered Triassic in age. Thus, the famous "Triassic" dinosaurian trackways first studied by Hitchcock (1858, 1865) from the Connecticut Valley are now dated as Early Jurassic. Significantly, it became apparent that certain tetrapod taxa, which, according to earlier authors, had disappeared well before the end of the Triassic, actually extended close or up to (or even beyond) the Triassic-Jurassic boundary. This, in turn, led to claims that a major extinction event among terrestrial vertebrates had taken place at this point in time.

A BIG BANG?

Proponents of a Late Triassic mass extinction among terrestrial animals and plants fall into two camps based on the timing of this event and two additional camps based on possible causes for the event. First, it has been argued that a rather sudden, substantial change in the composition of continental tetrapod communities occurred at the Triassic-Jurassic boundary (e.g., Olsen and Sues 1986; Olsen, Shubin, and Anders 1987; Olsen et al. 2002a, 2002b). Proponents of this hypothesis noted that certain taxa such as the leptopleuronine procolophonid *Hypsognathus* occur right up to the (palynologically identified) boundary

but not in younger Newark Supergroup strata whose ages are well constrained.

Olsen, Shubin, and Anders (1987) argued that the impact of a large extraterrestrial object at the Triassic-Jurassic boundary would have had the profound environmental impact that they saw in their reading of the fossil record. In this context, the Manicouagan impact feature in eastern Québec, Canada, received particular attention as a potential candidate for this agent of destruction. This structure may have originally had a diameter of more than 100 kilometers, but subsequent erosion (especially by Pleistocene glaciation) diminished its currently visible diameter to about 70 kilometers (figure 11.1). However, Hodych and Dunning (1992) removed Manicouagan from further consideration as the potential "smoking gun" when they redated the impact to 214 ± 1 Ma using U-Pb dates, which places the impact event long before the Triassic-Jurassic boundary. (Surprisingly, there are as yet no clear signs of a major biotic disturbance in the Newark Supergroup or elsewhere that coincide with the revised date for the impact.) Nevertheless other researchers have continued to present evidence pointing to an extraterrestrial impact at the Triassic-Jurassic boundary. Bice et al. (1992) reported what they considered shocked quartz from strata at the boundary in Italy, but this record has been questioned (Olsen et al. 2002a). Walkden, Parker, and Kelley (2002) interpreted spherules from Late Triassic sequences in southwestern Britain as impact ejecta. Olsen et al. (2002a, 2002b) presented three additional lines of evidence based on data from numerous localities in the Newark Supergroup in support of a possible end-Triassic impact. First, they reported (albeit only modestly) elevated levels of iridium in Triassic-Jurassic boundary strata. They also observed a spike in the abundance of fern spores in sedimentary rocks in the Newark basin of the Newark Supergroup in Pennsylvania

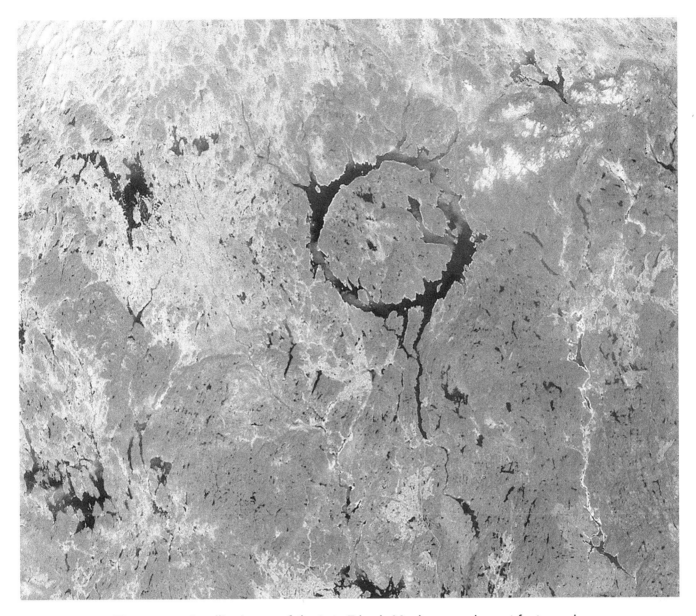

Figure 11.1.   Satellite image of the Late Triassic Manicouagan impact feature, visible as a large, ring-shaped lake, in eastern Québec. (Photograph courtesy of NASA [PIA-03434])

dated to the Triassic-Jurassic boundary. Ferns are known for their ability to proliferate rapidly immediately after major environmental disturbances (but see McElwain et al. 2007). For example, a "fern spike" has been reported from strata postdating the mass extinction at the Cretaceous-Paleogene boundary in western North America. Lastly, Olsen et al.

(2002b) documented a pronounced change in the taxonomic diversity of tetrapod trackways in Newark Supergroup strata at the Triassic-Jurassic boundary. In the tens of thousands of years immediately preceding the boundary, tracks produced by crurotarsan archosaurs declined markedly in relative abundance, while, over the same time interval, dinosaurs went from constituting 20 percent to making up 50 percent of the track makers. Moreover, the authors noted that the size of the makers of the theropod tracks increased dramatically. Olsen et al. (2002b) estimated that theropod dinosaurs might have on average doubled in body mass during this time interval. Lucas and Tanner (2007a) concurred that the disappearance of most crurotarsan archosaurs closely corresponds to the end of the Triassic but argued that the increases in dinosaurian abundance and body size had already occurred earlier during the Late Triassic.

END-TRIASSIC BLOWOUT?

The impact of some kind of extraterrestrial body on Earth is not the only potential cause for a mass extinction. Large-scale volcanic activity over a wide area of the globe's surface would produce similar results and might leave a similar geological signature, possibly even including elevated levels of certain rare elements such as iridium.

The most spectacular volcanic eruption during the last 15,000 years, that of Mount Tambora in Indonesia in 1815, released enormous quantities of ash into the atmosphere, greatly reducing solar radiation. During the following year, 1816, which came to be known as the "year without a summer," a drastic drop in annual temperatures in Europe and North America lead to widespread crop failures and famine. Multiply the force of such an event thousands of times and the resulting devastation becomes unimaginable. Yet the eruptions of the Siberian and Deccan traps represent just such cataclysmic volcanic events in Earth's history, at the end of the Permian and Cretaceous periods, respectively.

Marzoli et al. (1999) recognized a vast Central Atlantic Magmatic Province (CAMP) (figure 11.2), which formed around the Triassic-Jurassic boundary and appears to exceed even the Siberian Traps in extent. The evidence points to more or less synchronous eruptions of volcanic fissures from a common magma reservoir, possibly supplied by a mantle plume, along the pre-Atlantic rift zone. Tholeiitic flood basalts and associated igneous rocks of CAMP came to cover an area of as much as 11 million square kilometers on four present-day continents and could have had a combined volume of some 2 million cubic kilometers. The basalts have been most extensively studied in eastern North America (figure 11.3) and Morocco, where they are best exposed. Associated igneous rocks in Brazil, French Guyana, Surinam, and Guinea often occur in the form of dykes and sills (intrusions into older rocks) (figure 11.2). Significantly, the most recent radiometric ages of CAMP flood basalts tightly cluster together, with a peak around 199–200 Ma (Nomade et al. 2007). Such major volcanic activity over a relatively short period of time would have had dramatic environmental consequences through the release of vast quantities of carbon dioxide (CO_2) and sulfur dioxide (SO_2) as well as huge amounts of water vapor and ash into the atmosphere, resulting in significant cooling, "greenhouse" conditions, or both in succession, on a global scale (Olsen 1999; Guex et al. 2004; Schaltegger et al. 2008). McHone and Puffer (2003) estimated the volume of gaseous aerosols released during the formation of CAMP as well in excess of 10^{12} metric tons.

Even spread out over a period of 100,000 or even a million years, such a quantity of atmospheric emissions would have had a profound impact on global climates. Guex et al. (2004; see also Schaltegger et al. 2008) developed a detailed scenario in which the release of vast amounts of sulfur dioxide initially led

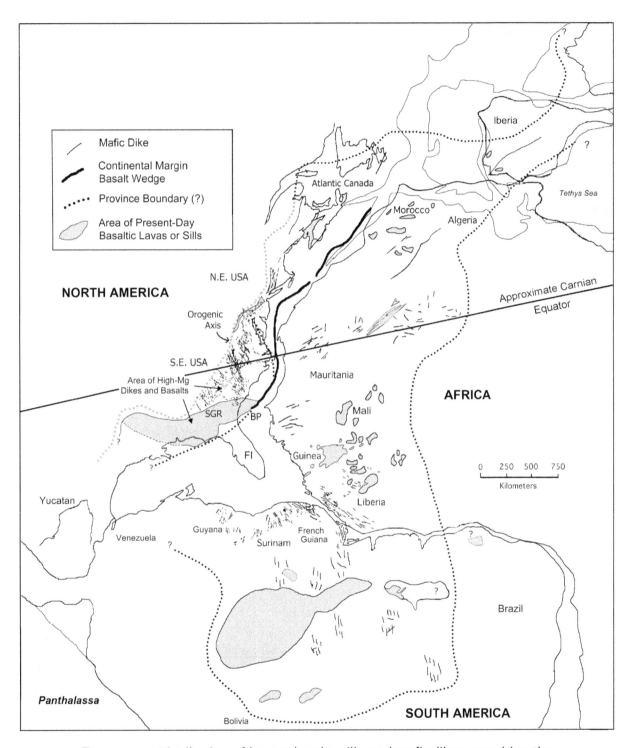

Figure 11.2. Distribution of known basalts, sills, and mafic dikes comprising the Central Atlantic Magmatic Province (CAMP). *BP*, Blake Plateau area, western Atlantic; *SGR*, South Georgia Rift terrane. (Map courtesy and copyright of J. G. McHone)

Figure 11.3. Triassic-Jurassic boundary section along the Bay of Fundy at Five Islands, Nova Scotia. The massive, cliff-forming North Mountain Basalt, which is part of CAMP, overlies Late Triassic lacustrine strata of the Blomidon Formation. (Photograph by H-DS)

to darkening and cooling of the atmosphere. Sulfur dioxide is generally considered a "greenhouse" gas contributing to increases in atmospheric temperatures. However, it also readily reacts with atmospheric water to form sulfate aerosols that absorb or reflect solar radiation, which would have led to a drop in temperatures. Eventually these aerosols were precipitated as acid rain. In the scenario outlined by Guex et al. (2004), this cooling phase was followed by long-term accumulation of vast quantities of carbon dioxide, which would have created "greenhouse" atmospheric conditions at the beginning of the Jurassic. Cleveland et al. (2008) even reported at least two episodes of dramatic increases in atmospheric carbon dioxide preceding the Triassic-Jurassic boundary and related them to volcanic outgassing. Thus, the appearance of the Central Atlantic Magmatic Province certainly provides a plausible causative agent for the end-Triassic mass extinction (Marzoli et al. 2004). However, Nomade et al. (2007) cautioned that such a link must remain tenuous due to the lack of a reliable Triassic-Jurassic boundary datum and difficulties in comparing the results of different methods for radiometric dating. Schaltegger et al. (2008) underscored the need for more precise dating of the early eruptions of CAMP to understand the chronology of events at the Triassic-Jurassic boundary.

McElwain, Beerling, and Woodward (1999) studied the cuticles on leaves from strata rich in plant fossils straddling the (floristically defined) Triassic-Jurassic boundary in the Jameson Land Basin of East Greenland. The fossil record from these beds documents a complete turnover in plant communities at the boundary (McElwain et al. 2007). Late Triassic high-diversity forests dominated by the broad-leaved conifer *Podozamites* and bennettitaleans (*Anomozamites*, *Pterophyllum*) gave way to lower-diversity forests with the ginkgophyte *Sphenobaeira*, the superficially ginkgo-like gymnosperm *Czekanowskia*, and the osmundaceous fern *Todites*. Interestingly, McElwain et al. (2007) observed a gradual decline in plant biodiversity at the genus and especially species level already below the Triassic-Jurassic boundary, which is inconsistent with a catastrophic extinction mechanism.

The cuticles of plant leaves bear many small openings (stomata) through which gas exchange with the surrounding atmosphere takes place. Experimental work on extant angiosperms and gymnosperms has established an inverse relationship between the stomatal index (which measures the density of stomata relative to epidermal cells) on plant leaves and increased levels of atmospheric carbon dioxide. McElwain, Beerling, and Woodward (1999) reported a marked decrease in stomatal density and the stomatal index (a number based on density) in plant fossils across the florally identified Triassic-Jurassic boundary in East Greenland. Based on these data, they inferred a marked increase in atmospheric carbon dioxide and resulting "greenhouse effect" with a sig-

nificant rise in global temperatures of perhaps 2–4 degrees Celsius across the Triassic-Jurassic boundary. However, various studies on extant plants indicate a need for caution when using stomatal density and index to assess atmospheric carbon dioxide levels. Kürschner (1997) demonstrated that even different species of the same plant genus (e.g., *Quercus*) have different stomatal indices, and Cantor et al. (2006) showed that species of *Quercus* raised under identical climatic and atmospheric gas conditions exhibited statistically significant differences in stomatal densities.

Impacts of extraterrestrial objects and cataclysmic volcanic activity are not necessarily mutually exclusive as causes for major extinction events, although it is not clear how the former might have triggered the latter (Olsen et al. 2002a). Certainly, the end-Cretaceous mass extinction more or less coincides temporally with both the impact that formed the giant Chicxulub crater on the present-day Yucatán Peninsula of México and the emplacement of the vast flood basalts forming the Deccan Traps in central India.

AN END-CARNIAN EVENT?

Not all authors have agreed that the main faunal changes occurred at the end of the Triassic. Benton (1986, 1991) argued that many of the tetrapod groups supposedly disappearing at the Triassic-Jurassic boundary had already gone into decline well before then. Nevertheless, he suggested that there was more than just a blip in extinctions of terrestrial vertebrates, and that a major extinction occurred among tetrapods at the end of the Carnian. Subsequent studies have found little if any evidence for an extinction event among continental tetrapods at the Carnian-Norian boundary. For example, the recent discovery of a new late Norian or Rhaetian tetrapod assemblage near Lisowice in southern Poland (Dzik, Sulej, and Niedzwiedzki 2008) has demonstrated that large kannemeyeriid dicynodonts persisted until the end of the Triassic and did not disappear at the end of the Carnian as previously assumed. The revised placement of the chronostratigraphic boundary between the Carnian and Norian at 227–228 Ma by Muttoni et al. (2004) has redated many "late Carnian" tetrapod-bearing strata as Norian, removing much of the evidence previously cited in support of an extinction at the Carnian-Norian transition (Mundil and Irmis 2008).

Is there any evidence for massive and sudden physical perturbations during the Late Triassic? That leads again to consideration of the Manicouagan impact. Spray, Kelley, and Rowley (1998) argued that this might represent not an isolated impact event but one of several in rapid succession. They noted that, in addition to Manicouagan, two other impact sites plot at virtually the same paleolatitude: St. Martin in western Canada and Rochechouart in France. Furthermore, Spray and his colleagues noted two additional impact sites at Obolon in Ukraine and at Red Wing in North Dakota, respectively. All five impact events are thought to have occurred sometime during the Late Triassic (although more precise dating of these impacts is still problematical), but the question is just when and whether they could conceivably have occurred within a short time, perhaps hours, of each other. Spray, Kelley, and Rowley (1998) believed that the answer to the latter question is affirmative, but other authors have categorically rejected this idea. Kent (1998) questioned that the Manicouagan and Rochechouart craters could have formed at the same time. He noted that the melt rocks at the two sites formed under opposing magnetic fields, and, because it takes a few thousand years for a geomagnetic reversal to occur, it was im-

possible for the two structures to have formed at the same time. Spray (1998) responded that there was a considerable difference in size between Rochechouart, with a diameter of 25 kilometers, and Manicouagan, with an original diameter of some 100 kilometers. He pointed out that this difference would have influenced the time required for the melt layers at the two sites to cool below their Curie points, when the magnetic minerals would acquire and lock in their magnetic polarities. Spray estimated that within 100 years after impact the Rochechouart melt layers would have passed through their Curie points, but that it may have taken perhaps 10,000 years in the case of Manicouagan. If this assumption is correct, then the different magnetic polarities of the melt layers would no longer necessarily be an issue, although it does mean that the impacts took place at one of those times when the polarity of Earth's magnetic field was changing.

A PAUCITY OF EVIDENCE

The continuing debate about the tempo and mode of Late Triassic biotic changes reflects to a large extent the scarcity of continuous fossiliferous sections and unresolved issues in chronostratigraphic dating. For example, although Late Triassic tetrapod-bearing strata are widespread throughout the American Southwest and represent a considerable range of depositional environments and geological ages, at least two major unconformities separate the various depositional sequences (chapter 8). Thus, any marked faunal differences observed between assemblages might reflect significant temporal gaps (along with environmental differences) rather than abrupt faunal change. Similarly, until the ages of the fossiliferous fissure fillings in southwestern Britain (chapter 6) can be constrained, these deposits can contribute only equivocal information to the debate.

Collectively, perhaps the most extensive and complete succession of Late Triassic continental strata is found in the Newark Supergroup (chapters 7 and 9). These sequences have had a long if quite sporadic history of exploration. The legendary rivals among nineteenth-century American vertebrate paleontologists, Edward Drinker Cope and Othniel Charles Marsh, were among the first to study tetrapod fossils from the Newark Supergroup, before directing their attention to the much greater paleontological riches of the Western Interior. Subsequently, many paleontologists reported occasional finds of tetrapod remains from the Newark Supergroup. Thus, it is not surprising that Benton would interpret this absence of records of Norian-age continental vertebrates as reflecting a genuine drop in diversity levels during that stage. He reasoned that, since the discovery of extensive assemblages of tetrapod trackways as early as the beginning of the nineteenth century, generation after generation of paleontologists had scrutinized the sedimentary formations of the Newark Supergroup without much in the way of extensive discoveries. Thus, perhaps these strata simply do not contain many tetrapod remains after all, and this, in turn, could indicate that Norian tetrapod communities were characterized by low diversity.

On the face of it, this is a reasonable assumption, but recent discoveries indicate that past generations of paleontologists had not realized the paleontological potential of the Newark Supergroup. Discoveries of reptilian remains in Norian-age strata of the Hartford basin in Connecticut in recent decades are a good case in point. These include the partial skull of an indeterminate sphenodontian (Sues and Baird 1993), skulls and other skeletal remains of the leptopleuronine procolophonid *Hypsognathus* (Sues et al. 2000), and a partial skull of the crocodile-like cruro-

tarsan *Erpetosuchus* (Olsen, Sues, and Norell 2001) from the middle and upper New Haven Formation. In recent years, early Norian strata in the Deep River basin in North Carolina have yielded an impressive diversity of tetrapod fossils (Sues et al. 2003; Peyer et al. 2008). It is becoming increasingly apparent that the Newark Supergroup may shed light on faunal changes during the Late Triassic and across the Triassic-Jurassic boundary after all. However, a more complete picture of these changes is still far in the future. At the present time, researchers must piece together the story from a limited number of snapshots provided by the various assemblages discussed in this book.

THE INSECT RECORD

Compared to insects, vertebrates comprise only a minor component of present-day continental ecosystems, yet they have commanded most of the attention in discussions concerning faunal changes during the Triassic. At the present time, the limitations of the known fossil record of insects and other terrestrial invertebrates make it difficult to offer meaningful comments concerning the pattern of extinction among these groups at the end of the Triassic.

Characteristic Triassic insects include the Titanoptera, which are best known from the Middle Triassic of New South Wales, Australia, and the Middle or Upper Triassic of Kyrgyzstan (Grimaldi and Engel 2005). Béthoux, Papier, and Nel (2005) considered titanopterans most closely related to the Pennsylvanian-age Geraridae and noted as yet undescribed titanopteran fossils from the Upper Permian of southern France and the Anisian-age Grès à Voltzia of eastern France. Gorochov (2007) recently described a titanopteran from the Upper Permian of Orenburg

Province in Russia. There is as yet no known post-Triassic record of this group, and it presumably did not survive the end of that period.

A few horizons yielding an abundance and diversity of Triassic insect fossils, such as the Grès à Voltzia of eastern France (chapter 3) and the Molteno Formation of southern Africa (chapter 4), have been known for many years, but ongoing work on existing collections and new collecting efforts in the field continue to generate significant new information. In addition, new localities, such as the exposures of the Cow Branch Formation in the Solite Quarry in Virginia (chapter 9), have started to yield a remarkable diversity of insects. For example, dipterans from the latter locality are particularly diverse. The earliest known records of belostomatid water bugs, thysanopterans, and staphylinid beetles come from Solite. If taken at face value, this record of insect diversity might hint at some major change at this point in geological time. However, considering the small number of known Early and Middle Triassic insect assemblages, this more likely reflects major gaps in the currently known fossil record of insects rather than actual faunal changes. If some of these gaps are filled by future discoveries, a rather different picture of the end-Triassic terrestrial faunal turnover is likely to emerge. Until that time, any assessment must be based on an inadequate fossil record.

The issues concerning the mode and tempo of terrestrial biotic turnover at the Triassic-Jurassic boundary are still far from being resolved. Yet, this fact should not prevent researchers from continuously reassessing the evidence for the various competing hypotheses. It is important to take stock of any controversy because it will help to focus the direction of future research efforts.

GLOSSARY

Acrodont A type of dentition in which the teeth are fused to the jawbone.

Amniota A clade of tetrapods, comprising reptiles, birds, mammals, and related forms. Amniotes have eggs that are characterized by the presence of an amnion, an embryonic membrane enclosing a fluid-filled chamber in which the embryo develops.

Amphibia Traditionally a grade comprising all extant and extinct tetrapods that are not referable to Amniota. Many authors now use this term only for a group comprising caecilians (Gymnophiona), frogs (Anura), salamanders (Caudata), and their closest relatives.

Apomorphic When different character-states in two taxa are compared and one is thought to have evolved from the other, the former is considered an apomorphic (derived) character-state. (Noun: apomorphy)

Arkose A sandstone containing significant quantities of feldspar (typically more than 25 percent).

Assemblage A set of fossils found together in the same stratum but not necessarily representing the actual composition of the original ancient biota.

Autapomorphic An apomorphic (derived) character-state unique to a particular taxon.

Basal taxon A taxon placed at the base of a phylogenetic tree.

Bennettitales A clade of gymnosperms (also known as Cycadeoidales) with simple or pinnate, cycad-like foliage, reproductive organs in "flower-like" structures in the trunk, and seeds surrounded by sterile scales. Grouped with cycads (Cycadales) in Cycadophyta.

Biostratigraphy The correlation of strata by their fossil content.

Biota The fauna, flora, and other organisms of a particular geographic area.

Bonebed A concentration of vertebrate bones and teeth (from multiple individuals) in a geographically and (or) stratigraphically restricted sedimentary unit.

Caliche Nodules of calcium carbonate that accumulate in the B-horizon of soils. Characteristic of environments with seasonally dry climatic conditions.

Carbonate platform A flat or gently sloping underwater surface comprising carbonate rocks and extending from the shore out to sea.

Carbonate rock A sedimentary rock primarily composed of carbonate minerals, typically calcite or dolomite.

Character An observable or quantifiable feature of the structure, behavior, biochemistry, ecology, or physiology of an organism or, more generally, a taxon.

Character-state The condition of a character found in an organism or, more generally, a taxon. It can denote presence or absence of a particular character or it can describe a specific expression of that character.

Chronostratigraphy The correlation of strata by their age.

Cladogram A branching diagram ("tree") representing a hypothesis of relationships among a number of taxa based on a cladistic analysis. Each branching point or node is supported by one or more synapomorphies.

Clastic rock A sedimentary rock made up of fragments of older rock that formed as a result of weathering of the latter.

Claystone A sedimentary rock primarily comprising particles that are smaller than 1/256 mm and typically are clay minerals. It is not fissile like shale, which it otherwise resembles in composition and texture.

Conglomerate   A sedimentary rock consisting of frequently rounded granules, pebbles, cobbles, or boulders cemented by more fine-grained material.

Coniferophyta   A clade of gymnosperms (also known as Pinophyta) with usually needle-like leaves, compound ovule-bearing and simple pollen cones, and dense (pycnoxylic) wood.

Corystospermales   A clade of gymnosperms with fern-like foliage typically comprising leaves with open, dichotomous venation, and with reproductive organs on branched axes.

Craton   A stable, usually ancient continental part of Earth's lithosphere.

Crown group   A group comprising the last common ancestor of an extant clade and all of its descendants.

Cursorial   Pertaining to running.

Diapsida   A clade of amniotes, which is characterized by the presence of two openings (fenestrae) behind the orbit on either side of the skull. Some derived diapsid taxa no longer retain one or even both openings. Extant Diapsida encompass birds, crocodylians, and lepidosaurs (lizards, snakes, and sphenodontians); some authors also consider turtles highly derived diapsids.

Digitigrade   A type of posture in which only the digits of the manus or pes contact the substrate and the wrist or ankle is raised off the ground.

Dike   A sheet-like body of igneous rock that cuts through adjacent sedimentary or crystalline rock.

Elytra   The hardened forewings forming a protective cover over the flight wings in beetles. (Singular: elytron)

Endemic   Living in a particular geographic area and nowhere else.

Eolian   Pertaining to the agency of wind.

Eon   The largest unit of the geological timescale. Example: Phanerozoic Eon.

Epicontinental   A term for typically shallow seas covering areas underlain by continental crust, as opposed to oceans.

Era   A unit of the geological timescale comprising two or more periods. Example: Mesozoic Era.

Evaporite   A sedimentary rock formed by precipitation of minerals from evaporating water. Example: gypsum.

Extant   Currently existing – the opposite of extinct.

Facies   The combined attributes of a sedimentary rock that reflect a particular depositional environment.

Fissile   Having the tendency to split along bedding planes.

Flood basalt   A basalt that erupted as sheets of lava from fissures, often covering vast areas.

Formation   The fundamental unit of rock stratigraphy (lithostratigraphy). A set of rocks characterized by a particular combination of physical attributes and given a formal name. Example: Ischigualasto Formation.

Fructification   A structure in plants that bears spores or seeds.

Ghost lineage   The temporal range of a lineage inferred from phylogenetic analysis but not yet documented by the fossil record.

Gondwana   The name of the ancient southern landmass that included much of present-day Africa, Antarctica, Australia, Madagascar, and South America as well as parts of Asia (especially India) and southern Europe.

Graben   An elongate downthrown block of crust that is bounded by normal faults along its sides. A half-graben is bounded by a normal fault only along one side.

Group   A lithostratigraphic unit comprising two or more formations. Example: Beaufort Group.

Gymnospermae   A group comprising all spermatophytes that have "naked" seeds that are not enclosed in modified leaves (carpels).

Halophyte   A plant tolerant of salty substrates.

Heterodont   A dentition in which the teeth vary in shape and/or size.

Holometaboly   A type of postembryonic development in insects where there are distinct larval, pupal, and adult stages as well as complete metamorphosis. Insect taxa with holometabolous development are often grouped together as Endopterygota or Holometabola.

Holotype   The individual specimen upon which the taxonomic name of a particular species is based, and which has been formally designated as such in the original published description. All other specimens are only referred to that species.

Homodont   A dentition in which all teeth are alike in shape and size.

Homology   The correspondence in origin, shape, or position of a feature in different taxa that is due to shared common ancestry.

Homoplasy   The appearance of a similar feature in different taxa that is not due to common ancestry. Also referred to as convergence.

Lacustrine Pertaining to lakes.

Lagerstätte A fossil deposit remarkable for the abundance and diversity or the quality of preservation of its fossils. There are two principal types—concentration Lagerstätten and conservation Lagerstätten. The former term refers to deposits that have a remarkable abundance and/or diversity of fossils. The latter denotes deposits in which the preservation of individual fossils is often exceptional. (Plural: Lagerstätten)

Laurasia The name of the ancient northern landmass that included North America and much of Asia and Europe.

Laurentia The name of an ancient landmass comprising North America and Greenland. Also known as the North American craton.

Lineage A group of organisms or taxa descended from a single common ancestor.

Lithostratigraphy The correlation of strata by the physical attributes of the rocks.

Marine Pertaining to the sea.

Marl A calcareous mudstone.

Mass extinction The (in geological time) sudden extinction of large numbers of taxa belonging to many different clades.

Member A lithostratigraphic unit of lower rank than a formation. Example: Tomahawk Member of the Vinita Formation.

Mesic Pertaining to conditions with a moderate amount of moisture.

Monophyletic A group of taxa that includes the last common ancestor of these taxa and all of its descendants. (Noun: monophyly) A strictly monophyletic taxon is referred to as a clade and is diagnosed by at least one synapomorphy.

Mudstone A sedimentary rock comprising clay- and silt-sized particles. Similar to shale in most attributes but not fissile.

Normal fault A type of fault that forms when the hanging wall of rock drops down relative to the footwall.

Paleosol A fossilized soil horizon.

Paludal Pertaining to marshes.

Pangaea The name of the supercontinent that formed through a series of collisions between landmasses during the late Paleozoic and that broke apart again during the Mesozoic. Divided into northern (Laurasia) and southern (Gondwana) portions.

Panthalassa The name of the vast ocean surrounding Pangaea.

Paraphyletic A group of taxa that includes the last common ancestor of these taxa but not all of the descendants of that ancestor. (Noun: paraphyly) Also often referred to as a grade. Most researchers now reject paraphyletic groupings as artificial.

Period The most common unit of the geological time scale, representing a subdivision of an era. Example: Triassic Period. Corresponding chronostratigraphic unit: system.

Pinna The primary division of a compound leaf or frond in plants.

Playa A flat, completely enclosed lake basin in an arid setting, composed of evenly stratified bands of clay, silt, and sand. After rains, water accumulates but quickly evaporates, leaving deposits of soluble salts.

Plesiomorphic When different character-states in two taxa are compared and one is thought to have evolved from the other, the latter is considered a plesiomorphic (primitive) character-state. (Noun: plesiomorphy)

Pleurodont A type of dentition in which the teeth are set in a groove in the jaw, with the outer wall of the groove higher than the inner wall and the teeth attached to the outer wall.

Pteridospermophyta A probably paraphyletic assemblage of Paleozoic and early Mesozoic spermatophytes ("seed ferns") with fern-like foliage and seeds with a small apical opening (micropyle) that did not seal following the capture of pollen.

Red beds Sedimentary strata of any grain size with predominantly red coloration due to the presence of iron oxide.

Sandstone A sedimentary rock primarily comprising particles that range in size from 1/16 mm to 2 mm.

Shale A sedimentary rock composed of clay. It is fissile and does not become plastic when wet.

Siliciclastic rock A sedimentary rock composed of grains of silicate minerals.

Siltstone A sedimentary rock primarily comprising particles that range in size from 1/256 mm to 1/16 mm.

Sister-taxon One of a pair of taxa that are more closely related to each other than to any other taxon.

Spermatophyta A clade comprising all plants in which the ovule is transformed into a seed after fertilization.

Stage A subdivision of strata below the level of system. Example: Norian stage. Corresponding time unit: age. The term "stage" is commonly used to refer to both the time interval and the strata deposited during that time interval.

Stem-group All extinct taxa more closely related to a crown group than to any other group but more basal than the last common ancestor of that crown group.

Stratum A distinct layer of rock. (Plural: strata)

Supergroup A lithostratigraphic unit comprising two or more groups. Example: Karoo Supergroup.

Symplesiomorphy A plesiomorphic character-state shared by two or more taxa.

Synapomorphy An apomorphic character-state shared by two or more taxa and uniting them into a clade.

Synapsida A clade of amniotes comprising mammals and their close relatives (traditionally known as "mammal-like reptiles"). Synapsids are characterized by the presence of a single opening behind the orbit on either side of the skull.

System The most common unit for succession of strata. Example: Triassic System. Corresponding time unit: period.

Taphonomy The study of the biological, chemical, and physical processes from the death of organisms to their preservation as fossils in rock.

Taxon A group of one or more organisms deemed to be a unit within a hierarchical biological classification. A taxon is typically given a formal name and assigned a rank. (Plural: taxa)

Terrestrial Pertaining to land.

Tethys The name of the vast sea that separated Gondwana and Laurasia in the east.

Tetrapod A member of the clade Tetrapoda, which comprises amphibians, reptiles, birds, mammals, and related extinct taxa. Tetrapods typically have four legs but some forms have modified or have lost their limbs.

Thecodont A dentition in which the teeth are set in deep sockets in the jawbone.

Unconformity A surface separating rock units that represents an interval of erosion or nondeposition.

Xeric Pertaining to conditions with little moisture.

Zone A body of strata characterized by the occurrence of a particular fossil or set of fossils (known as guide fossils or index fossils), which had wide geographic distribution but only short temporal ranges.

REFERENCES

Abdala, F., and N. P. Giannini. 2002. Chiniquodont cynodonts: Systematics and morphometric considerations. *Palaeontology* 45:1151–1170.

Abdala, F., P. J. Hancox, and J. Neveling. 2005. Cynodonts from the uppermost Burgersdorp Formation, South Africa, and their bearing on the biostratigraphy and correlation of the Triassic *Cynognathus* Assemblage Zone. *Journal of Vertebrate Paleontology* 25:192–199.

Abdala, F., J. Neveling, and J. Welman. 2006. A new trirachodontid cynodont from the lower levels of the Burgersdorp Formation (Lower Triassic) of the Beaufort Group, South Africa, and the cladistic relationships of Gondwanan gomphodonts. *Zoological Journal of the Linnean Society* 147:383–413.

Achilles, H. 1981. Die rätische und liassische Mikroflora Frankens. *Palaeontographica B* 179:1–86.

Achilles, H., and R. Schlatter. 1986. Palynostratigraphische Untersuchungen im Rhät-Bonebed von Hallau (Kt. Schaffhausen) mit einem Beitrag zur Ammonitenfauna im basalen Lias. *Eclogae geologicae Helvetiae* 79:149–179.

Agassiz, L. 1844. *Monographie des Poissons Fossiles du Vieux Grès Rouge ou Système Devonien (Old Red Sandstone) des Iles Britanniques et de Russie.* Neuchâtel: Jent et Gassmann.

Alberti, F. von. 1834. *Beitrag zu einer Monographie des bunten Sandsteins, Muschelkalks und Keupers, und die Verbindung dieser Gebilde zu einer Formation.* Stuttgart: Verlag der J. B. Cottaschen Buchhandlung.

Alberti, F. von. 1864. *Ueberblick über die Trias, mit Berücksichtigung ihres Vorkommens in den Alpen.* Stuttgart: Verlag der J. B. Cottaschen Buchhandlung.

Alcober, O. 2000. Redescription of the skull of *Saurosuchus galilei* (Archosauria: Rauisuchidae). *Journal of Vertebrate Paleontology* 20:302–316.

Alcober, O., and J. M. Parrish. 1997. A new poposaurid from the Upper Triassic of Argentina. *Journal of Vertebrate Paleontology* 17:548–556.

Anantharaman, S., G. P. Wilson, D. C. Das Sarma, and W. A. Clemens. 2006. A possible Late Cretaceous "haramiyidan" from India. *Journal of Vertebrate Paleontology* 26:488–490.

Anderson, J. M., and H. M. Anderson. 1993a. Terrestrial flora and fauna of the Gondwana Triassic: Part 1—Occurrences. In S. G. Lucas and M. Morales, eds., *The Nonmarine Triassic*, pp. 3–12. New Mexico Museum of Natural History & Science Bulletin 3. Albuquerque: New Mexico Museum of Natural History & Science.

Anderson, J. M., and H. M. Anderson. 1993b. Terrestrial flora and fauna of the Gondwana Triassic: Part 2—Co-evolution. In S. G. Lucas and M. Morales, eds., *The Nonmarine Triassic*, pp. 13–25. New Mexico Museum of Natural History & Science Bulletin 3. Albuquerque: New Mexico Museum of Natural History & Science.

Anderson, J. M., and H. M. Anderson. 2003. Heyday of the gymnosperms: Systematics and biodiversity of the Late Triassic Molteno fructifications. *Strelitzia* 15:1–398.

Anderson, J., H. Anderson, P. Fatti, and H. Sichel. 1996. The Triassic explosion(?): A statistical model for extrapolating biodiversity based on the terrestrial Molteno Formation. *Paleobiology* 22:318–328.

Andreis, R. R., G. E. Bossi, and D. K. Montardo. 1980. O Grupo Rosário do Sul (Triássico) no Rio Grande do

Sul. *Anais Congresso da Sociedade Brasileira de Geología* 31, *Camboriú* 2:659–673.

Arcucci, A. B. 1986. Nuevos materiales y reinterpretacion de *Lagerpeton chanarensis* Romer (Thecodontia, Lagerpetontidae nov.) del Triásico medio de La Rioja, Argentina. *Ameghiniana* 23:233–242.

Arcucci, A. B. 1987. Un nuevo Lagosuchidae (Thecodontia-Pseudosuchia) de la fauna de Los Chañares (Edad Reptil Chañarense, Triásico medio), La Rioja, Argentina. *Ameghiniana* 24:89–94.

Arcucci, A. B., C. A. Marsicano, and A. T. Caselli. 2004. Tetrapod association and palaeoenvironment of the Los Colorados Formation (Argentina): A significant sample from Western Gondwana at the end of the Triassic. *Geobios* 37:557–568.

Arnol'di, L. V., V. V. Zherikhin, L. M. Nikritin, and A. G. Ponomarenko. 1977. [Mesozoic Coleoptera.] *Trudy Paleontologicheskogo Instituta Akademiya Nauk SSSR* 161:1–204. [Russian.]

Arratia, G. 2001. The sister-group of Teleostei: Consensus and disagreements. *Journal of Vertebrate Paleontology* 21:767–773.

Ash, S. R. 1976. The systematic position of *Eoginkgoites*. *American Journal of Botany* 63:1327–1331.

Ash, S. R. 1980. Upper Triassic floral zones of North America. In D. L. Dilcher and T. N. Taylor, eds., *Biostratigraphy of Fossil Plants*, pp. 153–170. Stroudsburg, PA: Hutchinson and Ross.

Ash, S. R. 1986. Fossil plants and the Triassic-Jurassic boundary. In K. Padian, ed., *The Beginning of the Age of Dinosaurs: Faunal Change Across the Triassic-Jurassic Boundary*, pp. 21–30. New York: Cambridge University Press.

Ash, S. R., and G. Creber. 2000. The Late Triassic *Araucarioxylon arizonicum* trees of the Petrified Forest National Park, Arizona, USA. *Palaeontology* 43:15–28.

Ávila, J. N., F. Chemale Jr., G. Mallmann, K. Kawashita, and R. A. Armstrong. 2006. Combined stratigraphic and isotopic studies of Triassic strata, Cuyo Basin, Argentine Precordillera. *Geological Society of America Bulletin* 118:1088–1098.

Axsmith, B. J., T. N. Taylor, N. C. Fraser, and P. E. Olsen. 1997. An occurrence of the Gondwanan plant *Fraxinopsis* in the Upper Triassic of eastern North America. *Modern Geology* 21:299–308.

Bachmann, G. H., and H. W. Kozur. 2004. The Germanic Triassic: Correlations with the international chronostratigraphic scale, numerical ages and Milankovitch cyclicity. *Hallesches Jahrbuch für Geowissenschaften B* 26:17–62.

Bachmann, G. H., G. Beutler, H. Hagdorn, and N. Hauschke. 1999. Stratigraphie der Germanischen Trias. In N. Hauschke and V. Wilde, eds., *Trias—Eine ganz andere Welt*, pp. 81–104. Munich: Verlag Dr. Friedrich Pfeil.

Bain, A. G. 1845. On the discovery of fossil remains of bidental and other reptiles in South Africa. *Transactions of the Geological Society of London*, ser. 2, 7:53–59.

Baird, D. 1957. Triassic reptile footprint faunas from Milford, New Jersey. *Bulletin of the Museum of Comparative Zoology, Harvard College* 117:449–520.

Bandyopadhyay, S. 1988. A kannemeyeriid dicynodont from the Middle Triassic Yerrapalli Formation. *Philosophical Transactions of the Royal Society of London B* 320:185–233.

Bandyopadhyay, S. 1999. Gondwana vertebrate faunas of India. *Proceedings of the Indian National Science Academy A* 65:285–313.

Bandyopadhyay, S., T. K. Roy Chowdhury, and D. P. Sengupta. 2002. Taphonomy of some Gondwana vertebrate assemblages of India. *Sedimentary Geology* 147:219–245.

Barberena, M. C. 1978. A huge thecodont skull from the Triassic of Brazil. *Pesquisas, Porto Alegre* 9:62–75.

Barberena, M. C. 1981. Novos materiais de *Traversodon stahleckeri* da Formação Santa Maria (Triássico do Rio Grande do Sul). *Pesquisas, Porto Alegre* 14:149–162.

Barberena, M. C. 1982. Uma nova espécie de *Proterochampsa* (*P. nodosa*) do Triássico do Brasil. *Anais da Academia Brasileira de Ciências* 54:127–141.

Barberena, M. C., and J. E. F. Dornelles. 1998. A new morphological configuration of the skull and lower jaw of *Cerritosaurus binsfeldi* Price, 1946 after the elimination of distortions caused by taphonomic processes. *Anais da Academia Brasileira de Ciências* 70:469–476.

Barberena, M. C., D. C. Araújo, and E. L. Lavina. 1985. Late Permian and Triassic tetrapods of southern Brazil. *National Geographic Research and Exploration* 1:5–20.

Bartholomai, A. 1979. New lizard-like reptiles from the Early Triassic of Queensland. *Alcheringa* 3:225–234.

Battail, B., and M. V. Surkov. 2000. Mammal-like reptiles from Russia. In M. J. Benton, M. A. Shishkin, D. M. Unwin, and E. N. Kurochkin, eds., *The Age of Dinosaurs in Russia and Mongolia*, pp. 86–119. Cambridge: Cambridge University Press.

Belcher, C. M., and J. C. McElwain. 2008. Limits for combustion in low O_2 redefine paleoatmospheric predictions for the Mesozoic. *Science* 321:1197–1200.

Benton, M. J. 1983a. Dinosaur success in the Triassic: A non-competitive ecological model. *Quarterly Review of Biology* 58:29–55.

Benton, M. J. 1983b. The Triassic reptile *Hyperodapedon* from Elgin: Functional morphology and relationships. *Philosophical Transactions of the Royal Society of London B* 302:605–720.

Benton, M. J. 1985. Classification and phylogeny of the diapsid reptiles. *Zoological Journal of the Linnean Society* 84:97–164.

Benton, M. J. 1986. The Late Triassic tetrapod extinction events. In K. Padian, ed., *The Beginning of the Age of Dinosaurs: Faunal Change Across the Triassic-Jurassic Boundary*, pp. 303–320. New York: Cambridge University Press.

Benton, M. J. 1987. Mass extinctions among families of non-marine tetrapods: The data. *Mémoires de la Société Géologique de France*, n.s., 150:21–32.

Benton, M. J. 1990. The species of *Rhynchosaurus*, a rhynchosaur (Reptilia, Diapsida) from the Middle Triassic of England. *Philosophical Transactions of the Royal Society of London B* 328:213–306.

Benton, M. J. 1991. What really happened in the Late Triassic? *Historical Biology* 5:263–278.

Benton, M. J. 1993. Late Triassic terrestrial vertebrate extinctions: Stratigraphic aspects and the record of the Germanic Basin. In J. M. Mazin and G. Pinna, eds., *Evolution, Ecology and Biogeography of Triassic Reptiles*, pp. 19–38. Palaeontologia Lombarda, n.s., 2. Milan: Società Italiana di Scienze Naturali e Museo Civico di Storia Naturale.

Benton, M. J. 1994. Late Triassic to Middle Jurassic extinctions among continental tetrapods: Testing the pattern. In N. C. Fraser and H.-D. Sues, eds., *In the Shadow of the Dinosaurs: Early Mesozoic Tetrapods*, pp. 366–397. New York: Cambridge University Press.

Benton, M. J. 1999. *Scleromochlus taylori* and the origin of dinosaurs and pterosaurs. *Philosophical Transactions of the Royal Society of London B* 354:1423–1446.

Benton, M. J. 2005. *Vertebrate Palaeontology*. 3rd ed. Oxford: Blackwell Publishing.

Benton, M. J. 2006. *When Life Nearly Died: The Greatest Mass Extinction of All Time*. London: Thames & Hudson.

Benton, M. J., and J. M. Clark. 1988. Archosaur phylogeny and the relationships of the Crocodylia. In M. J. Benton, ed., *The Phylogeny and Classification of the Tetrapods*, vol. 1, *Amphibians, Reptiles, Birds*, pp. 295–338. Oxford: Clarendon Press.

Benton, M. J., and R. J. Twitchett. 2003. How to kill (almost) all life: The end-Permian extinction event. *Trends in Ecology and Evolution* 18:358–365.

Benton, M. J., and A. D. Walker. 1985. Palaeoecology, taphonomy and dating of Permo-Triassic reptiles from Elgin, north-east Scotland. *Palaeontology* 28:207–234.

Benton, M. J., and A. D. Walker. 2002. *Erpetosuchus*, a crocodile-like basal archosaur from the Late Triassic of Elgin, Scotland. *Zoological Journal of the Linnean Society* 136:25–47.

Benton, M. J., G. Warrington, A. J. Newell, and P. S. Spencer. 1994. A review of the British Middle Triassic tetrapod assemblages. In N. C. Fraser and H.-D. Sues, eds., *In the Shadow of the Dinosaurs: Early Mesozoic Tetrapods*, pp. 131–160. New York: Cambridge University Press.

Berman, D. S, and R. R. Reisz. 1992. *Dolabrosaurus aquitalis*, a small lepidosauromorph reptile from the Upper Triassic Chinle Formation of north-central New Mexico. *Journal of Paleontology* 66:1001–1009.

Berner, R. A. 2006. GEOCARBSULF: A combined model for Phanerozoic atmospheric O_2 and CO_2. *Geochimica et Cosmochimica Acta* 70:5653–5664.

Béthoux, O., F. Papier, and A. Nel. 2005. The Triassic radiation of the entomofauna. *Comptes Rendus Palevol* 4: 609–621.

Beutler, G., N. Hauschke, and E. Nitsch. 1999. Faziesentwicklung des Keupers im Germanischen Becken. In N. Hauschke and V. Wilde, eds., *Trias—Eine ganz andere Welt*, pp. 129–174. Munich: Verlag Dr. Friedrich Pfeil.

Bice, D., C. R. Newton, S. McCauley, P. W. Reiners, and C. A. McRoberts. 1992. Shocked quartz at the Triassic-Jurassic boundary in Italy. *Science* 255:443–446.

Bittencourt, J. S., and A. W. A. Kellner. 2009. The anatomy and phylogenetic position of the Triassic dinosaur *Staurikosaurus pricei* Colbert, 1970. *Zootaxa* 2079:1–56.

Blagoderov, V., D. A. Grimaldi, and N. C. Fraser. 2007. How time flies for flies: Diverse Diptera from the Triassic of Virginia and early radiation of the order. *American Museum Novitates* 3572:1–39.

Böhme, W. 1988. Zur Genitalmorphologie der Sauria: Funktionelle und stammesgeschichtliche Aspekte. *Bonner Zoologische Monographien* 27:1–176.

Bolt, J. R., and S. Chatterjee. 2000. A new temnospondyl amphibian from the Late Triassic of Texas. *Journal of Paleontology* 74:670–683.

Bonaparte, J. F. 1962. Descripción del cráneo y mandíbula de *Exaeretodon frenguelli*, Cabrera, y su comparación con

Diademodontidae, Tritylodontidae y los cinodontes sudamericanos. *Publicaciones del Museo Municipal de Ciencias Naturales y Tradicional de Mar del Plata* 1:135–202.

Bonaparte, J. F. 1963. *Promastodonsaurus bellmanni* n. gen. et n. sp., capitosáurido del Triásico medio de Argentina (Stereospondyli-Capitosauroidea). *Ameghiniana* 3:67–78.

Bonaparte, J. F. 1967. New vertebrate evidence for a southern transatlantic connexion during the Lower or Middle Triassic. *Palaeontology* 10:554–563.

Bonaparte, J. F. 1970. Annotated list of the South American Triassic tetrapods. In *Second Gondwana Symposium, South Africa, July to August 1970. Proceedings and Papers*, pp. 665–682. Pretoria: Council for Scientific and Industrial Research.

Bonaparte, J. F. 1972. Los tetrápodos del sector superior de la Formación Los Colorados, La Rioja, Argentina (Triásico superior). Parte I. *Opera Lilloana* 22:1–183.

Bonaparte, J. F. 1973. Edades/reptil para el Triásico de Argentina y Brasil. *Actas del Quinto Congreso Geológico Argentino* 3:93–129.

Bonaparte, J. F. 1975a. Nuevos materiales de *Lagosuchus talampayensis* Romer (Thecodontia-Pseudosuchia) y su significado en el origen de los Saurischia. Chañarense inferior, Triásico medio de Argentina. *Acta Geológica Lilloana* 13:1–90.

Bonaparte, J. F. 1975b. Sobre la presencia del laberintodonte *Pelorocephalus* en la Formación Ischigualasto y su significado estratigráfico (Brachyopoidea-Chigutisauridae). *Actas I Congreso Argentino de Paleontología y Estratigrafía, Tucumán* 1:537–544.

Bonaparte, J. F. 1976. *Pisanosaurus mertii* Casamiquela and the origin of the Ornithischia. *Journal of Paleontology* 50:808–820.

Bonaparte, J. F. 1978. El Mesozoico de América del Sur y sus tetrápodos. *Opera Lilloana* 26:1–596.

Bonaparte, J. F. 1980. El primer ictidosaurio (Reptilia-Therapsida) de América del Sur, *Chaliminia musteloides*, del Triásico superior de La Rioja, República Argentina. *Actas II Congreso Argentino de Paleontología y Bioestratigrafía, Buenos Aires 1978*, 1:123–133.

Bonaparte, J. F. 1981. Descripción de *Fasolasuchus tenax* y su significado en la sistemática y evolución de los Thecodontia. *Revista del Museo Argentino de Ciencas Naturales "Bernardino Rivadavia"* 3:55–101.

Bonaparte, J. F. 1982. Faunal replacement in the Triassic of South America. *Journal of Vertebrate Paleontology* 2: 362–371.

Bonaparte, J. F. 1997. *El Triásico de San Juan-La Rioja, Argentina, y sus Dinosaurios*. Buenos Aires: Museo Argentino de Ciencias Naturales "Bernardino Rivadavia."

Bonaparte, J. F., and M. C. Barberena. 1975. A possible mammalian ancestor from the Middle Triassic of Brazil (Therapsida-Cynodontia). *Journal of Paleontology* 49:931–936.

Bonaparte, J. F., and M. C. Barberena. 2001. On two advanced carnivorous cynodonts from the Late Triassic of southern Brazil. *Bulletin of the Museum of Comparative Zoology, Harvard University* 156:59–80.

Bonaparte, J. F., and J. A. Pumares. 1995. Notas sobre el primer cráneo de *Riojasaurus incertus* (Dinosauria, Prosauropoda, Melanorosauridae) del Triásico superior de La Rioja, Argentina. *Ameghiniana* 32:341–349.

Bonaparte, J. F., and H.-D. Sues. 2006. A new species of *Clevosaurus* (Lepidosauria: Rhynchocephalia) from the Upper Triassic of Rio Grande do Sul, Brazil. *Palaeontology* 49:917–923.

Bonaparte, J. F., J. Ferigolo, and A. M. Ribeiro. 1999. A new early Late Triassic saurischian dinosaur from Rio Grande do Sul State, Brazil. In Y. Tomida, T. H. Rich, and P. Vickers-Rich, eds., *Proceedings of the Second Gondwanan Dinosaur Symposium*, pp. 89–109. National Science Museum Monographs 15. Tokyo: National Science Museum.

Bonaparte, J. F., A. G. Martinelli, and C. L. Schultz. 2005. New information on *Brasilodon* and *Brasilitherium* (Cynodontia, Probainognathia) from the Late Triassic of southern Brazil. *Revista Brasileira de Paleontologia* 8:25–46.

Bonaparte, J. F., A. G. Martinelli, C. L. Schultz, and R. Rubert. 2003. The sister group of mammals: Small cynodonts from the Late Triassic of southern Brazil. *Revista Brasileira de Paleontologia* 5:5–27.

Bordy, E. M., P. J. Hancox, and B. S. Rubidge. 2004. Fluvial style variations in the Late Triassic–Early Jurassic Elliot Formation, main Karoo Basin, South Africa. *Journal of African Earth Sciences* 38:383–400.

Borsuk-Białynicka, M., and S. E. Evans. 2003. A basal archosauriform from the Early Triassic of Poland. *Acta Palaeontologica Polonica* 48:649–652.

Botha, J., and R. H. M. Smith. 2007. *Lystrosaurus* species composition across the Permo-Triassic boundary in the Karoo Basin of South Africa. *Lethaia* 40:125–137.

Bowring, S. A., D. H. Erwin, Y. Jin, M. W. Martin, K. Davidek, and W. Wang. 1998. U/Pb zircon geochronology of the end-Permian mass extinction. *Science* 280: 1039–1045.

Brack, P., H. Rieber, A. Nicora, and R. Mundil. 2005. The Global Boundary Stratotype Section and Point (GSSP) of the Ladinian Stage (Middle Triassic) at Bagolino (southern Alps, northern Italy) and its implications for the Triassic time-scale. *Episodes* 28:233–244.

Brink, A. S. 1963. On *Bauria cynops* Broom. *Palaeontologia Africana* 8:39–56.

Broili, F., and J. Schröder. 1934. Beobachungen an Wirbeltieren der Karooformation. Zur Osteologie des Kopfes von *Cynognathus. Sitzungsberichte der Bayerischen Akademie der Wissenschaften, mathematisch-naturwissenschaftliche Abteilung* 1934:163–177.

Broin, F. de. 1984. *Proganochelys ruchae* n.sp., chelonien du Trias supérieur de Thailande. *Studia Palaeocheloniologica* 1:87–97.

Broom, R. 1913. On the South African pseudosuchian *Euparkeria* and allied genera. *Proceedings of the Zoological Society of London* 1913:619–633.

Brusatte, S. L., M. J. Benton, M. Ruta, and G. T. Lloyd. 2008. Superiority, competition, and opportunism in the evolutionary radiation of dinosaurs. *Science* 321:1485–1488.

Brusatte, S. L., R. J. Butler, T. Sulej, and G. Niedzwiedzki. 2009. The taxonomy and anatomy of rauisuchian archosaurs from the Late Triassic of Germany and Poland. *Acta Palaeontologica Polonica* 54:221–230.

Buffetaut, E. 1983. Mesozoic vertebrates from Thailand: A review. *Acta Palaeontologica Polonica* 28:43–53.

Buffetaut, E. 1993. Phytosaurs in time and space. In J. M. Mazin and G. Pinna, eds., *Evolution, Ecology and Biogeography of Triassic Reptiles*, pp. 39–44. Palaeontologia Lombarda, n.s., 2. Milan: Società Italiana di Scienze Naturali e Museo Civico di Storia Naturale.

Buffetaut, E., and R. Ingavat. 1982. Phytosaur remains (Reptilia, Thecodontia) from the Upper Triassic of northeastern Thailand. *Geobios* 15:7–17.

Bürgin, T., O. Rieppel, P. M. Sander, and K. Tschanz. 1989. The fossils of Monte San Giorgio. *Scientific American* 260:74–81.

Burmeister, H. 1849. *Die Labyrinthodonten aus dem bunten Sandstein von Bernburg, zoologisch geschildert. Erste Abtheilung.* Trematosaurus. Berlin: Verlag von G. Reimer.

Butler, P. M., and G. T. MacIntyre. 1994. Review of the British Haramiyidae (?Mammalia, Allotheria), their molar occlusion and relationships. *Philosophical Transactions of the Royal Society of London B* 345:433–458.

Butler, R. J., R. M. H. Smith, and D. B. Norman. 2007. A primitive ornithischian dinosaur from the Late Triassic of South Africa, and the early evolution and diversification of Ornithischia. *Proceedings of the Royal Society B* 274:2041–2046.

Bystrov, A. P. 1935. Morphologische Untersuchungen der Deckknochen des Schädels der Wirbeltiere. I. Mitteilung. Schädel der Stegocephalen. *Acta Zoologica (Stockholm)* 16:65–141.

Bystrov, A. P., and I. A. Efremov. 1940. [*Benthosuchus sushkini* Efr.—a labyrinthodont from the Eotriassic of Sharzhenga River.] *Trudy Paleontologicheskogo Instituta Akademiya Nauk SSSR* 10:1–152. [Russian with English translation.]

Calzavara, M., G. Muscio, and R. Wild. 1981. *Megalancosaurus preonensis* n.g., n.sp., a new reptile from the Norian of Friuli, Italy. *Gortania* 2:49–64.

Camp, C. L. 1930. A study of the phytosaurs with description of new material from western North America. *Memoirs of the University of California* 10:1–161.

Camp, C. L. 1945. *Prolacerta* and the protorosaurian reptiles. Parts 1 and 2. *American Journal of Science* 243:17–32 and 84–101.

Camp, C. L., and S. P. Welles. 1956. Triassic dicynodont reptiles. Part I. The North American genus *Placerias. Memoirs of the University of California* 13:255–304.

Cantor, B. M., B. V. Aigler, D. W. Pace, S. B. Reid, C. Y. Thompson, and R. A. Gastaldo. 2006. Intra- and interspecific variation in stomatal proxies for *Quercus* and *Nyssa* in the subtropical southeastern USA. *Geological Society of America, Abstracts with Program* 38(7):487.

Cantrill, D. J., and J. A. Webb. 1998. Permineralised pleuromeid lycopsid remains from the Early Triassic Arcadia Formation, Queensland, Australia. *Review of Palaeobotany and Palynology* 102:189–211.

Carrier, D. R., and C. G. Farmer. 2000. The integration of ventilation and locomotion in archosaurs. *American Zoologist* 40:87–100.

Carroll, R. L. 1988. *Vertebrate Paleontology and Evolution.* New York: W. H. Freeman and Company.

Carroll, R. L. 2009. *The Rise of Amphibians.* Baltimore: Johns Hopkins University Press.

Carroll, R. L., and W. Lindsay. 1985. Cranial anatomy of the primitive reptile *Procolophon. Canadian Journal of Earth Sciences* 22:1571–1587.

Carroll, R. L., and R. Wild. 1994. Marine members of the diapsid order Sphenodontia. In N. C. Fraser and H.-D. Sues, eds., *In the Shadow of the Dinosaurs: Early Mesozoic Tetrapods*, pp. 70–83. New York: Cambridge University Press.

Carroll, R. L., E. S. Belt, D. L. Dineley, D. Baird, and D. C. McGregor. 1972. *Excursion A59: Vertebrate Paleontology of Eastern Canada.*, Montreal, Quebec: Twenty-Fourth International Geological Congress.

Casamiquela, R. M. 1960. Noticia preliminar sobre dos nuevos estagonolepoideos Argentinos. *Ameghiniana* 2:3–9.

Casamiquela, R. M. 1962. Dos nuevos estagonolepoideos Argentinos (de Ischigualasto, San Juan). *Revista de la Asociación Geológica Argentina* 16:143–203.

Casamiquela, R. M. 1967. Un nuevo dinosaurio ornitisquio triásico (*Pisanosaurus mertii*; Ornithopoda) de la Formación Ischigualasto, Argentina. *Ameghiniana* 5:47–64.

Case, E. C. 1922. New reptiles and stegocephalians from the Upper Triassic of western Texas. *Carnegie Institution of Washington Publication* 321:1–84.

Caselli, A. T., C. A. Marsicano, and A. B. Arcucci. 2001. Sedimentología y paleontología de la Formación Los Colorados, Triásico superior (provincias de La Rioja y San Juan, Argentina). *Revista de la Asociación Geológica Argentina* 56:171–188.

Casey, M. M., N. C. Fraser, and M. Kowalewski. 2007. Quantitative taphonomy of a Triassic reptile *Tanytrachelos ahynis* from the Cow Branch Formation, Dan River Basin, Solite Quarry, Virginia. *Palaios* 22:598–611.

Catuneanu, O., P. J. Hancox, and B. S. Rubidge. 1998. Reciprocal flexural behaviour and contrasting stratigraphies: A new basin development model for the Karoo retroarc foreland system, South Africa. *Basin Research* 10:417–439.

Catuneanu, O., H. Wopfner, P. G. Eriksson, B. Cairncross, B. S. Rubidge, R. M. H. Smith, and P. J. Hancox. 2005. The Karoo basins of south-central Africa. *Journal of African Earth Sciences* 43:211–253.

Chakraborty, C., N. Mandal, and S. K. Ghosh. 2003. Kinematics of the Gondwana basins of peninsular India. *Tectonophysics* 377:299–324.

Charig, A. J. 1984. Competition between therapsids and archosaurs during the Triassic period: A review and synthesis of current theories. In M. W. J. Ferguson, ed., *The Structure, Development and Evolution of Reptiles*, pp. 597–628. Zoological Society of London Symposia 52. London: Academic Press.

Chatterjee, S. 1974. A rhynchosaur from the Upper Triassic Maleri Formation of India. *Philosophical Transactions of the Royal Society of London B* 267:209–261.

Chatterjee, S. 1978. A primitive parasuchid (phytosaur) reptile from the Upper Triassic Maleri Formation of India. *Palaeontology* 21:83–127.

Chatterjee, S. 1980a. The evolution of rhynchosaurs. *Mémoires de la Société Géologique de France*, n.s., 139:57–65.

Chatterjee, S. 1980b. *Malerisaurus*, a new eosuchian reptile from the Late Triassic of India. *Philosphical Transactions of the Royal Society of London B* 291:163–200.

Chatterjee, S. 1982. A new cynodont reptile from the Triassic of India. *Journal of Paleontology* 56:203–214.

Chatterjee, S. 1983. An ictidosaur fossil from North America. *Science* 220:1151–1153.

Chatterjee, S. 1984. A new ornithischian dinosaur from the Triassic of North America. *Naturwissenschaften* 71:630–631.

Chatterjee, S. 1985. *Postosuchus*, a new thecodontian reptile from the Triassic of Texas and the origin of tyrannosaurs. *Philosophical Transactions of the Royal Society of London B* 309:395–460.

Chatterjee, S. 1991. Cranial anatomy and relationships of a new Triassic bird from Texas. *Philosophical Transactions of the Royal Society of London B* 332:277–342.

Chatterjee, S. 1993. *Shuvosaurus*, a new theropod. *National Geographic Research and Exploration* 9:274–285.

Chatterjee, S. 1999. *Protoavis* and the early evolution of birds. *Palaeontographica A* 254:1–100.

Chatterjee, S., and P. K. Majumdar. 1987. *Tikisuchus romeri*, a new rauisuchid reptile from the Late Triassic of India. *Journal of Paleontology* 61:787–793.

Christian, A., and H. Preuschoft. 1996. Deducing the body posture of extinct large vertebrates from the shape of the vertebral column. *Palaeontology* 39:801–812.

Cisneros, J. C., R. Damiani, C. Schultz, A. da Rosa, C. Schwanke, L. W. Neto, and P. L. P. Aurélio. 2004. A procolophonoid reptile with temporal fenestration from the Middle Triassic of Brazil. *Proceedings of the Royal Society of London B* 271:1541–1546.

Claessens, L. P. A. M., P. M. O'Connor, and D. M. Unwin. 2009. Respiratory evolution facilitated the origin of pterosaur flight and aerial gigantism. *PLoS One* 4:e4497.

Clark, J. M., H.-D. Sues, and D. S Berman. 2001. A new specimen of *Hesperosuchus* from the Upper Triassic of New Mexico and the interrelationships of basal crocodylomorph archosaurs. *Journal of Vertebrate Paleontology* 20:683–704.

Clemens, W. A. 1980. Rhaeto-Liassic mammals from Switzerland and West Germany. *Zitteliana* 5:51–92.

Cleveland, D. M., S. C. Atchley, and L. C. Nordt. 2007. Continental sequence-stratigraphy of the Late Triassic (Norian-Rhaetian) Chinle strata, northern New Mexico: Allo- and autocyclic origins of paleosol-bearing alluvial successions. *Journal of Sedimentary Research* 77: 909–924.

Cleveland, D. M., L. C. Nordt, S. I. Dworkin, and S. C. Atchley. 2008. Pedogenic carbonate isotopes as evidence for extreme climatic events preceding the Triassic-Jurassic boundary: Implications for the biotic crisis? *Geological Society of America Bulletin* 120:1408–1415.

Cluver, M. A. 1971. The cranial anatomy of the dicynodont genus, *Lystrosaurus. Annals of the South African Museum* 56:35–54.

Colbert, E. H. 1946. *Hypsognathus*, a Triassic reptile from New Jersey. *Bulletin of the American Museum of Natural History* 86:225–274.

Colbert, E. H. 1958. Tetrapod extinctions at the end of the Triassic. *Proceedings of the National Academy of Sciences U.S.A.* 44:973–977.

Colbert, E. H. 1965. A phytosaur from North Bergen, New Jersey. *American Museum Novitates* 2230:1–25.

Colbert, E. H. 1966. A gliding reptile from the Triassic of New Jersey. *American Museum Novitates* 2246:1–23.

Colbert, E. H. 1970a. The gliding Triassic reptile *Icarosaurus. Bulletin of the American Museum of Natural History* 143:85–142.

Colbert, E. H. 1970b. A saurischian dinosaur from the Triassic of Brazil. *American Museum Novitates* 2405:1–39.

Colbert, E. H. 1974. *Lystrosaurus* from Antarctica. *American Museum Novitates* 2535:1–44.

Colbert, E. H. 1989. The Triassic dinosaur *Coelophysis. Museum of Northern Arizona Bulletin* 57:1–160.

Colbert, E. H., and P. E. Olsen. 2001. A new and unusual aquatic reptile from the Lockatong Formation of New Jersey (Late Triassic, Newark Supergroup). *American Museum Novitates* 3334:1–24.

Colombi, C. E., and J. T. Parrish. 2008. Late Triassic environmental evolution in southwestern Pangea: Plant taphonomy of the Ischigualasto Formation. *Palaios* 23: 778–795.

Cope, E. D. 1887a. The dinosaurian genus *Coelurus. American Naturalist* 21:367–369.

Cope, E. D. 1887b. A contribution to the history of the Vertebrata of the Trias of North America. *Proceedings of the American Philosophical Society* 24:209–228.

Cope, E. D. 1889. On a new genus of Triassic Dinosauria. *American Naturalist* 23:626.

Cornet, B. 1993. Applications and limitations of palynology in age, climatic, and paleoenvironmental analysis of Triassic sequences in North America. In S. G. Lucas

and M. Morales, eds., *The Nonmarine Triassic*, pp. 75–93. New Mexico Museum of Natural History & Science Bulletin 3. Albuquerque: New Mexico Museum of Natural History & Science.

Cornet, B., and P. E. Olsen. 1990. *Early to Middle Carnian (Triassic) Flora and Fauna of the Richmond and Taylorsville Basins, Virginia and Maryland, U.S.A.* Virginia Museum of Natural History Guidebook, no. 1. Martinsville: Virginia Museum of Natural History.

Cornet, B., A. Traverse, and N. G. McDonald. 1973. Fossil spores, pollen, and fishes from Connecticut indicate Early Jurassic age for part of the Newark Group. *Science* 182:1243–1247.

Cosgriff, J. W. 1974. Lower Triassic Temnospondyli of Tasmania. *Geological Society of America Special Paper* 149: 1–131.

Cosgriff, J. W. 1984. The temnospondyl labyrinthodonts of the earliest Triassic. *Journal of Vertebrate Paleontology* 4:30–46.

Cox, C. B. 1965. New Triassic dicynodonts from South America, their origins and relationships. *Philosophical Transactions of the Royal Society of London B* 248:457–516.

Cox, C. B. 1969. Two new dicynodonts from the Triassic Ntawere Formation, Zambia. *Bulletin of the British Museum (Natural History), Geology* 17:257–294.

Cox, C. B. 1991. The Pangaean dicynodont *Rechnisaurus* and the comparative biostratigraphy of Triassic dicynodont faunas. *Palaeontology* 34:767–784.

Cox, C. B., and J. Li. 1983. A new genus of Triassic dicynodont from East Africa and its classification. *Palaeontology* 26:389–406.

Crompton, A. W. 1955. On some Triassic cynodonts from Tanganyika. *Proceedings of the Zoological Society of London* 125:617–669.

Crompton, A. W. 1972. Postcanine occlusion in cynodonts and tritylodonts. *Bulletin of the British Museum (Natural History), Geology* 21:27–71.

Cruickshank, A. R. I. 1967. A new dicynodont genus from the Manda Formation of Tanzania (Tanganyika). *Journal of Zoology, London* 153:163–208.

Cruickshank, A. R. I. 1972. The proterosuchian thecodonts. In K. A. Joysey and T. S. Kemp, eds., *Studies in Vertebrate Evolution*, pp. 89–119. Edinburgh: Oliver and Boyd.

Crush, P. J. 1984. A late Triassic sphenosuchid crocodilian from Wales. *Palaeontology* 34:131–157.

Dalla Vecchia, F. 2003. New morphological observations on Triassic pterosaurs. In E. Buffetaut and J.-M.

Mazin, eds., *Evolution and Palaeobiology of Pterosaurs*, pp. 23–44. Geological Society of London Special Publication 217. London: Geological Society of London.

Dalla Vecchia, F. 2006. The tetrapod fossil record from the Norian-Rhaetian of Friuli (northeastern Italy). In J. D. Harris, S. G. Lucas, J. A. Spielmann, M. G. Lockley, A. R. C. Milner, and J. I. Kirkland, eds., *The Triassic-Jurassic Terrestrial Transition*, pp. 432–444. New Mexico Museum of Natural History & Science Bulletin 37. Albuquerque: New Mexico Museum of Natural History & Science.

Damiani, R. J. 2001. A systematic revision of and phylogenetic analysis of the Triassic mastodonsauroids (Temnospondyli: Stereospondyli). *Zoological Journal of the Linnean Society* 133:379–482.

Damiani, R. J., J. Neveling, and P. J. Hancox. 2001. First record of a mastodonsaurid (Temnospondyli: Stereospondyli) from the Early Triassic *Lystrosaurus* Assemblage Zone (Karoo Basin) of South Africa. *Neues Jahrbuch für Geologie und Paläontologie, Abhandlungen* 221:133–144.

Damiani, R., S. Modesto, A. Yates, and J. Neveling. 2003. Earliest evidence of cynodont burrowing. *Proceedings of the Royal Society of London B* 270:1747–1751.

Damiani, R. J., R. R. Schoch, H. Hellrung, R. Werneburg, and S. Gastou. 2009. The plagiosaurid temnospondyl *Plagiosuchus pustuliferus* (Amphibia: Temnospondyli) from the Middle Triassic of Germany: Anatomy and functional morphology of the skull. *Zoological Journal of the Linnean Society* 155:348–373.

Damiani, R. J., C. Vasconcelos, A. Renaut, J. Hancox, and A. Yates. 2007. *Dolichuranus primaevus* (Therapsida: Amonodontia) from the Middle Triassic of Namibia and its phylogenetic relationships. *Palaeontology* 50:1531–1546.

Datta, P. M. 2005. Earliest mammal with transversely expanded upper molar from the Late Triassic (Carnian) Tiki Formation, South Rewa Gondwana Basin, India. *Journal of Vertebrate Paleontology* 25:200–207.

Datta, P. M., and D. P. Das. 1996. Discovery of the oldest mammal from India. *Indian Minerals* 50:217–222.

Datta, P. M., and S. Ray. 2006. Earliest lizard from the Late Triassic (Carnian) of India. *Journal of Vertebrate Paleontology* 26:795–800.

Datta, P. M., D. P. Das, and Z.-X. Luo. 2004. A Late Triassic dromatheriid (Synapsida: Cynodontia) from India. *Annals of Carnegie Museum* 73:72–84.

Delevoryas, T., and R. C. Hope. 1973. Fertile coniferophyte remains from the Late Triassic Deep River basin, North Carolina. *American Journal of Botany* 60:810–818.

Delevoryas, T., and R. C. Hope. 1975. *Voltzia andrewsii*, n. sp.: An Upper Triassic seed cone from North Carolina, U.S.A. *Review of Palaeobotany and Palynology* 20:67–74.

Demathieu, G. 1989. The appearance of the first dinosaur tracks in the French Middle Triassic and their probable significance. In D. D. Gillette and M. G. Lockley, eds., *Dinosaur Tracks and Traces*, pp. 201–207. New York: Cambridge University Press.

Demko, T. M., R. F. Dubiel, and J. T. Parrish. 1998. Plant taphonomy in incised valleys: Implications for interpreting paleoclimate from fossil plants. *Geology* 26:1119–1122.

Desojo, J. B., and A. M. Báez. 2007. Cranial morphology of the Late Triassic South American archosaur *Neoaetosauroides engaeus*: Evidence for aetosaurian diversity. *Palaeontology* 50:267–276.

Desojo, J. B., A. B. Arcucci, and C. A. Marsicano. 2002. Reassessment of *Cuyosuchus huenei*, a Middle-Late Triassic archosauriform from the Cuyo Basin, west-central Argentina. In A. B. Heckert and S. G. Lucas, eds., *Upper Triassic Stratigraphy and Paleontology*, pp. 143–148. New Mexico Museum of Natural History & Science Bulletin 21. Albuquerque: New Mexico Museum of Natural History & Science.

Deutsche Stratigraphische Kommission, ed. 2005. Stratigraphie von Deutschland. IV—Keuper. *Courier Forschungsinstitut Senckenberg* 253:1–296.

De Wit, M. J., and I. G. D. Ransome. 1992. Regional inversion tectonics along the southern margin of Gondwana. In M. J. de Wit and I. G. D. Ransome, eds., *Inversion Tectonics of the Cape Fold Belt, Karoo and Cretaceous Basins of Southern Africa*, pp. 15–22. Rotterdam: Balkema.

Dias-da-Silva, S., S. P. Modesto, and C. L. Schultz. 2007. New material of *Procolophon* (Parareptilia: Procolophonidae) from the Lower Triassic of Brazil, with remarks on the age of the Sanga do Cabral and Buena Vista formations of South America. *Canadian Journal of Earth Sciences* 43:1685–1693.

Dickinson, W. R., L. S. Beard, G. R. Brakenridge, J. L. Erjavec, R. C. Ferguson, K. F. Inman, R. A. Knepp, F. A. Lindberg, and P. T. Ryberg. 1983. Provenance of North American Phanerozoic sandstones in relation to tectonic setting. *Geological Society of America Bulletin* 94:222–235.

Dilkes, D. W. 1995. The rhynchosaur *Howesia browni* from the Lower Triassic of South Africa. *Palaeontology* 38:665–685.

Dilkes, D. W. 1998. The Early Triassic rhynchosaur *Mesosuchus browni* and the interrelationships of basal archosauromorph reptiles. *Philosophical Transactions of the Royal Society of London B* 353:501–541.

Dilkes, D. W., and H.-D. Sues. 2009. Redescription and phylogenetic relationships of *Doswellia kaltenbachi* (Diapsida: Archosauriformes) from the Upper Triassic of Virginia. *Journal of Vertebrate Paleontology* 29: 58–79.

DiMichele, W. A., H. Kerp, N. J. Tabor, and C. V. Looy. 2008. The so-called "Paleophytic-Mesophytic" transition in equatorial Pangea—multiple biomes and vegetational tracking of climate change through geological time. *Palaeogeography, Palaeoclimatology, Palaeoecology* 268:152–163.

DiMichele, W. A., S. H. Mamay, D. S. Chaney, R. W. Hook, and W. J. Nelson. 2001. An Early Permian flora with Late Permian and Mesozoic affinities from north-central Texas. *Journal of Paleontology* 75:449–460.

Dobruskina, I. A. 1994. *Triassic Floras of Eurasia.* Österreichische Akademie der Wissenschaften, Schriftenreihe der Erdwissenschaftlichen Kommissionen, Band 10. Vienna: Springer-Verlag.

Dobruskina, I. A. 1995. Keuper (Triassic) flora from Middle Asia (Madygen, Southern Fergana). *New Mexico Museum of Natural History & Science Bulletin* 5:1–49.

Dobruskina, I. A. 1998. Lunz flora in the Austrian Alps—a standard for Carnian floras. *Palaeogeography, Palaeoclimatology, Palaeoecology* 143:307–345.

Doyle, K. D., and H.-D. Sues. 1995. Phytosaurs (Reptilia: Archosauria) from the Upper Triassic New Oxford Formation of York County, Pennsylvania. *Journal of Vertebrate Paleontology* 15:545–553.

Dubiel, R.F., J. T. Parrish, J. M. Parrish, and S. C. Good. 1991. The Pangaean megamonsoon—evidence from the Upper Triassic Chinle Formation, Colorado Plateau. *Palaios* 6:347–370.

Dutuit, J.-M. 1976. Introduction à l'étude paléontologique du Trias continental marocain. Description des premiers Stégocephales recueillis dans le couloir d'Argana (Atlas occidental). *Mémoires du Muséum National d'Histoire Naturelle C* 36:1–253.

Dutuit, J.-M. 1989. Confirmation des affinités entre Trias supérieurs marocain et sud-américain: Découverté d'un troisième dicynodonte (Reptilia, Therapsida), *Azarifeneria robustus*, n. sp., de la formation d'Argana (Atlas occidental). *Comptes Rendus de l'Académie des Sciences*, sér. II, 309:1267–1270.

Dyke, G. J., R. L. Nudds, and J. M. V. Rayner. 2006. Flight of *Sharovipteryx mirabilis*: The world's first delta-winged glider. *Journal of Evolutionary Biology* 19:1040–1043.

Dzik, J. 2001. A new *Paleorhinus* fauna in the Late Triassic of Poland. *Journal of Vertebrate Paleontology* 21:625–627.

Dzik, J. 2003. A beaked herbivorous archosaur with dinosaur affinities from the early Late Triassic of Poland. *Journal of Vertebrate Paleontology* 23:556–574.

Dzik, J., and T. Sulej. 2007. A review of the early Late Triassic Krasiejów biota from Silesia, Poland. *Palaeontologia Polonica* 64:3–27.

Dzik, J., T. Sulej, and G. Niedzwiedzki. 2008. A dicynodont-theropod association in the latest Triassic of Poland. *Acta Palaeontologica Polonica* 53:733–738.

Ebel, C., F. Falkenstein, F. Haderer, and R. Wild. 1998. *Ctenosauriscus koeneni* (v. Huene) und der Rauisuchier von Waldshut—biomechanische Deutung der Wirbelsäule und Beziehungen zu *Chirotherium sickleri* Kaup. *Stuttgarter Beiträge zur Naturkunde B* 116:1–19.

Edwards, B., and S. E. Evans. 2006. A Late Triassic microvertebrate assemblage from Ruthin Quarry, Wales. In P. M. Barrett and S. E. Evans, eds., *Ninth International Symposium on Mesozoic Terrestrial Ecosystems and Biota, Abstracts and Proceedings*, pp. 33–35. London: Natural History Museum.

Ellenberger, P. 1970. Les niveaux paléontologiques de première apparition des mammifères primordiaux en Afrique du sud et leur ichnologie. Établissement de zones stratigraphiques détaillées dans le Stormberg du Lesotho (Afrique du sud) (Trias supérieur à Jurassique). In *Second Gondwana Symposium, South Africa, July to August 1970. Proceedings and Papers*, pp. 343–370. Pretoria: Council for Scientific and Industrial Research.

Ellenberger, P. 1972. Contributions à la classification des pistes de vertébrés du Trias: Les types du Stormberg d'Afrique du sud (I). *Palaeovertebrata, Mémoire Extraordinaire*:1–117.

Ellenberger, P. 1974. Contributions à la classification des pistes de vertébrés du Trias: Les types du Stormberg d'Afrique du sud (II). *Palaeovertebrata, Mémoire Extraordinaire*:1–141.

Embry, A. F. 1988. Triassic sea level changes: Evidence from the Canadian Arctic archipelago. *Society of Economic Paleontologists and Mineralogists Special Publication* 42: 249–259.

Emmons, E. 1856. *Geological Report of the Midland Counties of North Carolina.* New York: George P. Putnam & Co.

Emmons, E. 1857. *American Geology, Containing a State-ment of the Principles of the Science, with Full Illustrations of the Characteristic American Fossils. With an Atlas and a Geological Map of the United States.* Part VI. Albany: Sprague and Co.

Engel, M. S. 2000. A new interpretation of the oldest fossil bee (Hymenoptera: Apidae). *American Museum Novitates* 3296:1–11.

Erba, E. 2006. The first 150 million years history of calcareous nannoplankton: Biosphere-geosphere interactions. *Palaeogeography, Palaeoclimatology, Palaeoecology* 232: 237–250.

Erwin, D. H. 2006. *Extinction: How Life on Earth Nearly Ended 250 Million Years Ago.* Princeton, NJ: Princeton University Press.

Evans, S. E. 1980. The skull of a new eosuchian reptile from the Lower Jurassic of South Wales. *Zoological Journal of the Linnean Society* 70:203–264.

Evans, S. E. 1981. The postcranial skeleton of the Lower Jurassic eosuchian *Gephyrosaurus bridensis. Zoological Journal of the Linnean Society* 73:81–116.

Evans, S. E. 2003. At the feet of the dinosaurs: The early history and radiation of lizards. *Biological Reviews of the Cambridge Philosophical Society* 78:513–551.

Evans, S. E., and M. Borsuk-Białynicka. 1998. A stem-group frog from the Early Triassic of Poland. *Acta Palaeontologica Polonica* 43:573–580.

Evans, S. E., and K. A. Kermack. 1994. Assemblages of small tetrapods from the Early Jurassic in Great Britain. In N. C. Fraser and H.-D. Sues, eds., *In the Shadow of the Dinosaurs: Early Mesozoic Tetrapods,* pp. 271–283. New York: Cambridge University Press.

Ewer, R. F. 1965. The anatomy of the thecodont reptile *Euparkeria capensis* Broom. *Philosophical Transactions of the Royal Society of London B* 248:379–435.

Ezcurra, M. D., and F. E. Novas. 2007. Phylogenetic relationships of the Triassic theropod *Zupaysaurus rougieri* from NW Argentina. *Historical Biology* 19:35–72.

Faccini, U. F. 1989. O Permo-Triássico do Rio Grande do Sul: Uma ánalise sob o ponto de vista das seqüencias deposicionais. Thesis. Universidae Federal do Rio Grande do Sul, Porto Alegre.

Farrell, B. D. 1998. "Inordinate fondness" explained: Why are there so many beetles? *Science* 281:555–559.

Ferigolo, J., and M. C. Langer. 2007. A Late Triassic dinosauriform from south Brazil and the origin of the ornithischian predentary bone. *Historical Biology* 19: 23–33.

Fiorillo, A. R., K. Padian, and C. Musikasinthorn. 2000. Taphonomy and depositional setting of the *Placerias* Quarry (Chinle Formation: Late Triassic, Arizona). *Palaios* 15:373–386.

Flynn, J. J., J. M. Parrish, B. Rakotosamimanana, L. Ranivoharimanana, W. F. Simpson, and A. R. Wyss. 2000. New traversodontids (Synapsida: Cynodontia) from the Triassic of Madagascar. *Journal of Vertebrate Paleontology* 20:422–427.

Flynn, J. J., J. M. Parrish, B. Rakotosamimanana, W. F. Simpson, R. L. Whatley, and A. R. Wyss. 1999. A Triassic fauna from Madagascar, including early dinosaurs. *Science* 286:763–765.

Fontaine, W. M. 1883. Contributions to the knowledge of the older Mesozoic flora of Virginia. *Monographs of the United States Geological Survey* 6:1–144.

Fraas, E. 1889. Die Labyrinthodonten der schwäbischen Trias. *Palaeontographica* 36:1–158.

Fraas, E. 1896. *Die schwäbischen Trias-Saurier nach dem Material der Kgl. Naturalien-Sammlung in Stuttgart zusammengestellt. Mit Abbildungen der schönsten Schaustücke.* Stuttgart: E. Schweizerbart'sche Verlagshandlung (E. Koch).

Fraas, E. 1910. *Der Petrefaktensammler. Ein Leitfaden zum Sammeln und Bestimmen der Versteinerungen Deutschlands.* Stuttgart: K. G. Lutz Verlag.

Fraas, E. 1913. Neue Labyrinthodonten aus der schwäbischen Trias. *Palaeontographica* 60:275–294.

Fraas, O. 1877. *Aetosaurus ferratus* Fr. Die gepanzerte Vogel-Echse aus dem Stubensandstein bei Stuttgart. *Württembergische naturwissenschaftliche Jahreshefte* 33(3):1–21.

Francis, J. E. 1983. The dominant conifer of the Jurassic Purbeck Formation, England. *Palaeontology* 26:277–294.

Fraser, N. C. 1982. A new rhynchocephalian from the British Upper Trias. *Palaeontology* 25:709–725.

Fraser, N. C. 1985. Vertebrate faunas from Mesozoic fissure deposits of Southwest Britain. *Modern Geology* 9:273–300.

Fraser, N. C. 1986. New Triassic sphenodontids from southwest England and a review of their classification. *Palaeontology* 29:165–186.

Fraser, N. C. 1988a. Rare tetrapod remains from the Late Triassic fissure infillings of Cromhall Quarry, Gloucestershire. *Palaeontology* 31:567–576.

Fraser, N. C. 1988b. The osteology and relationships of *Clevosaurus* (Reptilia: Sphenodontida). *Philosophical Transactions of the Royal Society of London B* 321:125–178.

Fraser, N. C. 1993. A new sphenodontian from the early Mesozoic of England and North America: Implications for correlating early Mesozoic continental deposits. In S. G. Lucas and M. Morales, eds., *The Nonmarine Triassic*, pp. 135–139. New Mexico Museum of Natural History & Science Bulletin 3. Albuquerque: New Mexico Museum of Natural History & Science.

Fraser, N. C. 1994. Assemblages of small tetrapods from British Late Triassic fissure deposits. In N. C. Fraser and H.-D. Sues, eds., *In the Shadow of the Dinosaurs: Early Mesozoic Tetrapods*, pp. 214–226. New York: Cambridge University Press.

Fraser, N. C. 2006. *Dawn of the Dinosaurs: Life in the Triassic*. Bloomington: Indiana University Press.

Fraser, N. C., and M. J. Benton 1989. The Triassic reptiles *Brachyrhinodon* and *Polysphenodon* and the relationships of the sphenodontids. *Zoological Journal of the Linnean Society* 96:413–445.

Fraser, N. C., and D. A. Grimaldi. 2003. Late Triassic continental faunal change: New perspectives on Triassic insect diversity as revealed by a locality in the Danville basin, Virginia, Newark Supergroup. In P. M. LeTourneau and P. E. Olsen, eds., *The Great Rift Valleys of Pangaea in Eastern North America*, vol. 2, *Sedimentology, Stratigraphy, and Paleontology*, pp. 192–205. New York: Columbia University Press.

Fraser, N. C., and H.-D. Sues, eds. 1994. *In the Shadow of the Dinosaurs: Early Mesozoic Tetrapods*. New York: Cambridge University Press.

Fraser, N. C., and D. M. Unwin. 1990. Pterosaur remains from the Upper Triassic of Britain. *Neues Jahrbuch für Geologie und Paläontologie, Monatshefte* 1990:272–282.

Fraser, N. C., and G. M. Walkden. 1984. The postcranial skeleton of *Planocephalosaurus robinsonae*. *Palaeontology* 27:575–595.

Fraser, N. C., G. M. Walkden, and V. Stewart 1985. The first pre-Rhaetic therian mammal. *Nature* 314:161–163.

Fraser, N. C., D. A. Grimaldi, P. E. Olsen, and B. Axsmith. 1996. A Triassic Lagerstätte from eastern North America. *Nature* 380:615–619.

Fraser, N. C., P. E. Olsen, A. C. Dooley Jr., and T. R. Ryan. 2007. A new gliding tetrapod (Diapsida: ?Archosauromorpha) from the Upper Triassic (Carnian) of Virginia. *Journal of Vertebrate Paleontology* 27:261–265.

Fraser, N. C., K. Padian, G. M. Walkden, and A. L. M. Davis. 2002. Basal dinosauriform remains from Britain and the diagnosis of the Dinosauria. *Palaeontology* 45:79–95.

Frentzen, K. 1934. Über die Schachtelhalmgewächse des Keupers. *Aus der Heimat* 47:147–152.

Fröbisch, J. 2007. The cranial anatomy of *Kombuisia frerensis* Hotton (Synapsida, Dicynodontia) and a new phylogeny of anomodont therapsids. *Zoological Journal of the Linnean Society* 150:117–144.

Furin, S., N. Preto, M. Rigo, G. Roghi, P. Gianolla, J. L. Crowley, and S. A. Bowring. 2006. High-precision U-Pb zircon age from the Triassic of Italy: Implications for the Triassic time scale and the Carnian origin of calcareous nannoplankton and dinosaurs. *Geology* 34:1009–1012.

Gaffney, E. S. 1990. The comparative osteology of the Triassic turtle *Proganochelys*. *Bulletin of the American Museum of Natural History* 194:1–263.

Gall, J.-C. 1971. Faunes et paysages du Grès à Voltzia du Nord des Vosges. Essai paléoécologique sur le Buntsandstein supérieur. *Mémoires du Service de la Carte géologique d'Alsace-Lorraine* 34:1–318.

Gall, J.-C. 1985. Fluvial depositional environment evolving into deltaic setting with marine influences in the Buntsandstein of Northern Vosges (France). In D. Mader, ed., *Aspects of Fluvial Sedimentation in the Lower Buntsandstein of Europe*, pp. 449–477. Lecture Notes in Earth Sciences, vol. 4. Berlin: Springer-Verlag.

Gall, J.-C., and L. Grauvogel-Stamm. 1999. Die Paläoökologie des Oberen Buntsandsteins am Westrand des Germanischen Beckens. Der Voltziensandstein im nordöstlichen Frankreich als deltaische Bildung. In N. Hauschke and V. Wilde, eds., *Trias—Eine ganz andere Welt*, pp. 283–298. Munich: Verlag Dr. Friedrich Pfeil.

Galton, P. M. 1984. Cranial anatomy of the prosauropod dinosaur *Plateosaurus* from the Knollenmergel (Middle Keuper, Upper Triassic) of Germany. I. Two complete skulls from Trossingen/Württ. with comments on the diet. *Geologica et Palaeontologica* 18:139–171.

Galton, P. M. 1985a. The poposaurid thecodontian *Teratosaurus suevicus* v. Meyer, plus referred specimens mostly based on prosauropod dinosaurs, from the middle Stubensandstein (Upper Triassic) of Nordwürttemberg. *Stuttgarter Beiträge zur Naturkunde B* 116:1–29.

Galton, P. M. 1985b. Cranial anatomy of the prosauropod dinosaur *Plateosaurus* from the Knollenmergel (Middle Keuper, Upper Triassic) of Germany. II. All the cranial material and details of soft-part anatomy. *Geologica et Palaeontologica* 19:119–159.

Galton, P. M., and P. Upchurch. 2004. Prosauropoda. In D. B. Weishampel, P. Dodson, and H. Osmólska, eds., *The Dinosauria*, 2nd ed., pp. 232–258. Berkeley: University of California Press.

Galton, P. M., A. M. Yates, and D. Kermack. 2007. *Pantydraco* n. gen. for *Thecodontosaurus caducus* Yates, 2003, a basal sauropodomorph dinosaur from the Upper Triassic or Lower Jurassic of South Wales, UK. *Neues Jahrbuch für Geologie und Paläontologie, Abhandlungen* 243: 119–125.

Gans, C. 1983. Is *Sphenodon punctatus* a maladapted relict? In A. J. G. Rhodin and K. Miyata, eds., *Advances in Herpetology and Evolutionary Biology*, pp. 613–620. Cambridge, MA: Museum of Comparative Zoology, Harvard University.

Gans, C., I. Darevskii, and L. P. Tatarinov. 1987. *Sharovipteryx*, a reptilian glider? *Paleobiology* 13:415–426.

Gao, K., and N. H. Shubin. 2003. Earliest known crown-group salamanders. *Nature* 422:424–428.

Gauthier, J., A. G. Kluge, and T. Rowe. 1988. Amniote phylogeny and the importance of fossils. *Cladistics* 4:105–209.

Geyer, O. F., and M. P. Gwinner 1991. *Geologie von Baden-Württemberg*. 4th ed. Stuttgart: E. Schweizerbart'sche Verlagsbuchhandlung.

Getmanov, S. N. 1989. [Triassic amphibians of the Eastern European Platform (family Benthosuchidae Efremov).] *Trudy Paleontologicheskogo Instituta Akademiya Nauk SSSR* 236:1–102. [Russian.]

Gilmore, C. W. 1928. A new fossil reptile from New Jersey. *Proceedings of the United States National Museum* 73:1–8.

Godefroit, P., and B. Battail. 1997. Late Triassic cynodonts from Saint-Nicolas-de-Port (northeastern France). *Geodiversitas* 19:567–631.

Golonka, J. 2007. Late Triassic and Early Jurassic palaeogeography of the world. *Palaeogeography, Palaeoclimatology, Palaeoecology* 244:297–307.

Gorochov, A. V. 2007. [First representative of the Suborder Mesotitanina from the Paleozoic and notes on the system and evolution of the Order Titanoptera (Insecta: Polyneoptera).] *Paleontologicheskii Zhurnal* 2007(6): 31–35. [Russian.]

Gothan, W. 1914. Die unterliassische (rhätische) Flora der Umgebung von Nürnberg. *Abhandlungen der Naturhistorischen Gesellschaft Nürnberg* 19:89–186.

Gothan, W. 1935. Die Unterscheidung der Lias- und Rhätflora. *Zeitschrift der deutschen geologischen Gesellschaft* 87:692–695.

Gow, C. E. 1975. The morphology and relationships of *Youngina capensis* Broom and *Prolacerta broomi* Parrington. *Palaeontologia Africana* 18:89–131.

Gow, C. E. 1978. The advent of herbivory in certain reptilian lineages during the Triassic. *Palaeontologia Africana* 21:133–141.

Gower, D. J. 1999. The cranial and mandibular osteology of a new rauisuchian archosaur from the Middle Triassic of southern Germany. *Stuttgarter Beiträge zur Naturkunde B* 280:1–49.

Gower, D. J. 2000. Rauisuchian archosaurs (Reptilia, Diapsida): An overview. *Neues Jahrbuch für Geologie und Paläontologie, Abhandlungen* 218:447–488.

Gower, D. J. 2003. Osteology of the early archosaurian reptile *Erythrosuchus africanus*. *Annals of the South African Museum* 110:1–84.

Gower, D. J., and R. R. Schoch. 2009. Postcranial anatomy of the rauisuchian archosaur *Batrachotomus kupferzellensis*. *Journal of Vertebrate Paleontology* 29:103–122.

Gower, D. J., and A. G. Sennikov. 2000. Early archosaurs from Russia. In M. J. Benton, M. A. Shishkin, D. M. Unwin, and E. N. Kurochkin, eds., *The Age of Dinosaurs in Russia and Mongolia*, pp. 140–159. Cambridge: Cambridge University Press.

Gozzi, E., and S. Renesto. 2003. A complete specimen of *Mystriosuchus* (Reptilia, Phytosauria) from the Norian (Late Triassic) of Lombardy (northern Italy). *Rivista Italiana di Paleontologia e Stratigrafia* 109:475–493.

Gradstein, F. M., J. G. Ogg, and A. G. Smith, eds. 2004. *A Geologic Time Scale 2004*. Cambridge: Cambridge University Press.

Grauvogel, L. 1947. Note préliminaire sur la faune du Grès à Voltzia. *Comptes Rendus sommaires de la Société géologique de France* 1947:90–92.

Grauvogel-Stamm, L. 1978. La flore du Grès à Voltzia (Buntsandstein supérieur) des Vosges du Nord (France). Morphologie, anatomie, interprétations phylogénique et paléogéographique. *Sciences Géologiques, Mémoires* 50:1–225.

Grauvogel-Stamm, L. 1993. *Pleuromeia sternbergii* (Münster) Corda from the Lower Triassic of Germany—further observations and comparative morphology of its rooting organ. *Review of Palaeobotany and Palynology* 77:185–212.

Grauvogel-Stamm, L. 1999. *Pleuromeia sternbergii* (Münster) Corda, eine charakteristische Pflanze des deutschen Buntsandsteins. In N. Hauschke and V. Wilde, eds., *Trias—Eine ganz andere Welt*, pp. 271–282. Munich: Verlag Dr. Friedrich Pfeil.

Greb, S. F., W. A. DiMichele, and R. A. Gastaldo. 2006. Evolution and importance of wetlands in earth history. In S. F. Greb and W. A. DiMichele, eds., *Wetlands Through*

Time, pp. 1–40. Geological Society of America Special Paper 399. Boulder, CO: Geological Society of America.

Gregor, B. 1970. Denudation of the continents. *Nature* 228: 273–275.

Gregory, H. E. 1917. Geology of the Navajo Country: A reconnaissance of parts of Arizona, New Mexico, and Utah. *United States Geological Survey Professional Paper* 93:1–161.

Gregory, J. T. 1945. Osteology and relationships of *Trilophosaurus*. *University of Texas Special Publication* 4401: 273–359.

Grimaldi, D., and M. S. Engel. 2005. *Evolution of the Insects*. New York: Cambridge University Press.

Grimaldi, D., A. Shmakov, and N. C. Fraser. 2004. Mesozoic thrips and early evolution of the order Thysanoptera (Insecta). *Journal of Paleontology* 78:941–952.

Grimaldi, D., J. Zhang, N. C. Fraser, and A. Rasnitsyn. 2005. Revision of the bizarre Mesozoic scorpionflies in the Pseudopolycentropodidae (Mecopteroidea). *Insect Systematics and Evolution* 36:443–458.

Grine, F. E. 1977. Postcanine tooth function and jaw movement in the gomphodont cynodont *Diademodon* (Reptilia: Therapsida). *Palaeontologia Africana* 20:123–135.

Grine, F. E., C. A. Forster, M. A. Cluver, and J. A. Georgi. 2006. Cranial variability, ontogeny, and taxonomy of *Lystrosaurus* from the Karoo Basin of South Africa. In M. T. Carrano, T. J. Gaudin, R. W. Blob, and J. R. Wible, eds., *Amniote Paleobiology: Perspectives on the Evolution of Mammals, Birds, and Reptiles*, pp. 432–503. Chicago: University of Chicago Press.

Groenewald, G. H. 1991. Burrow casts from the *Lystrosaurus-Procolophon* Assemblage Zone, Karoo sequence, South Africa. *Koedoe* 34:13–22.

Groenewald, G. H., J. Welman, and J. A. MacEachern. 2001. Vertebrate burrow complexes from the Early Triassic *Cynognathus* Zone (Driekoppen Formation, Beaufort Group) of the Karoo Basin, South Africa. *Palaios* 16:148–160.

Guerra-Sommer, M., and M. C. Klepzig. 2000. The Triassic taphoflora from Paraná basin, southern Brazil. *Revista Brasileira de Geociências* 30:481–485.

Guex, J., A. Bartolini, V. Atudorei, and D. Taylor. 2004. High-resolution ammonite and carbon isotope stratigraphy across the Triassic-Jurassic boundary at New York Canyon (Nevada). *Earth and Planetary Science Letters* 225:29–41.

Hahn, G. 1973. Neue Zähne von Haramiyiden aus der deutschen Ober-Trias und ihre Beziehungen zu den Multituberculaten. *Palaeontographica A* 142:1–15.

Hahn, G., J.-C. Lepage, and G. Wouters. 1984. Cynodontier-Zähne aus der Ober-Trias von Medernach, Grossherzogtum Luxemburg. *Bulletin de la Société Belge de Géologique* 93:357–373.

Hahn, G., J.-C. Lepage, and G. Wouters. 1988. Traversodontiden-Zähne (Cynodontia) aus der Ober-Trias von Gaume (Süd-Belgien). *Bulletin de l'Institut Royal des Sciences Naturelles de Belgique, Sciences de la Terre* 58:177–186.

Hahn, G., R. Wild, and G. Wouters. 1987. Cynodontier-Zähne aus der Ober-Trias von Gaume (S-Belgien). *Mémoires pour servir à l'Explication des Cartes Géologiques et Minières de la Belgique* 24:1–32.

Hallam, A. 1992. *Phanerozoic Sea-Level Changes*. New York: Columbia University Press.

Hallam, A. 2002. How catastrophic was the end-Triassic mass extinction? *Lethaia* 35:147–157.

Halstead, L. B., and P. G. Nicoll. 1971. Fossilized caves of Mendip. *Studies in Speleology* 2:93–102.

Hancox, P. J. 2000. The continental Triassic of South Africa. *Zentralblatt für Geologie und Paläontologie, Teil I*, 1998(11–12):1285–1324.

Harris, J. D., and A. Downs. 2002. A drepanosaurid pectoral girdle from the Ghost Ranch (Whitaker) *Coelophysis* Quarry (Chinle Group, Rock Point Formation, Rhaetian), New Mexico. *Journal of Vertebrate Paleontology* 22:70–75.

Harris, T. M. 1958. Forest fire in the Mesozoic. *Journal of Ecology* 46:447–453.

Hasiotis, S. T. 2000. The invertebrate invasion and evolution of Mesozoic soil ecosystems: The ichnofossil record of ecological innovations. In R. A. Gastaldo and W. A. DiMichele, eds., *Phanerozoic Terrestrial Ecosystems*, pp. 141–169. The Paleontological Society Papers 6. New Haven, CT: Yale University.

Haubold, H. 1971. Die Tetrapodenfährten des Buntsandsteins in der Deutschen Demokratischen Republik und in Westdeutschland und ihre Äquivalente in der gesamten Trias. *Paläontologische Abhandlungen, A: Paläozoologie* 4(3):397–548.

Haubold, H., and E. Buffetaut. 1987. Une nouvelle interprétation de *Longisquama insignis*, reptile énigmatique du Trias supérieur d'Asie centrale. *Comptes Rendus de l'Académie des Sciences*, sér. II, 305:65–70.

Hauschke, N., and V. Wilde, eds. 1999. *Trias—Eine ganz andere Welt: Mitteleuropa im frühen Erdmittelalter*. Munich: Verlag Dr. Friedrich Pfeil.

Heckert, A. B. 2004. Late Triassic microvertebrates from the lower Chinle Group (Otischalkian-Adamanian:

Carnian), southwestern U.S.A. *New Mexico Museum of Natural History & Science Bulletin* 27:1–170.

Heckert, A. B., and S. G. Lucas. 1999. A new aetosaur (Reptilia: Archosauria) from the Upper Triassic of Texas and the phylogeny of aetosaurs. *Journal of Vertebrate Paleontology* 19:50–68.

Heckert, A. B., and S. G. Lucas. 2002. South American occurrences of the Adamanian (Late Triassic: latest Carnian) index taxon *Stagonolepis* (Archosauria: Aetosauria) and their biochronological significance. *Journal of Paleontology* 76:852–863.

Heckert, A. B., S. G. Lucas, L. F. Rinehart, and A. P. Hunt. 2008. A new genus and species of sphenodontian from the Ghost Ranch *Coelophysis* Quarry (Upper Triassic: Apachean), Rock Point Formation, New Mexico, USA. *Palaeontology* 51:827–845.

Heinrich, W.-D. 1999. First haramiyid (Mammalia, Allotheria) from the Mesozoic of Gondwana. *Mitteilungen aus dem Museum für Naturkunde Berlin, Geowissenschaftliche Reihe* 2:159–170.

Hellrung, H. 1987. Revision von *Hyperokynodon keuperinus* Plieninger (Amphibia: Temnospondyli) aus dem Schilfsandstein von Heilbronn (Baden-Württemberg). *Stuttgarter Beiträge zur Naturkunde B* 136:1–28.

Hellrung, H. 2003. *Gerrothorax pustuloglomeratus*, ein Temnospondyle (Amphibia) mit knöcherner Branchialkammer aus dem Unteren Keuper von Kupferzell (Süddeutschland). *Stuttgarter Beiträge zur Naturkunde B* 330:1–130.

Hesselbo, S. P., S. A. Robinson, F. Surlyk, and S. Piasecki. 2002. Terrestrial and marine extinction at the Triassic-Jurassic boundary synchronized with major carbon-cycle perturbation: A link to initiation of massive volcanism? *Geology* 30:251–254.

Heunisch, C. 1999. Die Bedeutung der Palynologie für Biostratigraphie und Fazies in der Germanischen Trias. In N. Hauschke and V. Wilde, eds., *Trias—Eine ganz andere Welt*, pp. 207–220. Munich: Verlag Dr. Friedrich Pfeil.

Hitchcock, E. 1858. *Ichnology of New England. A Report on the Sandstone of the Connecticut Valley, Especially Its Fossil Footmarks, Made to the Government of the Commonwealth of Massachusetts.* Boston: William White.

Hitchcock, E. 1865. *Supplement to the Ichnology of New England.* With an Appendix by C. H. Hitchcock. Boston: Wright and Potter.

Hodych, J. P., and G. R. Dunning. 1992. Did the Manicouagan impact trigger end-of-Triassic mass extinction? *Geology* 20:51–54.

Holz, M., and C. L. Schultz. 1998. Taphonomy of the south Brazilian Triassic herpetofauna: Fossilization mode and implications for morphological studies. *Lethaia* 31:335–345.

Hone, D. W. E., and M. J. Benton. 2008. A new genus of rhynchosaur from the Middle Triassic of south-west England. *Palaeontology* 51:95–115.

Hopson, J. A. 1984. Late Triassic traversodont cynodonts from Nova Scotia and southern Africa. *Palaeontologia Africana* 25:181–201.

Hopson, J. A., and J. W. Kitching. 1972. A revised classification of cynodonts (Reptilia, Therapsida). *Palaeontologia Africana* 14:71–85.

Hopson, J. A., and J. W. Kitching. 2001. A probainognathian cynodont from South Africa and the phylogeny of non-mammalian cynodonts. *Bulletin of the Museum of Comparative Zoology, Harvard University* 56:3–35.

Hopson, J. A., and H.-D. Sues. 2006. A traversodont cynodont from the Middle Triassic (Ladinian) of Baden-Württemberg (Germany). *Paläontologische Zeitschrift* 80:124–129.

Hotton, N., III. 1980. An alternative to dinosaurian endothermy: The happy wanderers. In R. D. K. Thomas and E. C. Olson, eds., *A Cold Look at the Warm-Blooded Dinosaurs*, pp. 311–350. American Association for the Advancement of Science Selected Symposium 28. Boulder, CO: Westview Press.

Hounslow, M. W., and G. McIntosh. 2003. Magnetostratigraphy of the Sherwood Sandstone Group (Lower and Middle Triassic), south Devon, U.K.: Detailed correlation of the marine and non-marine Anisian. *Palaeogeography, Palaeoclimatology, Palaeoecology* 193:325–348.

Huber, P., S. G. Lucas, and A. P. Hunt. 1993a. Vertebrate biochronology of the Newark Supergroup, Triassic, eastern North America. In S. G. Lucas and M. Morales, eds., *The Nonmarine Triassic*, pp. 179–186. New Mexico Museum of Natural History & Science Bulletin 3. Albuquerque: New Mexico Museum of Natural History & Science.

Huber, P., S. G. Lucas, and A. P. Hunt. 1993b. Late Triassic fish assemblages from the North American Western Interior. In M. Morales, ed., *Aspects of Mesozoic Geology and Paleontology of the Colorado Plateau*, pp. 51–66. Museum of Northern Arizona Bulletin 59. Flagstaff: Museum of Northern Arizona.

Hubert, J. F., and M. F. Forlenza. 1988. Sedimentology of braided-river deposits in the Wolfville redbeds, southern shore of Cobequid Bay, Nova Scotia, Canada. In W. Manspeizer, ed., *Triassic-Jurassic Rifting, Continental*

Breakup, and the Origin of the Atlantic Ocean and Passive Margins, Part A, pp. 231–248. Amsterdam: Elsevier.

Huene, F. von. 1907–8. Die Dinosaurier der europäischen Triasformation mit Berücksichtigung der aussereuropäischen Vorkommnisse. *Geologische und Palaeontologische Abhandlungen, Supplement-Band* 1:1–419.

Huene, F. von. 1910a. Ein primitiver Dinosaurier aus der mittleren Trias von Elgin. *Geologische und Palaeontologische Abhandlungen*, N.F. 8:315–322.

Huene, F. von. 1910b. Über einen echten Rhynchocephalen aus der Trias von Elgin, *Brachyrhinodon taylori*. *Neues Jahrbuch für Mineralogie, Geologie und Paläontologie* 1910: 29–62.

Huene, F. von. 1911. Beiträge zur Kenntnis und Beurteilung der Parasuchier. *Geologische und Palaeontologische Abhandlungen*, N.F. 10:67–122.

Huene, F. von. 1912. Die Cotylosaurier der Trias. *Palaeontographica* 59:69–102.

Huene, F. von. 1914. Beiträge zur Geschichte der Archosaurier. *Geologische und Palaeontologische Abhandlungen*, N.F. 13:1–53.

Huene, F. von. 1926. Vollständige Osteologie eines Plateosauriden aus dem schwäbischen Keuper. *Geologische und Palaeontologische Abhandlungen*, N.F. 15:139–179.

Huene, F. von. 1932. Die fossile Reptil-Ordnung Saurischia, ihre Entwicklung und Geschichte. *Monographien zur Geologie und Paläontologie* 1(4):1–361.

Huene, F. von. 1934. Ein neuer Coelurosaurier in der thüringischen Trias. *Palaeontologische Zeitschrift* 16:145–170.

Huene, F. von. 1935–42. *Die fossilen Reptilien des südamerikanischen Gondwanalandes. Ergebnisse der Sauriergrabungen in Südbrasilien 1928/29*. Munich: C. H. Becksche Verlagsbuchhandlung.

Huene, F. von. 1938a. *Stenaulorhynchus*, ein Rhynchosauride der ostafrikanischen Obertrias. *Nova Acta Leopoldina*, N. F. 6:83–121.

Huene, F. von. 1938b. Ein grosser Stagonolepide aus der jüngeren Trias Ostafrikas. *Neues Jahrbuch für Mineralogie, Geologie und Paläontologie, B, Beilage-Band* 80:264–278.

Huene, F. von. 1939. Die Lebensweise der Rhynchosauriden. *Palaeontologische Zeitschrift* 21:232–238.

Huene, F. von. 1940. The tetrapod fauna of the Upper Triassic Maleri beds. *Palaeontologia Indica*, n.s., 32:1–42.

Huene, F. von. 1960. Ein grosser Pseudosuchier aus der Orenburger Trias. *Palaeontographica A* 114:105–111.

Hungerbühler, A. 2002. The Late Triassic phytosaur *Mystriosuchus westphali*, with a revision of the genus. *Palaeontology* 45:377–418.

Hungerbühler, A., and A. P. Hunt. 2000. Two new phytosaur species (Archosauria, Crurotarsi) from the Upper Triassic of southwest Germany. *Neues Jahrbuch für Geologie und Paläontologie, Monatshefte* 2000:467–484.

Hungerbühler, A., S. Chatterjee, and T. S. Kutty. 2002. New phytosaurs from the Upper Triassic of India. *Journal of Vertebrate Paleontology* 22 (Supplement to 3):68A.

Hunt, A. P. 1989. A new ornithischian dinosaur from the Bull Canyon Formation (Upper Triassic) of east-central New Mexico. In S. G. Lucas and A. P. Hunt, eds., *Dawn of the Age of Dinosaurs in the American Southwest*, pp. 355–358. Albuquerque: New Mexico Museum of Natural History.

Hunt, A. P. 1993. A revision of the Metoposauridae (Amphibia: Temnospondyli) of the Late Triassic with description of a new genus from the western United States. In M. Morales, ed., *Aspects of the Mesozoic Geology and Paleontology of the Colorado Plateau*, pp. 67–97. Museum of Northern Arizona Bulletin 59. Flagstaff: Museum of Northern Arizona.

Hunt, A. P. 2001. The vertebrate fauna, biostratigraphy and biochronology of the type Revueltian land-vertebrate faunachron, Bull Canyon Formation (Upper Triassic), east-central New Mexico. *New Mexico Geological Society, Guidebook* 52:123–151.

Hunt, A. P., and S. G. Lucas. 1990. Re-evaluation of "*Typothorax*" *meadei*, a Late Triassic aetosaur from the United States. *Paläontologische Zeitschrift* 64:317–328.

Hunt, A. P., and S. G. Lucas. 1993. A new phytosaur (Reptilia: Archosauria) genus from the uppermost Triassic of the western United States and its biochronological significance. In S. G. Lucas and M. Morales, eds., *The Nonmarine Triassic*, pp. 193–196. New Mexico Museum of Natural History & Science Bulletin 3. Albuquerque: New Mexico Museum of Natural History & Science.

Hunt, A. P., and S. G. Lucas. 1994. Ornithischian dinosaurs from the Upper Triassic of the United States. In N. C. Fraser and H.-D. Sues, eds., *In the Shadow of the Dinosaurs: Early Mesozoic Tetrapods*, pp. 227–241. New York: Cambridge University Press.

Hunt, A. P., A. B. Heckert, S. G. Lucas, and A. Downs. 2002. The distribution of the enigmatic reptile *Vancleavea* in the Upper Triassic Chinle Group of the western United States. In A. B. Heckert and S. G. Lucas, eds., *Upper Triassic Stratigraphy and Paleontology*, pp. 269–274. New Mexico Museum of Natural History & Science Bulletin 21. Albuquerque: New Mexico Museum of Natural History & Science.

Hurley, I. A., R. L. Mueller, K. A. Dunn, E. J. Schmidt, M. Friedman, R. K. Ho, V. E. Prince, Z. Yang, M. G. Thomas, and M. I. Coates. 2007. A new time-scale for ray-finned fish evolution. *Proceedings of the Royal Society B* 274:489–498.

Huxley, T. H. 1865. On a collection of vertebrate fossils from the Panchet rocks, near Ranigunj, Bengal. *Memoirs of the Geological Survey of India, Palaeontologia Indica* (4)1(1):1–24.

Huxley, T. H. 1869. On *Hyperodapedon. Quarterly Journal of the Geological Society of London* 25:138–157.

Huxley, T. H. 1875. On *Stagonolepis Robertsoni*, and on the evolution of the Crocodilia. *Quarterly Journal of the Geological Society of London* 31:423–438.

Huxley, T. H. 1877. The crocodilian remains found in the Elgin Sandstones, with remarks on the ichnites of Cummingstone. *Memoirs of the Geological Survey U. K., Monograph* 3:1–52.

Ingavat, R., and P. Janvier. 1981. *Cyclotosaurus* cf. *posthumus* Fraas (Capitosauridae, Stereospondyli) from the Huai Hin Lat Formation (Upper Triassic), northeastern Thailand, with a note on capitosaurid biogeography. *Geobios* 14:711–725.

Irmis, R. B. 2005. The vertebrate fauna of the Upper Triassic Chinle Formation in northern Arizona. In S. J. Nesbitt, W. G. Parker, and R. B. Irmis, eds., *Guidebook to the Triassic Formations of the Colorado Plateau in Northern Arizona: Geology, Paleontology, History*, pp. 63–88. Mesa Southwest Museum Bulletin 9. Mesa, AZ: Mesa Southwest Museum.

Irmis, R. B., S. J. Nesbitt, K. Padian, N. D. Smith, A. H. Turner, D. Woody, and A. Downs. 2007. A Late Triassic dinosauromorph assemblage from New Mexico and the rise of dinosaurs. *Science* 317:358–361.

Ivakhnenko, M. F. 1978. [Tailed amphibians from the Triassic and Jurassic of Middle Asia.] *Paleontologicheskii Zhurnal* 1978(3):84–89. [Russian.]

Ivakhnenko, M. F. 1979. [Permian and Triassic procolophons of the Russian Platform.] *Trudy Paleontologicheskogo Instituta Akademiya Nauk SSSR* 164:1–80. [Russian.]

Jadoul, F. 1986. Stratigrafia e paleogeografia del Norico nelle Prealpi Bergamasche Occidentali. *Rivista Italiana di Paleontologia e Stratigrafia* 91:479–512.

Jadoul, F., D. Masetti, S. Cirilli, F. Berra, M. Claps, and S. Frisia. 1994. Norian-Rhaetian stratigraphy and palaeogeographical evolution of the Lombardy Basin (Bergamase Alps). In G. Carannante and R. Tonielli, eds., *Excursion B, 15th International Association of Sedimentologists Regional Meeting, Ischia, Italy, 1994*, pp. 5–38. Oxford: International Association of Sedimentologists.

Jaeger, G. F. 1828. *Über die fossile Reptilien, welche in Würtemberg aufgefunden worden sind.* Stuttgart: Verlag der J. B. Metzler'schen Buchhandlung.

Jalil, N.-E. 1996. Les Vertébrés permiens et triasiques de la Formation d'Argana (Haut Atlas occidental): liste faunique préliminaire et implications stratigraphiques. In F. Medina, ed., *Le Permien et le Trias du Maroc: État des Connaissances*, pp. 227–250. Marrakech: Éditions PUMAG.

Jalil, N.-E., and K. Peyer. 2007. A new rauisuchian (Archosauria, Suchia) from the Upper Triassic of the Argana Basin, Morocco. *Palaeontology* 50:417–430.

Jeffries, R. P. S. 1979. The origin of chordates—a methodological essay. In M. R. House, ed., *The Origin of Major Invertebrate Groups*, pp. 443–477. Systematics Association Special Volume 12. London: Academic Press.

Jenkins, F. A., Jr. 1970. The Chañares (Argentina) Triassic reptile fauna. VII. The postcranial skeleton of the traversodontid *Massetognathus pascuali* (Therapsida, Cynodontia). *Breviora* 352:1–28.

Jenkins, F. A., Jr. 1971. The postcranial skeleton of African cynodonts. *Bulletin of the Peabody Museum of Natural History, Yale University* 36:1–216.

Jenkins, F. A., Jr., S. M. Gatesy, N. H. Shubin, and W. W. Amaral. 1997. Haramyids and Triassic mammalian evolution. *Nature* 385:715–718.

Jenkins, F. A., Jr., and N. H. Shubin. 1998. *Prosalirus bitis* and the anuran caudopelvic mechanism. *Journal of Vertebrate Paleontology* 18:495–510.

Jenkins, F. A., Jr., D. M. Walsh, and R. L. Carroll. 2007. Anatomy of *Eocaecilia micropodia*, a limbed gymnophionan of the Early Jurassic. *Bulletin of the Museum of Comparative Zoology, Harvard University* 158:285–365.

Jenkins, F. A., Jr., N. H. Shubin, S. M. Gatesy, and A. Warren. 2008. *Gerrothorax pulcherrimus* from the Upper Triassic Fleming Fjord Formation of East Greenland and a reassessment of head lifting in temnospondyl feeding. *Journal of Vertebrate Paleontology* 28:935–950.

Jenkins, F. A., Jr., N. H. Shubin, W. W. Amaral, S. M. Gatesy, C. R. Schaff, L. B. Clemmensen, W. R. Downs, A. R. Davidson, N. Bonde, and F. Osbaeck. 1994. Late Triassic continental vertebrates and depositional environments of the Fleming Fjord Formation, Jameson Land, East Greenland. *Meddelelser om Grønland, Geoscience* 32:1–25.

Johnson, M. R., and N. Hiller. 1990. Burgersdorp Formation. In M. R. Johnson, ed., *Catalogue of South African Lithostratigraphic Units*, pp. 2-9–2-10. South African Committee for Stratigraphy. Pretoria: Council for Geoscience.

Jones, M. E. H. 2006. The Jurassic clevosaurs from China (Diapsida: Lepidosauria). In J. D. Harris, S. G. Lucas, J. A. Spielmann, M. G. Lockley, A. R. C. Milner, and J. I. Kirkland, eds., *The Triassic-Jurassic Terrestrial Transition*, pp. 548–562. New Mexico Museum of Natural History & Science Bulletin 37. Albuquerque: New Mexico Museum of Natural History & Science.

Jones, T. D., J. A. Ruben, L. D. Martin, E. N. Kurochkin, A. Feduccia, P. F. A. Maderson, W. J. Hillenius, N. R. Geist, and V. Alifanov. 2000. Nonavian feathers in a Late Triassic archosaur. *Science* 288:2202–2205.

Joyce, W. G. 2007. Phylogenetic relationships of Mesozoic turtles. *Bulletin of the Peabody Museum of Natural History, Yale University* 48:3–102.

Joyce, W. G., S. G. Lucas, T. M. Scheyer, A. B. Heckert, and A. P. Hunt. 2009. A thin-shelled reptile from the Late Triassic of North America and the origin of the turtle shell. *Proceedings of the Royal Society B* 276:507–513.

Kalandadze, N. N. 1970. [New Triassic kannemeyeriids from southern Cisuralia.] In K. K. Flerov, ed., [*Materials on the Evolution of Terrestrial Vertebrates*], pp. 51–57. Moscow: Nauka. [Russian.]

Kalandadze, N. N. 1975. [First find of a lystrosaur on the territory of the European part of the USSR.] *Paleontologicheskii Zhurnal* 1975(4):140–142. [Russian.]

Kammerer, C., J. J. Flynn, L. Ranivoharimanana, and A. R. Wyss. 2008. New material of *Menadon besairiei* (Cynodontia: Traversodontidae) from the Triassic of Madagascar. *Journal of Vertebrate Paleontology* 28:445–462.

Kamphausen, D. 1989. Der Schädel von *Eocyclotosaurus woschmidti* ORTLAM (Amphibia, Stegocephalia) aus dem Oberen Buntsandstein (Trias) des Schwarzwaldes (SW-Deutschland). *Stuttgarter Beiträge zur Naturkunde B* 149:1–65.

Kaye, F. T., and K. Padian. 1994. Microvertebrates from the *Placerias* Quarry: A window on Late Triassic vertebrate diversity in the American Southwest. In N. C. Fraser and H.-D. Sues, eds., *In the Shadow of the Dinosaurs: Early Mesozoic Tetrapods*, pp. 171–196. New York: Cambridge University Press.

Kelber, K.-P. 1999. Neue Befunde über die Schachtelhalme des Keupers. In N. Hauschke and V. Wilde, eds., *Trias—*

Eine ganz andere Welt, pp. 355–370. Munich: Verlag Dr. Friedrich Pfeil.

Kemp, T. S. 1982. *Mammal-like Reptiles and the Origin of Mammals*. New York: Academic Press.

Kemp, T. S. 2005. *The Origin and Evolution of Mammals*. Oxford: Oxford University Press.

Kent, D. V. 1998. Impacts on Earth in the Late Triassic: Discussion. *Nature* 395:126.

Kent, D. V., and P. E. Olsen. 1997. Paleomagnetism of Upper Triassic continental sedimentary rocks from the Dan River–Danville rift basin (eastern North America). *Geological Society of America Bulletin* 109:366–377.

Kent, D. V., P. E. Olsen, and W. K. Witte. 1995. Late Triassic–earliest Jurassic geomagnetic polarity and paleolatitudes from drill cores in the Newark rift basin, eastern North America. *Journal of Geophysical Research* 100:14965–14998.

Kent, D. V., and L. Tauxe. 2005. Corrected Late Triassic latitudes for continents adjacent to the North Atlantic. *Science* 307:240–247.

Kerp, H. 2000. The modernization of landscapes during the late Paleozoic–early Mesozoic. In R. A. Gastaldo and W. A. DiMichele, eds., *Phanerozoic Terrestrial Ecosystems*, pp. 79–113. The Paleontological Society Papers 6. New Haven, CT: Yale University.

Kerp, H., A. A. Hamad, B. Vörding, and K. Bandel. 2006. Typical Triassic Gondwanan floral elements in the Upper Permian of the paleotropics. *Geology* 34:265–268.

Keyser, A. W. 1973. A new Triassic vertebrate fauna from South West Africa. *Palaeontologia Africana* 16:1–15.

Kielan-Jaworowska, Z., R. L. Cifelli, and Z.-X. Luo. 2004. *Mammals from the Age of Dinosaurs*. New York: Columbia University Press.

King, G. M. 1991. The aquatic *Lystrosaurus*: A palaeontological myth. *Historical Biology* 4:285–321.

King, G. M., and M. A. Cluver. 1991. The aquatic *Lystrosaurus*: An alternative lifestyle. *Historical Biology* 4:323–341.

Kiparisova, L. D., and Y. N. Popov. 1956. [Subdivision of the lower series of the Triassic system into stages.] *Doklady Akademii Nauk SSSR, Seriya Geologicheskaya* 109:842–845. [Russian.]

Kischlat, E.-E., and S. G. Lucas. 2003. A phytosaur from the Upper Triassic of Brazil. *Journal of Vertebrate Paleontology* 23:464–467.

Kitching, J. W., and M. A. Raath. 1984. Fossils from the Elliot and Clarens formations (Karoo sequence) of the northeastern Cape, Orange Free State and Lesotho, and a suggested biozonation based on tetrapods. *Palae-ontologia Africana* 25:111–125.

Klavins, S. D., D. W. Kellogg, M. Krings, E. L. Taylor, and T. N. Taylor. 2005. Coprolites in a Middle Triassic cycad pollen cone: Evidence for insect pollination in early cycads? *Evolutionary and Ecological Research* 7:479–488.

Knoll, A. H. 1984. Patterns of extinction in the fossil record of vascular plants. In M. H. Nitecki, ed., *Extinctions*, pp. 21–68. Chicago: University of Chicago Press.

Kokogian, D., L. A. Spalletti, E. M. Morel, A. E. Artabe, R. N. Martínez, O. A. Alcober, J. P. Milana, and A. M. Zavattieri. 2001. Estratigrafía del Triásico argentino. In A. E. Artabe, E. M. Morel, and A. B. Zamuner, eds., *El Sistema Triásico en la Argentina*, pp. 23–54. La Plata: Fundación Museo de La Plata "Francisco Pascasio Moreno."

Kozur, H. W., and R. E. Weems. 2007. Upper Triassic conchostracan biostratigraphy of the continental rift basins of eastern North America: Its importance for correlating the Newark Supergroup events with the Germanic basin and the international geologic timescale. In S. G. Lucas and J. A. Spielmann, eds., *The Global Triassic*, pp. 137–188. New Mexico Museum of Natural History & Science Bulletin 41. Albuquerque: New Mexico Museum of Natural History & Science.

Krebs, B. 1965. Die Triasfauna der Tessiner Kalkalpen. XIX. *Ticinosuchus ferox* nov. gen. nov. sp. Ein neuer Pseudosuchier aus der Trias des Monte San Giorgio. *Schweizerische Paläontologische Abhandlungen* 81:1–140.

Krebs, B. 1969. *Ctenosauriscus koeneni* (v. Huene), die Pseudosuchia und die Buntsandstein-Reptilien. *Eclogae geologicae Helvetiae* 62:697–714.

Krebs, B. 1974. Die Archosaurier. *Naturwissenschaften* 61:17–24.

Kreuser, T. 1990. Permo-Trias im Ruhuhu-Becken (Tansania) und anderen Karoo-Becken von SE-Afrika. *Geologisches Institut der Universitaet zu Koeln, Sonderveroeffentlichungen* 75:1–131.

Krzemiński, W., and E. Krzemińska. 2003. Triassic Diptera: Descriptions, revisions and phylogenetic relations. *Acta Zoologica Cracoviensia* 46 (Supplement):153–184.

Krzemiński, W., E. Krzemińska, and F. Papier. 1994. *Grauvogelia arzvilleriana* sp. n.—the oldest Diptera species (Lower/Middle Triassic of France). *Acta Zoologica Cracoviensia* 37:95–99.

Kuhn, O. 1932. Labyrinthodonten und Parasuchier aus dem mittleren Keuper von Ebrach in Oberfranken. *Neues Jahrbuch für Mineralogie, Geologie und Paläontologie, Beilage-Band B* 69:94–144.

Kuhn, O. 1936. Weitere Parasuchier und Labyrinthodonten aus dem Blasensandstein des mittleren Keuper von Ebrach. *Palaeontographica A* 83:61–98.

Kuhn-Schnyder, E. 1974. Die Triasfauna der Tessiner Kalkalpen. *Neujahrsblatt der naturforschenden Gesellschaft in Zürich* 176:1–119.

Kürschner, W. M. 1997. The anatomical diversity of recent and fossil leaves of the durmast oak (*Quercus petraea* Lieblein/*Q. pseudocastanea* Goeppert): Implications for their use as biosensors of palaeoatmospheric CO_2 levels. *Review of Palaeobotany and Palynology* 96:1–30.

Kurtz, F. 1921. Atlas de plantas fósiles de la República Argentina. *Actas de la Academia Nacional de Ciencias en Córdoba (Rep. Argentina)* 7:129–153.

Kutty, T. S., S. Chatterjee, P. M. Galton, and P. Upchurch. 2007. Two basal sauropodomorphs (Dinosauria: Saurischia) from the Lower Jurassic of India: Their anatomy and relationships. *Journal of Paleontology* 81:1218–1240.

Kutty, T. S., and D. P. Sengupta. 1989. The Late Triassic formations of the Pranhita-Godavari Valley and their vertebrate faunal succession: A reappraisal. *Indian Journal of Earth Sciences* 16:189–206.

Labandeira, C. C. 2006. The four phases of plant-arthropod associations in deep time. *Geologica Acta* 4:409–438.

Labandeira, C. C., J. Kvaček, and M. B. Mostovski. 2007. Pollination drops, pollen, and insect pollination of Mesozoic gymnosperms. *Taxon* 56:663–695.

Langer, M. C. 2005. Studies on continental Late Triassic tetrapod biochronology. II. The Ischigualastian and a Carnian global correlation. *Journal of South American Earth Sciences* 19:219–239.

Langer, M. C., F. Abdala, M. Richter, and M. J. Benton. 1999. A sauropodomorph dinosaur from the Upper Triassic (Carnian) of southern Brazil. *Comptes Rendus de l'Académie des Sciences Paris, Sciences de la Terre et des Planètes* 329:511–517.

Langer, M. C., and M. J. Benton. 2006. Early dinosaurs: A phylogenetic study. *Journal of Systematic Palaeontology* 4:309–358.

Langer, M. C., A. M. Ribeiro, C. L. Schultz, and J. Ferigolo. 2007. The continental tetrapod-bearing Triassic of south Brazil. In S. G. Lucas and J. A. Spielmann, eds., *The Global Triassic*, pp. 201–218. New Mexico Museum of Natural History & Science Bulletin 41. Albuquerque: New Mexico Museum of Natural History & Science.

Langer, M. C., and C. L. Schultz. 2000. A new species of the Late Triassic rhynchosaur *Hyperodapedon* from the

Santa Maria Formation of South Brazil. *Palaeontology* 43:633–652

Laurin, M., and R. R. Reisz. 1995. A reevaluation of early amniote phylogeny. *Zoological Journal of the Linnean Society* 113:165–223.

Lea, I. 1851. Remarks on the bones of a fossil reptilian quadruped. *Proceedings of the Academy of Natural Sciences of Philadelphia* 5:171–172.

Leal, L. A., S. A. K. Azevedo, A. W. A. Kellner, and A. A. S. Da Rosa. 2004. A new early dinosaur (Sauropodomorpha) from the Caturrita Formation (Late Triassic), Paraná Basin, Brazil. *Zootaxa* 690:1–24.

Lee, M. S. Y. 1995. Historical burden in systematics and the interrelationships of 'parareptiles'. *Biological Reviews of the Cambridge Philosophical Society* 70:459–547.

Lee, M. S. Y., T. W. Reeder, J. B. Slowinski, and R. Lawson. 2004. Resolving reptile relationships: Molecular and morphological markers. In J. Cracraft and M. J. Donoghue, eds., *Assembling the Tree of Life*, pp. 451–467. New York: Oxford University Press.

Lefebvre, F., A. Nel, F. Papier, L. Grauvogel-Stamm, and J.-C. Gall. 1998. The first "cicada-like Homoptera" from the Triassic of the Vosges, France. *Palaeontology* 41:1195–1200.

Lehman, T., and S. Chatterjee. 2005. Depositional setting and vertebrate biostratigraphy of the Triassic Dockum Group of Texas. *Journal of Earth System Science* 114:325–351.

LeTourneau, P. M. 1999. Depositional history and tectonic evolution of Late Triassic age rifts of the U.S. Central Atlantic Margin: Results of an integrated stratigraphic, structural, and paleomagnetic analysis of the Taylorsville and Richmond basins. Ph.D. diss., Columbia University, New York.

LeTourneau, P. M. 2003. Tectonic and climatic controls on the stratigraphic architecture of the Late Triassic Taylorsville basin, Virginia and Maryland. In P. M. LeTourneau and P. E. Olsen, eds., *The Great Rift Valleys of Pangea in Eastern North America*, vol. 2, *Sedimentology, Stratigraphy, and Paleontology*, pp. 12–58. New York: Columbia University Press.

Li, C., X.-C. Wu, O. Rieppel, L.-T. Wang, and L.-J. Zhao. 2008. An ancestral turtle from the Late Triassic of southwestern China. *Nature* 456:497–501.

Li, C., L.-J. Zhao, and L.-T. Wang. 2007. A new species of *Macrocnemus* (Reptilia: Protorosauria) from the Middle Triassic of southwestern China and its palaeogeographical implication. *Science in China, Series D: Earth Sciences* 50:1601–1605.

Li, J. 1983. Tooth replacement in a new genus of procolophonid from the Early Triassic of China. *Palaeontology* 26:567–583.

Litwin, R. J., S. R. Ash, and A. Traverse. 1991. Preliminary palynological zonation of the Chinle Formation, southwestern U.S.A., and its correlation to the Newark Supergroup (eastern U.S.A.). *Review of Palaeobotany and Palynology* 68:269–287.

Long, R. A., and K. L. Ballew. 1985. Aetosaur dermal armor from the Late Triassic of southwestern North America, with special reference to material from the Chinle Formation of Petrified Forest National Park. *Museum of Northern Arizona Bulletin* 47:45–68.

Long, R. A. and P. A. Murry. 1995. Late Triassic (Carnian and Norian) tetrapods from the southwestern United States. *New Mexico Museum of Natural History & Science Bulletin* 4:1–254.

Looy, C. V., W. A. Brugman, D. L. Dilcher, and H. Visscher. 1999. The delayed resurgence of equatorial forests after the Permian-Triassic ecologic crisis. *Proceedings of the National Academy of Sciences U.S.A.* 96:13857–13862.

Lourenço, W. R., and J.-C. Gall. 2004. Fossil scorpions from the Buntsandstein (Early Triassic) of France. *Comptes Rendus Palevol* 3:369–378.

Lucas, S. G. 1993. The Chinle Group: Revised stratigraphy and chronology of the Upper Triassic nonmarine strata in the western United States. In M. Morales, ed., *Aspects of Mesozoic Geology and Paleontology of the Colorado Plateau*, pp. 27–50. Museum of Northern Arizona Bulletin 59. Flagstaff: Museum of Northern Arizona.

Lucas, S. G. 1994. The beginning of the age of dinosaurs in Wyoming. *Wyoming Geological Association Guidebook* 44:105–113.

Lucas, S. G. 1997. Upper Triassic Chinle Group, western United States: A nonmarine standard for Late Triassic time. In J. M. Dickens, Z. Yang, H. Yin, S. G. Lucas, and S. K. Acharyya, eds., *Late Palaeozoic and Early Mesozoic Circum-Pacific Events and Their Global Correlation*, pp. 209–228. Cambridge: Cambridge University Press.

Lucas, S. G. 1998. Global Triassic tetrapod biostratigraphy and biochronology. *Palaeogeography, Palaeoclimatology, Palaeoecology* 143:347–384.

Lucas, S. G. 1999. Tetrapod-based correlation of the nonmarine Triassic. *Zentralblatt für Geologie und Paläontologie, Teil I*, 1998(7–8):497–521.

Lucas, S. G. 2001. *Chinese Fossil Vertebrates*. New York: Columbia University Press.

Lucas, S. G., and P. Huber. 2003. Vertebrate biostratigraphy and biochronology of the nonmarine Late Triassic. In P. M. LeTourneau and P. E. Olsen, eds., *The Great Rift Valleys of Pangea in Eastern North America*, vol. 2, *Sedimentology, Stratigraphy, and Paleontology*, pp. 143–191. New York: Columbia University Press.

Lucas, S. G., and A. P. Hunt, eds. 1989. *Dawn of the Age of Dinosaurs in the American Southwest*. Albuquerque: New Mexico Museum of Natural History.

Lucas, S. G., and A. P. Hunt. 1993. Tetrapod biochronology of the Chinle Group (Upper Triassic), western United States. In S. G. Lucas and N. Morales, eds., *The Nonmarine Triassic*, pp. 327–329. New Mexico Museum of Natural History & Science Bulletin 3. Albuquerque: New Mexico Museum of Natural History & Science.

Lucas, S. G., and Z. Luo. 1993. *Adelobasileus* from the Upper Triassic of West Texas: The oldest mammal. *Journal of Vertebrate Paleontology* 13:309–334.

Lucas, S. G., and J. E. Marzolf. 1993. Stratigraphy and sequence stratigraphic interpretation of Upper Triassic strata in Nevada. In G. C. Dunne and K. A. McDougall, eds., *Mesozoic Paleogeography of the Western United States*, vol. 2, pp. 375–388. Los Angeles: Society of Economic Paleontologists and Mineralogists, Pacific Section.

Lucas, S. G., and R. R. Schoch. 2002. Triassic temnospondyl biostratigraphy, biochronology and correlation of the German Buntsandstein and North American Moenkopi Formation. *Lethaia* 35:97–106.

Lucas, S. G., and L. H. Tanner. 2007a. The nonmarine Triassic-Jurassic boundary in the Newark Supergroup of eastern North America. *Earth-Science Reviews* 84:1–20.

Lucas, S. G., and L. H. Tanner. 2007b. Tetrapod biostratigraphy and biochronology of the Triassic-Jurassic transition on the southern Colorado Plateau, USA. *Palaeogeography, Palaeoclimatology, Palaeoecology* 244:242–256.

Lucas, S. G., and R. Wild. 1995. A Middle Triassic dicynodont from Germany and the biochronology of Triassic dicynodonts. *Stuttgarter Beiträge zur Naturkunde B* 220:1–16.

Lucas, S. G., A. P. Hunt, and J. A. Spielmann. 2006. *Rioarribasuchus*, a new name for an aetosaur from the Upper Triassic of north-central New Mexico. In J. D. Harris, S. G. Lucas, J. A. Spielmann, M. G. Lockley, A. R. C. Milner, and J. I. Kirkland, eds., *The Triassic-Jurassic Terrestrial Transition*, pp. 581–582. New Mexico Museum of Natural History & Science Bulletin 37. Albuquerque: New Mexico Museum of Natural History & Science.

Lucas, S. G., R. Wild, and A. P. Hunt. 1998. *Dyoplax* O. Fraas, a Triassic sphenosuchian from Germany. *Stuttgarter Beiträge zur Naturkunde B* 263:1–13.

Lucas, S. G., A. B. Heckert, N. C. Fraser, and P. Huber. 1999. *Aetosaurus* from the Upper Triassic of Great Britain and its biochronological significance. *Neues Jahrbuch für Geologie und Paläontologie, Monatshefte* 1999:568–576.

Lucas, S. G., A. P. Hunt, A. B. Heckert, and J. A. Spielmann. 2007. Global Triassic tetrapod biostratigraphy and biochronology: 2007 status. In S. G. Lucas and J. A. Spielmann, eds., *The Global Triassic*, pp. 229–240. New Mexico Museum of Natural History & Science Bulletin 41. Albuquerque: New Mexico Museum of Natural History & Science.

Lyell, C. 1847. On the structure and probable age of the coal-field of the James River, near Richmond. *Quarterly Journal of the Geological Society of London* 3:261–280.

Maheswari, H. K., and K. P. N. Kumaran. 1979. Upper Triassic Sporae Dispersae from the Tiki Formation. 1: Miospores from the Son River section between Tharipathar and Ghiar, South Rewa Gondwana Basin. *Palaeontographica B* 71:1–126.

Mancuso, A. C., and C. A. Marsicano. 2008. Paleoenvironments and taphonomy of a Triassic lacustrine system (Los Rastros Formation, central-western Argentina). *Palaios* 23:535–547.

Mancuso, A. C., O. F. Gallego, and R. G. Martins-Neto. 2007. The Triassic insect fauna from the Los Rastros Formation (Bermejo Basin), La Rioja Province (Argentina): Its context, taphonomy and paleobiology. *Ameghiniana* 44:337–348.

Marchal-Papier, F. 1998. Les insectes du Buntsandstein des Vosges (NE de la France). Biodiversité et contribution aux modalités de la crise biologique du Permo-Trias. Thèse, Université Louis Pasteur, Strasbourg.

Marchal-Papier, F., A. Nel, and L. Grauvogel-Stamm. 2000. Nouveaux Orthoptères (Ensifera, Insecta) du Trias des Vosges (France). *Acta Geologica Hispanica* 35:5–18.

Markwort, S. 1991. Sedimentation und Diagenese der untertriadischen Karoo-Schichten des Ruhuhu-Beckens, SW Tansania. *Geologisches Institut der Universitaet zu Koeln, Sonderveroeffentlichungen* 80:1–118.

Marsh, O. C. 1893. Restoration of *Anchisaurus*. *American Journal of Science*, ser. 3, 45:169–170.

Marsh, O. C. 1896. A new belodont reptile (*Stegomus*) from the Connecticut River Sandstone. *American Journal of Science*, ser. 4, 2:59–62.

Marshall, J. E. A., and D. I. Whiteside. 1980. Marine influence in the Triassic 'uplands'. *Nature* 287:627–628.

Marsicano, C. A. 1999. Chigutisaurid amphibians from the Upper Triassic of Argentina and their phylogenetic relationships. *Palaeontology* 42:1–21.

Martinez, R. N., and O. Alcober. 2009. A basal sauropodomorph (Dinosauria: Saurischia) from the Ischigualasto Formation (Triassic, Carnian) and the early evolution of Sauropodomorpha. *PLoS One* 4:e4397.

Martinez, R. N., and C. A. Forster. 1996. The skull of *Probelesodon sanjuanensis* sp. nov. from the Late Triassic Ischigualasto Formation of Argentina. *Journal of Vertebrate Paleontology* 16:285–291.

Martinez, R. N., C. L. May, and C. A. Forster. 1996. A new carnivorous cynodont from the Ischigualasto Formation (Late Triassic) with comments on eucynodont phylogeny. *Journal of Vertebrate Paleontology* 16:271–284.

Martins-Neto, R. G., O. F. Gallego, and R. N. Melchor. 2003. The Triassic insect fauna from South America (Argentina, Brazil and Chile): A checklist (except Blattoptera and Coleoptera) and descriptions of new taxa. *Acta Zoologica Cracoviensia* 46 (Supplement):229–256.

Martins-Neto, R. G., O. F. Gallego, and A. M. Zavattieri. 2007. A new Triassic insect fauna from Cerro Bayo, Potrerillos (Mendoza Province, Argentina) with descriptions of new taxa (Insecta: Blattoptera and Coleoptera). *Alcheringa* 31:199–213.

Marzoli, A., H. Bertrand, K. B. Knight, S. Cirilli, N. Buratti, C. Vérati, S. Nomade, P. R. Renne, N. Youbi, R. Martini, K. Allenbach, R. Neuwerth, C. Rapaille, L. Zaninetti, and G. Bellieni. 2004. Synchrony of the Central Atlantic Magmatic Province and the Triassic-Jurassic boundary climatic and biotic crisis. *Geology* 32:973–976.

Marzoli, A., P. R. Renne, E. M. Piccirillo, M. Ernesto, G. Bellieni, and A. De Min. 1999. Extensive 200-million-year-old continental flood basalts of the Central Atlantic Magmatic Province. *Science* 284:616–618.

McElwain, J. C., and S. W. Punyasena. 2007. Mass extinction events and the plant fossil record. *Trends in Ecology and Evolution* 22:548–557.

McElwain, J. C., D. J. Beerling, and F. I. Woodward. 1999. Fossil plants and global warming at the Triassic-Jurassic boundary. *Science* 285:1386–1390.

McElwain, J. C., M. E. Popa, S. P. Hesselbo, M. Haworth, and F. Surlyk. 2007. Macroecological responses of terrestrial vegetation to climatic and atmospheric change across the Triassic/Jurassic boundary in East Greenland. *Paleobiology* 33:547–573.

McGregor, J. H. 1906. The Phytosauria, with especial reference to *Mystriosuchus* and *Rhytidodon*. *Memoirs of the American Museum of Natural History* 9:27–100.

McHone, J. G., and J. H. Puffer. 2003. Flood basalt provinces of the Pangean Atlantic Rift: Regional extent and environmental significance. In P. M. LeTourneau and P. E. Olsen, eds., *The Great Rift Valleys of Pangea in Eastern North America*, vol. 1, *Tectonics, Structure, and Volcanism*, pp. 141–154. New York: Columbia University Press.

McKeown, K. C. 1937. New fossil insect wings (Protohemiptera, family Mesotitanidae). *Records of the Australian Museum* 20:31–37.

Medina, F. 2000. Structural styles of the Moroccan Triassic basins. *Zentralblatt für Geologie und Paläontologie, Teil I*, 1998 (9–10):1167–1192.

Meyer, H. von. 1847–55. Zur *Fauna der Vorwelt. Die Saurier des Muschelkalkes, mit Rücksicht auf die Saurier aus Buntem Sandstein und Keuper*. Frankfurt am Main: Verlag von Heinrich Keller.

Meyer, H. von. 1861. Reptilien aus dem Stubensandstein des oberen Keupers. *Palaeontographica* 7:253–346.

Meyer, H. von. 1863. Der Schädel des *Belodon* aus dem Stubensandstein des oberen Keuper. *Palaeontographica* 10:227–246.

Meyer, H. von, and T. Plieninger. 1844. *Beiträge zur Paläontologie Württemberg's, enthaltend die fossilen Wirbelthierreste aus den Triasgebilden mit besonderer Rücksicht auf die Labyrinthodonten des Keupers*. Stuttgart: E. Schweizerbart'sche Verlagsbuchhandlung.

Meyertons, C. T. 1963. Triassic formations of the Danville basin. *Virginia Division of Mineral Resources, Report of Investigations* 6:1–65.

Milana, J. P., and O. Alcober. 1994. Modelo tectosedimentario de la cuenca triásica de Ischigualasto (San Juan, Argentina). *Revista de la Asociación Geológica Argentina* 49:217–235.

Miller, C. N. 1977. Mesozoic conifers. *Botanical Review* 43:218–271.

Milner, A. R., B. G. Gardiner, N. C. Fraser, and M. A. Taylor. 1990. Vertebrates from the Middle Triassic Otter Sandstone Formation of Devon. *Palaeontology* 33:873–892.

Mitter, C., B. Farrell, and B. Wiegmann. 1988. The phylogenetic study of adaptive zones: Has phytophagy promoted insect diversification? *American Naturalist* 132:107–128.

Modesto, S. P., and R. J. Damiani. 2003. Taxonomic status of *Thelegnathus browni* Broom, a procolophonid reptile from the South African Triassic. *Annals of Carnegie Museum* 72:53–64.

Modesto, S. P., and H.-D. Sues. 2004. The skull of the Early Triassic archosauromorph reptile *Prolacerta broomi* and its phylogenetic significance. *Zoological Journal of the Linnean Society* 140:335–351.

Modesto, S. P., H.-D. Sues, and R. J. Damiani. 2001. A new Triassic procolophonoid reptile and its implications for procolophonoid survivorship during the Permo-Triassic extinction event. *Proceedings of the Royal Society of London B* 268:2047–2052.

Morales, M. 1987. Terrestrial fauna and flora from the Triassic Moenkopi Formation of the southwestern United States. *Journal of the Arizona-Nevada Academy of Science* 22:1–20.

Moser, M. 2003. *Plateosaurus engelhardti* Meyer, 1837 (Dinosauria: Sauropodomorpha) aus dem Feuerletten (Mittelkeuper; Obertrias) von Bayern. *Zitteliana B* 24:3–186.

Mueller, B. D., and W. G. Parker. 2006. A new species of *Trilophosaurus* (Diapsida: Archosauromorpha) from the Sonsela Member (Chinle Formation) of Petrified Forest National Park, Arizona. In W. G. Parker, S. R. Ash, and R. B. Irmis, eds., *A Century of Research in Petrified Forest National Park. Geology and Paleontology*, pp. 119–125. Museum of Northern Arizona Bulletin 62. Flagstaff: Museum of Northern Arizona.

Müller, J., S. Renesto, and S. E. Evans. 2005. The marine diapsid reptile *Endennasaurus* from the Upper Triassic of Italy. *Palaeontology* 48:15–30.

Mundil, R., and R. Irmis. 2008. New U-Pb age constraints for terrestrial sediments in the Late Triassic: Implications for faunal evolution and correlations with marine environments. Abstracts, 33rd International Geological Congress, Oslo 2008, HPF-16: Correlation Between Marine and Terrestrial Ecosystems. Oslo: 33rd International Geological Congress.

Mundil, R., K. R. Ludwig, I. Metcalfe, and P. R. Renne. 2004. Age and timing of the end Permian mass extinctions: U/Pb geochronology on closed-system zircons. *Science* 305:1760–1763.

Murry, P. A. 1987. New reptiles from the Upper Triassic Chinle Formation of Arizona. *Journal of Paleontology* 61:773–786.

Muttoni, G., D. V. Kent, S. Meco, M. Balini, A. Nicora, R. Rettori, M. Gaetani, and L. Krystyn. 1998. Toward a better definition of the Middle Triassic magnetostratigraphy and biostratigraphy in the Tethyan realm. *Earth and Planetary Science Letters* 164:285–302.

Muttoni, G., D. V. Kent, P. E. Olsen, P. DiStefano, W. Lowrie, S. M. Bernasconi, and F. M. Hernández. 2004. Tethyan magnetostratigraphy from Pizzo Mondello (Sicily) and correlation to the Late Triassic Newark astrochronological polarity timescale. *Geological Society of America Bulletin* 116:1043–1058.

Nesbitt, S. J. 2003. *Arizonasaurus* and its implications for archosaur divergence. *Proceedings of the Royal Society of London B* 270 (Supplement 2):S234–S237.

Nesbitt, S. J. 2005. Osteology of the Middle Triassic pseudosuchian archosaur *Arizonasaurus babbitti*. *Historical Biology* 17:19–47.

Nesbitt, S. J. 2007. The anatomy of *Effigia okeeffeae* (Archosauria, Suchia), theropod-like convergence, and the distribution of related taxa. *Bulletin of the American Museum of Natural History* 302:1–84.

Nesbitt, S. J. 2009. The early evolution of archosaurs: Relationships and the origin of major clades. Ph.D. diss., Columbia University, New York.

Nesbitt, S. J., and K. D. Angielczyk. 2002. Fossil evidence of large dicynodonts in the upper Moenkopi Formation (Middle Triassic) of northern Arizona. *PaleoBios* 22(2):10–17.

Nesbitt, S. J., and S. Chatterjee. 2008. Late Triassic dinosauriforms from the Post Quarry and surrounding areas, west Texas, U.S.A. *Neues Jahrbuch für Geologie und Paläontologie, Abhandlungen* 249:143–156.

Nesbitt, S. J., and M. A. Norell. 2006. Extreme convergence in the body plans of an early suchian (Archosauria) and ornithomimid dinosaurs (Theropoda). *Proceedings of the Royal Society of London B* 273:1045–1048.

Nesbitt, S. J., and R. L. Whatley. 2004. The first discovery of a rhynchosaur from the upper Moenkopi Formation (Middle Triassic) of Arizona. *PaleoBios* 24(3):1–10.

Nesbitt, S. J., R. B. Irmis, and W. G. Parker. 2007. A critical re-evaluation of the Late Triassic dinosaur taxa of North America. *Journal of Systematic Palaeontology* 5:209–243.

Nesbitt, S. J., A. H. Turner, G. M. Erickson, and M. A. Norell. 2006. Prey choice and cannibalistic behaviour in the theropod *Coelophysis*. *Biology Letters* 2:611–614.

Neveling, J., P. J. Hancox, and B. S. Rubidge. 2005. Biostratigraphy of the lower Burgersdorp Formation (Beaufort Group; Karoo Supergroup) of South Africa—implications for the stratigraphic ranges of early Triassic tetrapods. *Palaeontologia Africana* 41:81–87.

Newton, E. T. 1894. Reptiles from the Elgin Sandstone.—Description of two new genera. *Philosophical Transactions of the Royal Society of London B* 185:573–607.

Nomade, S., K. B. Knight, E. Beutel, P. R. Renne, C. Vérati, G. Féraud, A. Marzoli, N. Youbi, and H. Bertrand. 2007. Chronology of the Central Atlantic Magmatic Province: Implications for central Atlantic rifting processes and the Triassic-Jurassic biotic crisis. *Palaeogeography, Palaeoclimatology, Palaeoecology* 244:326–344.

Nosotti, S. 2007. *Tanystropheus longobardicus* (Reptilia, Protorosauria): Re-interpretations of the anatomy based on new specimens from the Middle Triassic of Besano (Lombardy, northern Italy). *Memorie della Società Italiana di Scienze Naturali e del Museo Civico di Storia Naturale di Milano* 35(3):1–88.

Novas, F. E. 1994. New information on the systematics and postcranial skeleton of *Herrerasaurus ischigualastensis* (Theropoda: Herrerasauridae) from the Ischigualasto Formation (Upper Triassic) of Argentina. *Journal of Vertebrate Paleontology* 13:400–423.

Novas, F. E. 1996. Dinosaur monophyly. *Journal of Vertebrate Paleontology* 16:723–741.

Novikov, I. V., and M. A. Shishkin. 2000. Triassic chroniosuchians (Amphibia, Anthracosauromorpha) and the evolution of trunk dermal scutes in bystrowianids. *Paleontological Journal* 34 (Supplement 2):S165–S178.

Novikov, I. V., and H.-D. Sues. 2004. Cranial osteology of *Kapes* (Parareptilia: Procolophonidae) from the Lower Triassic of Orenburg Province, Russia. *Neues Jahrbuch für Geologie und Paläontologie, Abhandlungen* 232:267–281.

Ochev, V. G. 1958. [New data concerning the pseudosuchians of the USSR.] *Doklady Akademii Nauk SSSR* 123:749–751. [Russian.]

Ochev, V. G. 1966. [*Systematics and Phylogeny of Capitosauroid Labyrinthodonts.*] Saratov: Izdatel'stvo Saratovoskogo Gosuniversiteta. [Russian.]

Ochev, V. G. 1972. [*Capitosauroid Labyrinthodonts from the South-Eastern European Part of the USSR.*] Saratov: Izdatel'stvo Saratovoskogo Universiteta. [Russian.]

Ochev, V. G., and M. A. Shishkin. 1989. On the principles of global correlation of the continental Triassic on the tetrapods. *Acta Palaeontologica Polonica* 34:149–173.

O'Connor, P. M., and L. P. A. M. Claessens. 2005. Basic avian pulmonary design and flow-through ventilation in non-avian theropod dinosaurs. *Nature* 436:253–256.

Olsen, P. E. 1978. On the use of the term Newark for Triassic and Early Jurassic rocks of eastern North America. *Newsletter of Stratigraphy* 7:90–95.

Olsen, P. E. 1979. A new aquatic eosuchian from the Newark Supergroup (Late Triassic-Early Jurassic) of North Carolina and Virginia. *Postilla* 176:1–14.

Olsen, P. E. 1986. A 40-million year lake record of early Mesozoic orbital climatic forcing. *Science* 234:842–848.

Olsen, P. E. 1988. Paleontology and paleoecology of the Newark Supergroup (early Mesozoic, eastern North America). In W. Manspeizer, ed., *Triassic-Jurassic Rifting, Continental Breakup, and the Origin of the Atlantic Ocean and Passive Margins,* Part A, pp. 185–230. New York: Elsevier.

Olsen, P. E. 1997. Stratigraphic record of the early Mesozoic breakup of Pangea in the Laurasia-Gondwana rift system. *Annual Review of Earth and Planetary Sciences* 25:337–401.

Olsen, P. E. 1999. Giant lava flows, mass extinctions and mantle plumes. *Science* 284:604–605.

Olsen, P. E., and P. M. Galton. 1977. Triassic-Jurassic tetrapod extinctions: Are they real? *Science* 197:983–986.

Olsen, P. E., and A. K. Johansson. 1994. Field guide to Late Triassic tetrapod sites in Virginia and North Carolina. In N. C. Fraser and H.-D. Sues, eds., *In the Shadow of the Dinosaurs: Early Mesozoic Tetrapods,* pp. 408–430. New York: Cambridge University Press.

Olsen, P. E., and D. V. Kent. 2000. High-resolution early Mesozoic Pangean climatic transect in lacustrine environments. *Zentralblatt für Geologie und Paläontologie, Teil I,* 1998 (11–12):1475–1496.

Olsen, P. E., and H.-D. Sues. 1986. Correlation of continental Late Triassic and Jurassic sediments, and the Triassic-Jurassic tetrapod transition. In K. Padian, ed., *The Beginning of the Age of Dinosaurs: Faunal Change Across the Triassic-Jurassic Boundary,* pp. 321–351. New York: Cambridge University Press

Olsen, P. E., R. W. Schlische, and P. J. W. Gore. 1989. *Tectonic, Depositional, and Paleoecological History of Early Mesozoic Rift Basins, Eastern North America.* 28th International Geological Congress, Guidebook for Field Trip T351. Washington, DC: American Geophysical Union.

Olsen, P. E., N. H. Shubin, and M. H. Anders. 1987. New Early Jurassic tetrapod assemblages constrain Triassic-Jurassic tetrapod extinction event. *Science* 237:1025–1029.

Olsen, P. E., H.-D. Sues, and M. A. Norell. 2001. First record of *Erpetosuchus* (Reptilia: Archosauria) from the Late Triassic of North America. *Journal of Vertebrate Paleontology* 20:633–636.

Olsen, P. E., C. L. Remington, B. Cornet, and K. S. Thomson. 1978. Cyclic change in Late Triassic lacustrine communities. *Science* 201:729–733.

Olsen, P. E., C. Koeberl, H. Huber, A. Montanari, S. J. Fowell, M. Et-Touhami, and D. V. Kent. 2002a. Continental Triassic-Jurassic boundary in central Pangaea:

Recent progress and discussion of an Ir anomaly. In C. Koeberl and K. G. MacLeod, eds., *Catastrophic Events and Mass Extinctions: Impacts and Beyond*, pp. 505–522. Geological Society of America Special Paper 356. Boulder, CO: Geological Society of America.

Olsen, P. E., D. V. Kent, H.-D. Sues, C. Koeberl, H. Huber, A Montanari, E. C. Rainforth, S. J. Fowell, M. J. Szajna, and B. W. Hartline. 2002b. Ascent of dinosaurs linked to an iridium anomaly at the Triassic-Jurassic boundary. *Science* 296:1305–1307.

Ortlam, D. 1970. *Eocyclotosaurus woschmidti* n.g. n.sp.—ein neuer Capitosauride aus dem Oberen Buntsandstein des nördlichen Schwarzwaldes. *Neues Jahrbuch für Geologie und Paläontologie, Monatshefte* 1970:568–580.

Padian, K. 1984. The origin of pterosaurs. In F. Westphal and W.-E. Reif, eds., *Third Symposium on Terrestrial Ecosystems, Short Papers*, pp. 163–168. Tübingen: Attempto Verlag.

Padian, K., ed. 1986a. *The Beginning of the Age of Dinosaurs: Faunal Change Across the Triassic-Jurassic Boundary*. New York: Cambridge University Press.

Padian, K. 1986b. Introduction. In K. Padian, ed., *The Beginning of the Age of Dinosaurs: Faunal Change Across the Triassic-Jurassic Boundary*, pp. 1–7. New York: Cambridge University Press.

Pálfy, J., J. K. Mortensen, E. S. Carter, P. L. Smith, R. M. Friedman, and H. W. Tipper. 2000. Timing the end-Triassic mass extinction: First on land, then in the sea? *Geology* 28:39–42.

Pálfy, J., and R. Mundil. 2006. The age of the Triassic/Jurassic boundary: New data and their implications for the extinction and recovery. *Volumina Jurassica* 4:294.

Papier, F., and L. Grauvogel-Stamm. 1995. Les Blattodea du Trias: le genre *Voltziablatta* n. gen. du Buntsandstein supérieur des Vosges (France). *Palaeontographica A* 235: 141–162.

Papier, F., and A. Nel. 2001. Les Subioblattidae (Blattodea, Insecta) du Trias d'Asie Centrale. *Paläontologische Zeitschrift* 74:533–542.

Papier, F., L. Grauvogel-Stamm, and A. Nel. 1994. *Subioblatta undulata* n. sp., une nouvelle blatte (Subioblattidae Schneider) du Buntsandstein supérieur (Anisien) des Vosges (France). Morphologie, systématique et affinities. *Neues Jahrbuch für Geologie und Paläontologie, Monatshefte* 1994:277–290.

Papier, F., L. Grauvogel-Stamm, and A. Nel. 1996a. Deux nouveaux Mecopteroidea du Buntsandstein supérieur (Trias) des Vosges (France). In J.-C. Gall, ed., *Triassic Insects of Western Europe*, pp. 37–45. Palaeontologia Lombarda, n.s. 5. Milan: Società Italiana di Scienze Naturali e Museo Civico di Storia Naturale.

Papier, F., L. Grauvogel-Stamm, and A. Nel. 1996b. Nouveaux Blattodea du Buntsandstein supérieur (Trias de Vosges, France). In J.-C. Gall, ed., *Triassic Insects of Western Europe*, pp. 47–60. Palaeontologia Lombarda, n.s. 5. Milan: Società Italiana di Scienze Naturali e Museo Civico di Storia Naturale.

Papier, F., A. Nel, L. Grauvogel-Stamm, and J.-C. Gall. 1997. La plus ancienne sauterelle Tettigoniidae, Orthoptera (Trias, NE France): Mimétisme ou exaptation? *Paläontologische Zeitschrift* 71:71–77.

Papier, F., A. Nel, L. Grauvogel-Stamm, and J.-C. Gall. 2005. La diversité des Coleoptera (Insecta) du Trias dans le nord-est de la France. *Geodiversitas* 27:181–199.

Parker, W. G. 2005. Faunal review of the Upper Triassic Chinle Formation of Arizona. In R. D. McCord, ed., *Vertebrate Paleontology of Arizona*, pp. 34–54. Mesa Southwest Museum Bulletin 11. Mesa, AZ: Mesa Southwest Museum.

Parker, W. G. 2006. The stratigraphic distribution of major fossil localities in Petrified Forest National Park, Arizona. In W. G. Parker, S. R. Ash, and R. B. Irmis, eds., *A Century of Research at Petrified Forest National Park 1906–2006: Geology and Paleontology*, pp. 46–61. Museum of Northern Arizona Bulletin 62. Flagstaff: Museum of Northern Arizona.

Parker, W. G. 2007. Reassessment of the aetosaur *"Desmatosuchus" chamaensis* with a reanalysis of the phylogeny of the Aetosauria (Archosauria: Pseudosuchia). *Journal of Systematic Palaeontology* 5:41–68.

Parker, W. G. 2008. Description of new material of the aetosaur *Desmatosuchus spurensis* (Archosauria: Suchia) from the Chinle Formation of Arizona and a revision of the genus *Desmatosuchus*. *PaleoBios* 28:1–40.

Parker, W. G., and R. B. Irmis. 2006. A new species of the Late Triassic phytosaur *Pseudopalatus* (Archosauria: Pseudosuchia) from Petrified Forest National Park, Arizona. In W. G. Parker, S. R. Ash, and R. B. Irmis, eds., *A Century of Research at Petrified Forest National Park 1906–2006: Geology and Paleontology*, pp. 126–143. Museum of Northern Arizona Bulletin 62. Flagstaff: Museum of Northern Arizona.

Parker, W. G., R. B. Irmis, S. J. Nesbitt, J. W. Martz, and L. S. Browne. 2005. The Late Triassic pseudosuchian

Revueltosaurus callenderi and its implications for the diversity of early ornithischian dinosaurs. *Proceedings of the Royal Society B* 272:963–969.

Parker, W. G., M. R. Stocker, and R. B. Irmis. 2008. A new desmatosuchine aetosaur (Archosauria: Suchia) from the Upper Triassic Tecovas Formation (Dockum Group) of Texas. *Journal of Vertebrate Paleontology* 28:692–701.

Parks, P. 1969. Cranial anatomy and mastication of the Triassic reptile *Trilophosaurus*. M.A. thesis, University of Texas at Austin.

Parrington, F. R. 1935. On *Prolacerta broomi*, gen. et sp. n., and the origin of lizards. *Annals and Magazine of Natural History*, ser. 10, 16:197–205.

Parrish, J. M., J. T. Parrish, and A. M. Ziegler. 1986. Permian-Triassic paleogeography and paleoclimatology and implications for therapsid distributions. In N. Hotton III, P. D. MacLean, J. J. Roth, and E. C. Roth, eds., *The Biology and Ecology of Mammal-like Reptiles*, pp. 109–132. Washington, DC: Smithsonian Institution Press.

Parrish, J. T. 1993. Climate of the supercontinent Pangea. *Journal of Geology* 101:215–233.

Parrish, J. T. 1998. *Interpreting Pre-Quaternary Climate from the Geologic Record*. New York: Columbia University Press.

Parrish, J. T. 1999. Pangaea und das Klima der Trias. In N. Hauschke and V. Wilde, eds., *Trias—Eine ganz andere Welt*, pp. 37–42. Munich: Verlag Dr. Friedrich Pfeil.

Paul, J. 1999. Fazies und Sedimentstrukturen des Buntsandsteins. In N. Hauschke and V. Wilde, eds., *Trias—Eine ganz andere Welt*, pp. 105–114. Munich: Verlag Dr. Friedrich Pfeil.

Peabody, F. E. 1948. Reptile and amphibian trackways from the Lower Triassic Moenkopi formation of Arizona and Utah. *University of California Publications, Bulletin of the Department of Geological Sciences* 27:295–468.

Peabody, F. E. 1956. Ichnites from the Triassic Moenkopi formation of Arizona and Utah. *Journal of Paleontology* 30:731–740.

Pearson, H. S. 1924. A dicynodont reptile reconstructed. *Proceedings of the Zoological Society of London* 1924:827–855.

Peng, J. 1991. [A new genus of Proterosuchia from the Lower Triassic of Shaanxi, China.] *Vertebrata PalAsiatica* 29:95–107. [Chinese with English summary.]

Péron, S., S. Bourquin, F. Fluteau, and F. Guillocheau. 2005. Paleoenvironment reconstructions and climate simulations of the Early Triassic: Impact of the water and sediment supply on the preservation of fluvial systems. *Geodinamica Acta* 18:431–446.

Peyer, B. 1937. Die Triasfauna der Tessiner Kalkalpen. XII. *Macrocnemus bassanii* Nopcsa. *Abhandlungen der Schweizerischen Paläontologischen Gesellschaft* 59:1–140.

Peyer, B. 1944. *Die Reptilien vom Monte San Giorgio. 1924–1944*. Neujahrsblatt auf das Jahr 1944. Zürich: Naturforschende Gesellschaft.

Peyer, B. 1956. Über Zähne von Haramiyden, Triconodonten und wahrscheinlich synapsiden Reptilien aus dem Rhät von Hallau, Kt. Schaffhausen, Schweiz. *Abhandlungen der Schweizerischen Paläontologischen Gesellschaft* 72:1–72.

Peyer, K., J. G. Carter, H.-D. Sues, S. E. Novak, and P. E. Olsen. 2008. A new suchian archosaur from the Upper Triassic of North Carolina. *Journal of Vertebrate Paleontology* 28:363–381.

Pickford, M. 1995. Karoo Supergroup palaeontology of Namibia and brief description of a thecodont from Omingonde. *Palaeontologia Africana* 32:51–66.

Pinna, G. 1980. *Drepanosaurus unguicaudatus*, nuovo genere e nuova specie di Lepidosauro del Trias Alpino (Reptilia). *Atti della Società Italiana di Scienze Naturali e del Museo Civico di Storia Naturale di Milano* 121:181–192.

Pinna, G. 1984. Osteologia di *Drepanosaurus unguicaudatus*, Lepidosauro triassico del Sottordine Lacertilia. *Memoria della Società Italiana di Scienze Naturali e del Museo Civico di Storia Naturale di Milano* 24:7–28.

Pinna, G., and S. Nosotti. 1989. Anatomia, morfologia funzionale e paleoecologia del rettile placodonte *Psephoderma alpinum* Meyer, 1858. *Memoria della Società Italiana di Scienze Naturali e del Museo Civico di Storia Naturale di Milano* 25:17–49.

Pipiringos, G. N., and R. B. O'Sullivan. 1978. Principal unconformities in Triassic and Jurassic rocks, western United States. *U.S. Geological Survey Professional Paper* 1035A:A-1–A-29.

Ponomarenko, A. G. 1969. [Historical development of archostematan beetles.] *Trudy Paleontologicheskogo Instituta Akademiya Nauk SSSR* 125:1–240. [Russian.]

Pott, C., C. C. Labandeira, M. Krings, and H. Kerp. 2008. Fossil insect eggs and ovipositional damage on bennettitalean leaf cuticles from the Carnian (Upper Triassic) of Austria. *Journal of Paleontology* 82:778–789.

Prochnow, S. J., L. C. Nordt, S. C. Atchley, and M. R. Hudec. 2006. Multi-proxy paleosol evidence for Middle and Late Triassic climate trends in eastern Utah. *Palaeogeography, Palaeoclimatology, Palaeoecology* 232:53–72.

Rage, J.-C., and Z. Roček. 1989. Redescription of *Triadobatrachus massinoti* (Piveteau, 1936), an anuran amphibian from the early Triassic. *Palaeontographica A* 206:1–16.

Rasnitsyn, A. P., and D. L. J. Quicke, eds. 2002. *History of Insects*. Dordrecht: Kluwer Academic Publishers.

Raup, D. M., and J. J. Sepkoski Jr. 1982. Mass extinction in the fossil record. *Science* 215:1501–1503.

Ray, S. 2005. *Lystrosaurus* (Therapsida, Dicynodontia) from India: Taxonomy, relative growth and cranial dimorphism. *Journal of Systematic Palaeontology* 3:203–221.

Rayfield, E. J., P. M. Barrett, R. McDonnell, and K. J. Willis. 2005. A Geographical Information System (GIS) study of Triassic vertebrate biochronology. *Geological Magazine* 142:327–354.

Rayfield, E. J., P. M. Barrett, and A. R. Milner. 2009. Utility and validity of Middle and Late Triassic 'land vertebrate faunachrons.' *Journal of Vertebrate Paleontology* 29:80–87.

Redfield, W. C. 1856. On the relations of fossil fishes of the sandstone of the Connecticut and other Atlantic States to the Liassic and Oolitic periods. *American Journal of Science*, ser. 2, 22:357–363.

Rees, P. M. 2002. Land-plant diversity and the end-Permian mass extinction. *Geology* 30:827–830.

Reif, W.-E. 1971. Zur Genese des Muschelkalk/Keuper-Grenzbonebeds in Südwestdeutschland. *Neues Jahrbuch für Geologie und Paläontologie, Abhandlungen* 139:369–404.

Reig, O. A. 1959. Primeros datos descriptivos sobre nuevos reptiles arcosaurios del Triásico de Ischigualasto (San Juan, Argentina). *Revista del la Asociación Geológica Argentina* 13:257–270.

Reig, O. A. 1963. La presencia de dinosaurios saurisquios en los "Estratos de Ischigualasto" (Mesotriásico superior) de las provincias de San Juan y La Rioja (República Argentina). *Ameghiniana* 3:3–20.

Reisz, R. R., and H.-D. Sues. 2000a. Herbivory in late Paleozoic and Triassic tetrapods. In H.-D. Sues, ed., *Evolution of Herbivory in Terrestrial Vertebrates: Perspectives from the Fossil Record*, pp. 9–41. Cambridge: Cambridge University Press.

Reisz, R. R., and H.-D. Sues. 2000b. The "feathers" of *Longisquama*. *Nature* 408:428.

Renaut, A. J., and P. J. Hancox. 2001. Cranial redescription and taxonomic re-evaluation of *Kannemeyeria argentinensis* (Therapsida: Dicynodontia). *Palaeontologia Africana* 37:81–91.

Renesto, S. 1984. A new lepidosaur (Reptilia) from the Norian beds of the Bergamo Prealps. *Rivista Italiana di Paleontologia e Stratigrafia* 90:156–167.

Renesto, S. 1994. A new prolacertiform reptile from the Late Triassic of northern Italy. *Rivista Italiana di Paleontologia e Stratigrafia* 100:285–306.

Renesto, S. 1995. A sphenodontid from the Norian (Late Triassic) of Lombardy (northern Italy): A preliminary note. *Modern Geology* 20:149–158.

Renesto, S. 2000. Bird-like head on a chameleon body: New specimens of the enigmatic diapsid reptile *Megalancosaurus* from the Late Triassic of northern Italy. *Rivista Italiana di Paleontologia e Stratigrafia* 106:157–180.

Renesto, S. 2006. A reappraisal of the diversity and biogeographic significance of the Norian (Late Triassic) reptiles from the Calcare di Zorzino. In J. D. Harris, S. G. Lucas, J. A. Spielmann, M. G. Lockley, A. R. C. Milner, and J. I. Kirkland, eds., *The Triassic-Jurassic Terrestrial Transition*, pp. 445–456. New Mexico Museum of Natural History & Science Bulletin 37. Albuquerque: New Mexico Museum of Natural History & Science.

Renesto, S., and G. Binelli. 2006. *Vallesaurus cenensis* Wild, 1991, a drepanosaurid (Reptilia, Diapsida) from the Late Triassic of northern Italy. *Rivista Italiana di Paleontologia e Stratigrafia* 112:77–94.

Renesto, S., and F. M. Dalla Vecchia. 2000. The unusual dentition and feeding habits of the prolacertiform reptile *Langobardisaurus* (Late Triassic, northern Italy). *Journal of Vertebrate Paleontology* 20:622–627.

Renesto, S., and N. C. Fraser. 2003. Drepanosaurid (Reptilia: Diapsida) remains from a Late Triassic fissure infilling at Cromhall Quarry (Avon, Great Britain). *Journal of Vertebrate Paleontology* 23:703–705.

Renesto, S., and C. Lombardo. 1999. Structure of the tail of a phytosaur (Reptilia, Archosauria) from the Norian (Late Triassic) of Lombardy (northern Italy). *Rivista Italiana di Paleontologia e Stratigrafia* 105:135–144.

Renne, P. R., Z. Zhang, M. A. Richardson, M. T. Black, and A. R. Basu. 1995. Synchrony and causal relations between Permo-Triassic boundary crises and Siberian flood volcanism. *Science* 269:1413–1416.

Retallack, G. J. 1997. Earliest Triassic origin of *Isoetes* and quillwort evolutionary radiation. *Journal of Paleontology* 71:500–521.

Retallack, G. J., and W. R. Hammer. 1998. Paleoenvironment of the Triassic therapsid *Lystrosaurus* in the central Transantarctic Mountains, Antarctica. *U.S. Antarctic Journal* 31(for 1996):33–35.

Retallack, G. J., R. M. H. Smith, and P. D. Ward. 2003. Vertebrate extinction across Permian-Triassic boundary in Karoo Basin, South Africa. *Geological Society of America Bulletin* 115:1133–1152.

Retallack, G. J., J. J. Veevers, and R. Morante. 1996. Global coal gap between Permian-Triassic extinction and Middle Triassic recovery of peat-forming plants. *Geological Society of America Bulletin* 108:195–207.

Reynoso, V.-H. 2005. Possible evidence of a venom apparatus in a Middle Jurassic sphenodontian from the Huizachal Red Beds of Tamaulipas, México. *Journal of Vertebrate Paleontology* 25:646–654.

Riabinin, A. N. 1930. [A labyrinthodont stegocephalian *Wetlugasaurus angustifrons* nov. gen., nov. sp. from the Lower Triassic of Vetluga Land in northern Russia.] *Ezhegodnik Vserossiisskogo Paleontologicheskogo Obshchestva* 1930(8):49–76. [Russian.]

Ricqlès, A. de, K. Padian, and J. R. Horner. 2003. On the histology of some Triassic pseudosuchian archosaurs and related taxa. *Annales de Paléontologie* 89:67–101.

Riek, E. F. 1974. Upper Triassic insects from the Molteno Formation, South Africa. *Palaeontologia Africana* 17:19–31.

Riek, E. F. 1976. An unusual mayfly (Insecta: Ephemeroptera) from the Triassic of South Africa. *Palaeontologia Africana* 19:149–151.

Rieppel, O. 1989. The hind limb of *Macrocnemus bassanii* (Reptilia, Diapsida): Development and functional anatomy. *Journal of Vertebrate Paleontology* 9:373–387.

Rieppel, O. 2000. *Sauropterygia I. Handbuch der Paläoherpetologie*, Teil 12A. Munich: Verlag Dr. Friedrich Pfeil.

Rieppel, O., and R. R. Reisz. 1999. The origin and early evolution of turtles. *Annual Review of Ecology and Systematics* 30:1–22.

Riggs, N. R., T. M. Lehman, G. E. Gehrels, and W. R. Dickinson. 1996. Detrital zircon link between headwaters and terminus of the Upper Triassic Chinle-Dockum paleoriver system. *Science* 273:97–100.

Robinson, P. L. 1957a. The Mesozoic fissures of the Bristol Channel area and their vertebrate faunas. *Journal of the Linnean Society (Zoology)* 43:260–282.

Robinson, P. L. 1957b. An unusual sauropsid dentition. *Journal of the Linnean Society (Zoology)* 43:283–293.

Robinson, P. L. 1962. Gliding lizards from the Upper Keuper of Great Britain. *Proceedings of the Geological Society of London* 1601:137–146.

Robinson, P. L. 1967. Triassic vertebrates from lowland and upland. *Science and Culture* 33:169–173.

Robinson, P. L. 1971. A problem of faunal replacement on Permo-Triassic continents. *Palaeontology* 14:131–153.

Robinson, P. L. 1973. A problematic reptile from the British Upper Trias. *Journal of the Geological Society of London* 129:457–479.

Rogers, R. R., A. B. Arcucci, F. Abdala, P. C. Sereno, C. A. Forster, and C. L. May. 2001. Paleoenvironment and taphonomy of the Chañares Formation tetrapod assemblage (Middle Triassic), northwestern Argentina: Spectacular preservation in volcanigenic concretions. *Palaios* 16:461–481.

Rogers, R. R., C. C. Swisher III, P. C. Sereno, A. M. Monetta, C. A. Forster, and R. N. Martinez. 1993. The Ischigualasto tetrapod assemblage (Late Triassic, Argentina) and $^{40}Ar/^{39}Ar$ dating of dinosaur origins. *Science* 260:794–797.

Rohdendorf, B. B., and A. P. Rasnitsyn, eds. 1980. [Historical development of the class Insecta.] *Trudy Paleontologicheskii Instituta Akademiya Nauk SSSR* 175:1–269. [Russian.]

Romer, A. S. 1960. Vertebrate-bearing continental Triassic strata in Mendoza Region, Argentina. *Geological Society of America Bulletin* 71:1279–1294.

Romer, A. S. 1962. La evolución explosiva de los rhynchosaurios del Triásico. *Revista del Museo Argentino de Ciencias Naturales "Bernardino Rivadavia", Ciencias Zoológicas* 8:1–14.

Romer, A. S. 1966. The Chañares (Argentina) Triassic reptile fauna. I. Introduction. *Breviora* 247:1–14.

Romer, A. S. 1967. The Chañares (Argentina) Triassic reptile fauna. III. Two new gomphodonts, *Massetognathus pascuali* and *M. teruggii*. *Breviora* 264:1–25.

Romer, A. S. 1969. The Chañares (Argentina) Triassic reptile fauna. V. A new chiniquodont cynodont, *Probelesodon lewisi*—cynodont ancestry. *Breviora* 333:1–24.

Romer, A. S. 1970. The Chañares (Argentina) Triassic reptile fauna. VI. A chiniquodontid cynodont with an incipient squamosal-dentary jaw articulation. *Breviora* 344:1–18.

Romer, A. S. 1971a. The Chañares (Argentina) Triassic reptile fauna. X. Two new but incompletely known long-limbed pseudosuchians. *Breviora* 378:1–10.

Romer, A. S. 1971b. The Chañares (Argentina) Triassic reptile fauna. XI. Two new long-snouted thecodonts, *Chanaresuchus* and *Gualosuchus*. *Breviora* 379:1–22.

Romer, A. S. 1972a. The Chañares (Argentina) Triassic reptile fauna. XII. The postcranial skeleton of the thecodont *Chanaresuchus*. *Breviora* 385:1–21.

Romer, A. S. 1972b. The Chañares (Argentina) Triassic reptile fauna. XIII. An early ornithosuchid pseudosuchian *Gracilisuchus stipanicicorum* gen. et sp. nov. *Breviora* 389:1–24.

Romer, A. S. 1973. The Chañares (Argentina) Triassic reptile fauna. XX. Summary. *Breviora* 413:1–20.

Rothwell, G. W., L. Grauvogel-Stamm, and G. Mapes. 2000. An herbaceous fossil conifer: Gymnospermous ruderals in the evolution of Mesozoic vegetation. *Palaeogeography, Palaeoclimatology, Palaeoecology* 156:139–145.

Rougier, G. W., M. S. de la Fuente, and A. B. Arcucci. 1995. Late Triassic turtles from South America. *Science* 268:855–858.

Rowe, T. 1988. Definition, diagnosis, and the origin of Mammalia. *Journal of Vertebrate Paleontology* 8:241–264.

Roy Chowdhury, T. 1965. A new metoposaurid amphibian from the Upper Triassic Maleri Formation of Central India. *Philosophical Transactions of the Royal Society of London B* 250:1–52.

Roy Chowdhury, T. 1970. Two new dicynodonts from the Yerrapalli Formation of central India. *Palaeontology* 13:132–144.

Rubidge, B. S., ed. 1995. *Biostratigraphy of the Beaufort Group (Karoo Supergroup).* South African Committee for Stratigraphy, Biostratigraphic Series 1:1–46. Pretoria: Council for Geoscience.

Rubidge, B. S. 2005. 27th DuToit Memorial Lecture. Reuniting lost continents—fossil reptiles from the ancient Karoo and their wanderlust. *South African Journal of Geology* 108:135–172.

Rubidge, B. S., and C. A. Sidor. 2001. Evolutionary patterns among Permo-Triassic therapsids. *Annual Review of Ecology and Systematics* 32:449–480.

Ruta, M., and M. J. Benton. 2008. Calibrated diversity, tree topology and the mother of mass extinctions: The lesson of temnospondyls. *Palaeontology* 51:1261–1288.

Ruta, M., and J. R. Bolt. 2008. The brachyopoid *Hadrokkosaurus bradyi* from the early Middle Triassic of Arizona, and a phylogenetic analysis of lower jaw characters in temnospondyl amphibians. *Acta Palaeontologica Polonica* 53:579–592.

Säilä, L. K. 2005. A new species of the sphenodontian reptile *Clevosaurus* from the Lower Jurassic of Wales. *Palaeontology* 48:817–831.

Säilä, L. K. 2008. The osteology and affinities of *Anomoiodon liliensterni*, a procolophonid reptile from the Lower Triassic Bundsandstein of Germany. *Journal of Vertebrate Paleontology* 28:1199–1205.

Sander, P. M. 1992. The Norian *Plateosaurus* bonebeds of central Europe and their taphonomy. *Palaeogeography, Palaeoclimatology, Palaeoecology* 93:255–299.

Sawin, H. J. 1947. The pseudosuchian reptile *Typothorax meadei*, new species. *Journal of Paleontology* 21:201–238.

Schaeffer, B. 1967. Late Triassic fishes from the western United States. *Bulletin of the American Museum of Natural History* 135:285–342.

Schaeffer, B., and N. G. McDonald. 1978. Redfieldiid fishes from the Triassic-Liassic Newark Supergroup of eastern North America. *Bulletin of the American Museum of Natural History* 159:129–174.

Schaltegger, U., J. Guex, A. Bartolini, B. Schoene, and M. Ovtcharova. 2008. Precise U-Pb constraints for end-Triassic mass extinction, its correlation to volcanism and Hettangian post-extinction recovery. *Earth and Planetary Science Letters* 267:266–275.

Schenk, A. 1867. *Die fossile Flora der Grenzschichten des Keupers und Lias Frankens.* Wiesbaden: C. W. Kreidel's Verlag.

Schimper, W. P. 1880. Palaeophytologie. (Fortgesetzt und vollendet von A. Schenk.) K. A. v. Zittel, ed., *Handbuch der Palaeontologie.* II. *Abtheilung.* Munich: Oldenbourg.

Schlische, R. W., and P. E. Olsen, 1988. Structural evolution of the Newark basin. In J. M. Husch and M. J. Hozik, eds., *Geology of the Central Newark Basin: Fifth Annual Meeting of the New Jersey Geological Association,* pp. 43–65. Lawrenceville, NJ: Rider College.

Schmidt, M. 1928. *Die Lebewelt unserer Trias.* Oehringen: Hohenlohe'sche Buchhandlung Ferdinand Rau.

Schmidt, M. 1938. *Die Lebewelt unserer Trias. Nachtrag 1938.* Oehringen: Hohenlohe'sche Buchhandlung Ferdinand Rau.

Schoch, R. R. 1997. A new capitosaur from the Upper Lettenkeuper (Triassic: Ladinian) of Kupferzell (southern Germany). *Neues Jahrbuch für Geologie und Paläontologie, Abhandlungen* 203:239–272.

Schoch, R. R. 1999. Comparative osteology and reconstruction of *Mastodonsaurus giganteus* (Jaeger, 1828) from the Middle Triassic (Longobardian: Lettenkeuper) of Germany (Baden-Württemberg, Bayern, Thüringen). *Stuttgarter Beiträge zur Naturkunde B* 278:1–170.

Schoch, R. R. 2000. The status and osteology of two new capitosaurid amphibians from the upper Moenkopi Formation of Arizona (Amphibia: Temnospondyli; Middle Triassic). *Neues Jahrbuch für Geologie und Paläontologie, Abhandlungen* 216:387–411.

Schoch, R. R. 2006a. A complete trematosaurid amphibian from the Middle Triassic of Germany. *Journal of Vertebrate Paleontology* 26:29–43.

Schoch, R. R. 2006b. Kupferzell: Saurier aus den Keupersümpfen. *Stuttgarter Beiträge zur Naturkunde C* 61: 1–79.

Schoch, R. R. 2007. Osteology of the small archosaur *Aetosaurus* from the Upper Triassic of Germany. *Neues Jahrbuch für Geologie und Paläontologie, Abhandlungen* 246:1–35.

Schoch, R. R. 2008. A new stereospondyl from the German Middle Triassic, and the origin of the Metoposauridae. *Zoological Journal of the Linnean Society* 152:79–113.

Schoch, R. R., and A. R. Milner. 2000. Stereospondyli. *Handbuch der Paläoherpetologie*, Part 3B. Munich: Verlag Dr. Friedrich Pfeil.

Schoch, R. R., and B. S. Rubidge. 2005. The amphibamid *Micropholis* from the *Lystrosaurus* Assemblage Zone of South Africa. *Journal of Vertebrate Paleontology* 25: 502–522.

Schoch, R. R., and R. Wild. 1999a. Die Wirbeltiere des Muschelkalks unter besonderer Berücksichtigung Süddeutschlands. In N. Hauschke and V. Wilde, eds., *Trias—Eine ganz andere Welt*, pp. 331–342. Munich: Verlag Dr. Friedrich Pfeil.

Schoch, R. R., and R. Wild. 1999b. Die Wirbeltier-Fauna im Keuper von Süddeutschland. In N. Hauschke and V. Wilde, eds., *Trias—Eine ganz andere Welt*, pp. 395–408. Munich: Verlag Dr. Friedrich Pfeil.

Schröder, B. 1982. Entwicklung des Sedimentbeckens und Stratigraphie der klassischen Germanischen Trias. *Geologische Rundschau* 71:783–794.

Schroeder, H. 1913. Ein Stegocephalen-Schädel von Helgoland. *Jahrbuch der Königlich Preussischen Geologischen Landesanstalt zu Berlin* (für das Jahr 1912) 33:232–264.

Schultze, H.-P., and J. Kriwet. 1999. Die Fische der Germanischen Trias. In N. Hauschke and V. Wilde, eds., *Trias—Eine ganz andere Welt*, pp. 239–250. Munich: Verlag Dr. Friedrich Pfeil.

Schwartz, H. L., and D. D. Gillette. 1994. Geology and taphonomy of the *Coelophysis* Quarry, Upper Triassic Chinle Formation, Ghost Ranch, New Mexico. *Journal of Paleontology* 68:1118–1130.

Schweitzer, H.-J. 1978. Die räto-jurassischen Floren des Iran und Afghanistans. 5. *Todites princeps, Thaumatopteris brauniana* und *Phlebopteris polypodiodes*. *Palaeontographica B* 168:17–60.

Scotese, C. R. 2000. *PALEOMAP Project*. Arlington, TX: University of Texas at Arlington.

Scott, A. C., J. M. Anderson, and H. M. Anderson. 2004. Evidence of plant-insect interactions in the Upper Triassic Molteno Formation of South Africa. *Journal of the Geological Society of London* 161:401–410.

Selden, P. A., and J.-C. Gall. 1992. A Triassic mygalomorph spider from the northern Vosges, France. *Palaeontology* 35:211–235.

Selden, P. A., H. M. Anderson, and J. M. Anderson. 2009. A review of the fossil record of spiders (Araneae) with special reference to Africa, and description of a new specimen from the Triassic Molteno Formation of South Africa. *African Invertebrates* 50:105–116.

Selden, P. A., J. M. Anderson, H. M. Anderson, and N. C. Fraser. 1999. Fossil araneomorph spiders from the Triassic of South Africa and Virginia. *Journal of Arachnology* 27:401–414.

Selwood, B. W., and P. J. Valdes. 2006. Mesozoic climates: General circulation models and the rock record. *Sedimentary Geology* 190:269–287.

Sen, K. 2003. *Pamelaria dolichotrachela*, a new prolacertid reptile from the Middle Triassic of India. *Journal of Asian Earth Sciences* 21:663–681.

Sen, K. 2005. A new rauisuchian archosaur from the Middle Triassic of India. *Palaeontology* 48:185–196.

Sengupta, D. P. 1995. Chigutisaurid temnospondyls from the Late Triassic of India and a review of the family Chigutisauridae. *Palaeontology* 38:313–339.

Sennikov, A. G. 1995. [Primitive thecodonts of Eastern Europe.] *Trudy Paleontologicheskogo Instituta RAN* 263:1–141. [Russian.]

Sereno, P. C. 1991. Basal archosaurs: Phylogenetic relationships and functional implications. *Society of Vertebrate Paleontology Memoir* 2:1–53.

Sereno, P. C. 1994. The pectoral girdle and forelimb of the basal theropod *Herrerasaurus ischigualastensis*. *Journal of Vertebrate Paleontology* 13:425–450.

Sereno, P. C. 1997. The origin and evolution of dinosaurs. *Annual Review of Earth and Planetary Sciences* 25:435–489.

Sereno, P. C. 2007. The phylogenetic relationships of early dinosaurs: A comparative report. *Historical Biology* 19:145–155.

Sereno, P. C., and A. B. Arcucci. 1994a. Dinosaurian precursors from the Middle Triassic of Argentina: *Lagerpeton chanarensis*. *Journal of Vertebrate Paleontology* 13:385–399.

Sereno, P. C., and A. B. Arcucci. 1994b. Dinosaurian precursors from the Middle Triassic of Argentina: *Marasuchus lilloensis*, gen. nov. *Journal of Vertebrate Paleontology* 14:53–73.

Sereno, P. C., and F. E. Novas. 1994. The skull and neck of the basal theropod *Herrerasaurus ischigualastensis*. *Journal of Vertebrate Paleontology* 13:451–476.

Sereno, P. C., and R. Wild. 1992. *Procompsognathus*: Theropod, "thecodont" or both? *Journal of Vertebrate Paleontology* 12:435–458.

Sereno, P. C., C. A. Forster, R. R. Rogers, and A. M. Monetta. 1993. Primitive dinosaur skeleton from Argentina and the early evolution of Dinosauria. *Nature* 361:64–66.

Sharov, A. G. 1968. [Phylogeny of orthopteroid insects.] *Trudy Paleontologicheskogo Instituta Akademiya Nauk SSSR* 118:1–216. [Russian.]

Sharov, A. G. 1970. [An unusual reptile from the Lower Triassic of Fergana.] *Paleontologicheskii Zhurnal* 1970(1): 127–131. [Russian.]

Sharov, A. G. 1971. [New flying reptiles from the Mesozoic of Kazakhstan and Kirghizia.] *Trudy Paleontologicheskogo Instituta Akademiya Nauk SSSR* 130:104–113. [Russian.]

Shcherbakov, D. E. 2008a. Madygen, Triassic Lagerstätte number one, before and after Sharov. *Alavesia* 2:113–124.

Shcherbakov, D. E. 2008b. Insect recovery after the Permian/Triassic crisis. *Alavesia* 2:125–131.

Shear, W. A., P. A. Selden, and J.-C. Gall. 2009. Millipedes from the Grès à Voltzia, Triassic of France, with comments on Mesozoic millipedes (Diplopoda: Helminthomorpha: Eugnatha). *International Journal of Myriapodology* 1:1–13.

Sheehan, P. M. 1977. Species diversity in the Phanerozoic: A reflection of labor by systematists? *Paleobiology* 3: 325–328.

Shishkin, M. A. 1987. [Evolution of ancient amphibians (Plagiosauroidea).] *Trudy Paleontologicheskogo Instituta Akademiya Nauk SSSR* 225:1–144. [Russian.]

Shishkin, M. A., V. G. Ochev, V. R. Lozovskii, and I. V. Novikov. 2000. Tetrapod biostratigraphy of the Triassic of Eastern Europe. In M. J. Benton, M. A. Shishkin, D. M. Unwin, and E. N. Kurochkin, eds., *The Age of Dinosaurs in Russia and Mongolia*, pp. 120–139. Cambridge: Cambridge University Press.

Shubin, N. H., A. W. Crompton, H.-D. Sues, and P. E. Olsen. 1991. New fossil evidence on the sister-group of mammals and early Mesozoic faunal distributions. *Science* 251:1063–1065.

Shukla, U. K., and G. H. Bachmann. 2007. Estuarine sedimentation in the Stuttgart Formation (Carnian, Late Triassic), South Germany. *Neues Jahrbuch für Geologie und Paläontologie, Abhandlungen* 243:305–323.

Sickler, F. K. L. 1834. *Sendschreiben an Dr. J. F. Blumenbach über die höchst merkwürdigen, vor einigen Monaten erst entdeckten Reliefs der Fährten urweltlicher, grosser und unbekannter Thiere in den Hessberger Sandsteinbrüchen bei der Stadt Hildburghausen*. Hildburghausen: Schulprogramm des herzoglichen Gymnasiums zu Hildburghausen.

Sidor, C. A., and P. J. Hancox. 2006. *Elliotherium kersteni*, a new tritheledontid from the Lower Elliot Formation (Upper Triassic) of South Africa. *Journal of Paleontology* 80:333–342.

Sidor, C. A., M. F. Miller, and J. L. Isbell. 2008. Tetrapod burrows from the Triassic of Antarctica. *Journal of Vertebrate Paleontology* 28:277–284.

Sigogneau-Russell, D. 1983a. A new therian mammal from the Rhaetic locality of Saint-Nicolas-de-Port (France). *Zoological Journal of the Linnean Society* 78:175–186.

Sigogneau-Russell, D. 1983b. Nouveaux taxons de mammifères rhétiens. *Acta Palaeontologica Polonica* 28:233–249.

Sigogneau-Russell, D. 1989. Haramiyidae (Mammalia, Allotheria) en provenance du Trias supérieur de Lorraine (France). *Palaeontographica A* 206:137–198.

Sigogneau-Russell, D., and G. Hahn. 1994. Late Triassic microvertebrates from central Europe. In N. C. Fraser and H.-D. Sues, eds., *In the Shadow of the Dinosaurs: Early Mesozoic Tetrapods*, pp. 197–213. New York: Cambridge University Press.

Sigogneau-Russell, D., and A. Sun. 1979. A brief review of Chinese synapsids. *Geobios* 14:215–219.

Sill, W. D. 1967. *Proterochampsa barrionuevoi* and the early evolution of the Crocodilia. *Bulletin of the Museum of Comparative Zoology, Harvard University* 135:415–446.

Sill, W. D. 1974. The anatomy of *Saurosuchus galilei* and the relationship of the rauisuchid thecodonts. *Bulletin of the Museum of Comparative Zoology, Harvard University* 146:317–362.

Silvestri, S. M., and M. J. Szajna. 1993. Biostratigraphy of vertebrate footprints in the Late Triassic section of the Newark basin, Pennsylvania: Reassessment of stratigraphic ranges. In S. G. Lucas and M. Morales, eds., *The Nonmarine Triassic*, pp. 439–445. New Mexico Museum of Natural History & Science Bulletin 3. Albuquerque: New Mexico Museum of Natural History & Science.

Simms, M. J. 1990. Triassic palaeokarst in Britain. *Cave Science* 17:93–101.

Simms, M. J., A. H. Ruffell, and A. L. A. Johnson. 1994. Biotic and climatic changes in the Carnian (Triassic) of Europe and adjacent areas. In N. C. Fraser and H.-D. Sues, eds., *In the Shadow of the Dinosaurs: Early Mesozoic Tetrapods*, pp. 352–365. New York: Cambridge University Press.

Simpson, G. G. 1926. Mesozoic Mammalia. V. *Dromatherium* and *Microconodon*. *American Journal of Science*, ser. 5, 12:87–108.

Sinclair, W. J. 1918. A large parasuchian from the Triassic of Pennsylvania. *American Journal of Science*, ser. 4, 45:457–462.

Small, B. J. 1989. Aetosaurs from the Upper Triassic Dockum Formation, Post Quarry, West Texas. In S. G. Lucas and A. P. Hunt, eds., *Dawn of the Age of Dinosaurs in the American Southwest*, pp. 301–308. Albuquerque: New Mexico Museum of Natural History.

Small, B. J. 2002. Cranial anatomy of *Desmatosuchus haplocerus* (Reptilia: Archosauria: Stagonolepididae). *Zoological Journal of the Linnean Society* 136:97–111.

Smith, R. M. H. 1990. A review of stratigraphy and sedimentary environments of the Karoo Basin of South Africa. *Journal of African Earth Sciences* 10:117–137.

Smith, R. M. H. 1995. Changing fluvial environments across the Permian-Triassic boundary in the Karoo Basin, South Africa and possible causes of tetrapod extinctions. *Palaeogeography, Palaeoclimatology, Palaeoecology* 117:81–104.

Smith, R. M. H., and R. Swart. 2002. Changing fluvial environments and vertebrate taphonomy in response to climatic drying in a Mid-Triassic rift valley fill: The Omingonde Formation (Karoo Supergroup) of Central Namibia. *Palaios* 17:249–267.

Soergel, W. 1925. *Die Fährten der Chirotheria. Eine paläobiologische Studie.* Jena: Verlag von Gustav Fischer.

Spalletti, L., A. Artabe, E. Morel, and M. Brea. 1999. Biozonación paleoflorística y cronoestratigrafía del Triásico argentino. *Ameghiniana* 36:419–451.

Spalletti, L. A., E. M. Morel, A. E. Artabe, A. M. Zavattieri, and D. Ganuza. 2005. Estratigrafía, facies y paleoflora de la sucesión triásica de Potrerillos, Mendoza, República Argentina. *Revista Geológica de Chile* 32:249–272.

Spray, J. G. 1998. [Impacts on Earth in the Late Triassic: Discussion.] *Nature* 395:126.

Spray, J. G., S. P. Kelley, and D. B. Rowley. 1998. Evidence for a late Triassic multiple impact event on Earth. *Nature* 392:171–173.

Stanley, G. D., Jr. 2003. The evolution of modern corals and their early history. *Earth-Science Reviews* 60: 195–225.

Stein, K., C. Palmer, P. G. Gill, and M. J. Benton. 2008. The aerodynamics of the British Late Triassic Kuehneosauridae. *Palaeontology* 51:967–981.

Stewart, J. H., F. G. Poole, and R. F. Wilson. 1972. Stratigraphy and origin of the Triassic Moenkopi Formation in the Colorado Plateau region. *U.S. Geological Survey Professional Paper* 691:1–195.

Stollhofen, H., G. H. Bachmann, J. Barnasch, U. Bayer, G. Beutler, M. Franz, M. Kästner, B. Legler, J. Mutterlose, and D. Radies. 2008. Upper Rotliegend to Early Cretaceous basin development. In R. Littke, U. Bayer, D. Gajewski, and S. Nelskamp, eds., *Dynamics of Complex Intracontinental Basins: The Central European Basin System*, pp. 181–210. Berlin: Springer-Verlag.

Stur, D. 1885. Die obertriadische Flora der Lunzer-Schichten und des bituminösen Schiefers von Raibl. *Denkschriften der Kaiserlichen Akademie der Wissenschaften, Wien* 3: 93–103.

Sues, H.-D. 1991. Venom-conducting teeth in a Triassic reptile. *Nature* 351:141–143.

Sues, H.-D. 1992. A remarkable new armored archosaur from the Upper Triassic of Virginia. *Journal of Vertebrate Paleontology* 12:142–149.

Sues, H.-D. 1996. A reptilian tooth with apparent venom canals from the Chinle Group (Upper Triassic) of Arizona. *Journal of Vertebrate Paleontology* 16:571–572.

Sues, H.-D. 2001. On *Microconodon*, a Late Triassic cynodont from the Newark Supergroup of eastern North America. *Bulletin of the Museum of Comparative Zoology, Harvard University* 156:37–48.

Sues, H.-D. 2003. An unusual new archosauromorph reptile from the Upper Triassic Wolfville Formation of Nova Scotia. *Canadian Journal of Earth Sciences* 40:635–649.

Sues, H.-D., and D. Baird. 1993. A skull of a sphenodontian lepidosaur from the New Haven Arkose (Upper Triassic) of Connecticut. *Journal of Vertebrate Paleontology* 13:370–372.

Sues, H.-D., and D. Baird. 1998. Procolophonidae (Reptilia: Parareptilia) from the Upper Triassic Wolfville Formation of Nova Scotia. Canada. *Journal of Vertebrate Paleontology* 18:525–532.

Sues, H.-D., J. A. Hopson, and N. H. Shubin. 1992. Affinities of *?Scalenodontoides plemmyridon* Hopson, 1984 (Synapsida: Cynodontia) from the Upper Triassic of Nova Scotia. *Journal of Vertebrate Paleontology* 12:168–171.

Sues, H.-D., and P. E. Olsen 1990. Triassic vertebrates of Gondwanan aspect from the Richmond basin of Virginia. *Science* 249:1020–1023.

Sues, H.-D., and R. R. Reisz. 1995. First record of the early Mesozoic sphenodontian *Clevosaurus* (Lepidosauria: Rhynchocephalia) from the Southern Hemisphere. *Journal of Paleontology* 69:123–126.

Sues, H.-D., and R. R. Reisz. 2008. Anatomy and phylogenetic relationships of *Sclerosaurus armatus* (Amniota: Parareptilia) from the Buntsandstein (Triassic) of Europe. *Journal of Vertebrate Paleontology* 28:1031–1042.

Sues, H.-D., P. E. Olsen, and P. A. Kroehler. 1994. Small tetrapods from the Upper Triassic of the Richmond basin (Newark Supergroup), Virginia. In N. C. Fraser and H.-D. Sues, eds., *In the Shadow of the Dinosaurs: Early Mesozoic Tetrapods*, pp. 161–170. New York: Cambridge University Press.

Sues, H.-D., N. H. Shubin, and P. E. Olsen. 1994. A new sphenodontian (Lepidosauria: Rhynchocephalia) from the McCoy Brook Formation (Lower Jurassic) of Nova Scotia, Canada. *Journal of Vertebrate Paleontology* 14:327–340.

Sues, H.-D., P. E. Olsen, J. G. Carter, and D. M. Scott. 2003. A new crocodylomorph reptile from the Upper Triassic of North Carolina. *Journal of Vertebrate Paleontology* 23:329–343.

Sues, H.-D., P. E. Olsen, D. M. Scott, and P. S. Spencer. 2000. Cranial osteology of *Hypsognathus fenneri*, a latest Triassic procolophonid reptile from the Newark Supergroup of eastern North America. *Journal of Vertebrate Paleontology* 20:275–284.

Sulej, T. 2002. Species discrimination of the Late Triassic temnospondyl amphibian *Metoposaurus diagnosticus*. *Acta Palaeontologica Polonica* 47:535–546.

Sulej, T. 2005. A new rauisuchian reptile (Diapsida: Archosauria) from the Late Triassic of Poland. *Journal of Vertebrate Paleontology* 25:78–86.

Sulej, T. 2007. Osteology, variability and evolution of *Metoposaurus*, a temnospondyl from the Late Triassic of Poland. *Palaeontologia Polonica* 64:29–139.

Sulej, T., and D. Majer. 2005. The temnospondyl amphibian *Cyclotosaurus* from the Late Triassic of Poland. *Palaeontology* 48:157–170.

Sun, A.-L. 1963. The Chinese kannemeyerids. *Palaeontologia Sinica*, n.s., C 17:1–109. [Chinese with English summary.]

Surkov, M. V., and M. J. Benton. 2008. Head kinematics and feeding adaptations of the Permian and Triassic dicynodonts. *Journal of Vertebrate Paleontology* 28:1120–1129.

Swinton, W. E. 1939. A new Triassic rhynchocephalian from Gloucestershire. *Annals and Magazine of Natural History*, ser. 11, 4:591–594.

Sytchevskaya, E. K. 1999. Freshwater fish fauna from the Triassic of Northern Asia. In G. Arratia and H.-P. Schultze, eds., *Mesozoic Fishes 2—Systematics and Fossil Record*, pp. 445–468. Munich: Verlag Dr. Friedrich Pfeil.

Tatarinov, L. P. 2005. [A new cynodont (Reptilia, Theriodontia) from the Madygen Svita of Fergana, Kyrgyzstan.] *Paleontologicheskii Zhurnal* 2005(2):81–87. [Russian.]

Taylor, T. N., E. L. Taylor, and M. Krings. 2009. *Paleobotany: The Biology and Evolution of Fossil Plants.* 2nd ed. Boston: Academic Press.

Throckmorton, G. S., J. A. Hopson, and P. Parks. 1981. A redescription of *Toxolophosaurus cloudi* Olson, a Lower Cretaceous herbivorous sphenodontid reptile. *Journal of Paleontology* 55:586–597.

Thulborn, R. A. 1979. A proterosuchian thecodont from the Rewan Formation of Queensland. *Memoirs of the Queensland Museum* 19:331–355.

Thulborn, R. A. 1983. A mammal-like reptile from Australia. *Nature* 303:330–331.

Tintori, A. 1992. Fish taphonomy and Triassic anoxic basins from the Alps: A case history. *Rivista Italiana di Paleontologia e Stratigrafia* 97:393–408.

Tourani, A., J. J. Lund, N. Benaouiss, and R. Gaupp. 2000. Stratigraphy of Triassic syn-rift deposition in Western Morocco *Zentralblatt für Geologie und Paläontologie, Teil I*, 1998(9–10):1193–1215.

Tozer, E. T. 1984. The Triassic and its ammonoids: The evolution of a geological timescale. *Geological Survey of Canada, Miscellaneous Reports* 35:1–171.

Tripathi, S., and P. P. Satsangi. 1963. *Lystrosaurus* fauna from the Panchet Series of the Raniganj Coalfield. *Palaeontologia Indica*, n.s., 37:1–53.

Tverdokhlebov, V. P., G. I. Tverdokhlebova, M. V. Surkov, and M. J. Benton. 2002. Tetrapod localities from the Triassic of the SE of European Russia. *Earth-Science Reviews* 60:1–66.

Unwin, D. M. 2006. *The Pterosaurs from Deep Time.* New York: Pi Press.

Unwin, D. M., V. R. Alifanov, and M. J. Benton. 2000. Enigmatic small reptiles from the Middle-Late Triassic of Kirgizstan. In M. J. Benton, M. A. Shishkin, D. M. Unwin, and E. N. Kurochkin, eds., *The Age of Dinosaurs in Russia and Mongolia*, pp. 177–186. Cambridge: Cambridge University Press.

Veevers, J. J., C. McA. Powell, J. W. Collinson, and O. R. Gamundí. 1994. Synthesis. In J. J. Veevers and C. McA. Powell, eds., *Permian-Triassic Basins and Foldbelts Along the Panthalassan Margin of Gondwanaland*, pp. 331–353. Geological Society of America Memoir 184. Boulder, CO: Geological Society of America.

Vishnyakova, V. N. 1998. [Cockroaches (Insecta, Blattodea) from the Triassic Madygen locality, Central Asia.] *Paleontologicheskii Zhurnal* 1998(5):69–76. [Russian.]

Visscher, H., and C. J. van der Zwan. 1980. Palynology of the circum-Mediterranean Triassic: Phytogeographical and palaeoclimatological implications. *Geologische Rundschau* 70:625–634.

Voigt, S., M. Buchwitz, J. Fischer, D. Krause, and R. Georgi. 2009. Feather-like development of Triassic diapsid skin appendages. *Naturwissenschaften* 96:81–86.

Voigt, S., H. Haubold, S. Meng, D. Krause, J. Buchantschenko, K. Ruckwied, and A. E. Götz. 2006. Die Fossil-Lagerstätte Madygen: Ein Beitrag zur Geologie und Paläontologie der Madygen-Formation (Mittel-bis Ober-Trias, SW-Kirgisistan, Zentralasien). *Hallesches Jahrbuch für Geowissenschaften, Beiheft* 22:85–119.

Vršanský, P. 2003. *Phylloblatta grimaldii* sp. nov.—a new Triassic cockroach (Insecta: Blattaria) from Virginia. *Entomological Problems* 33:51–53.

Walkden, G. M., and N. C. Fraser 1993. Late Triassic fissure sediments and vertebrate faunas: Environmental change and faunal succession at Cromhall, South West Britain. *Modern Geology* 18:511–535.

Walkden, G. M., J. Parker, and S. Kelley. 2002. A Late Triassic impact ejecta layer in southwestern Britain. *Science* 298:2185–2188.

Walker, A. D. 1961. Triassic reptiles from the Elgin area: *Stagonolepis, Dasygnathus* and their allies. *Philosophical Transactions of the Royal Society of London B* 244:103–204.

Walker, A. D. 1964. Triassic reptiles from the Elgin area: *Ornithosuchus* and the origin of carnosaurs. *Philosophical Transactions of the Royal Society of London B* 248:53–134.

Walker, A. D. 1970. A revision of the Jurassic crocodile *Hallopus victor* (Marsh), with remarks on the classification of crocodiles. *Philosophical Transactions of the Royal Society of London B* 257:323–372.

Wall, G. R. T., and H. C. Jenkyns. 2004. The age, origin and tectonic significance of Mesozoic sediment-filled fissures in the Mendip Hills (SW England): Implications for extension models and Jurassic sea-level curves. *Geological Magazine* 141:471–504.

Wang, Z. S., E. T. Rasbury, G. N. Hanson, and W. J. Meyers. 1998. Using the U-Pb system of calcretes to date the time of sedimentation of clastic sedimentary rocks. *Geochimica et Cosmochimica Acta* 62:2823–2835.

Ward, L. F. 1900. The older Mesozoic. In *Status of the Mesozoic Floras of the United States*, pp. 213–748. U.S. Geological Survey 20th Annual Report (for 1898–1899), Part 2. Washington, DC: U.S. Geological Survey.

Ward, P. D. 2006. *Out of Thin Air: Dinosaurs, Birds, and Earth's Ancient Atmosphere*. Washington, DC: Joseph Henry Press.

Ward, P. D., D. R. Montgomery, and R. M. H. Smith. 2000. Altered river morphology in South Africa related to the Permian-Triassic extinction. *Science* 289:1740–1743.

Warren, A. A. 1999. Karoo tupilakosauroid: A relict from Gondwana. *Transactions of the Royal Society of Edinburgh: Earth Sciences* 89:145–160.

Warren, A. A. 2000. Secondarily aquatic temnospondyls of the Upper Permian and Mesozoic. In H. Heatwole and R. L. Carroll, eds., *Amphibian Biology*, vol. 4, *Palaeontology: The Evolutionary History of Amphibians*, pp. 1121–1149. Chipping Norton, NSW: Surrey Beatty.

Warren, A. A., and R. Damiani. 1999. Stereospondyl amphibians from the Elliot Formation of South Africa. *Palaeontologia Africana* 35:45–54.

Warren, A. A., and C. Marsicano. 2000. A phylogeny of Brachyopoidea (Temnospondyli, Stereospondyli). *Journal of Vertebrate Paleontology* 20:462–483.

Warren, A. A., R. Damiani, and A. M. Yates. 2006. The South African stereospondyl *Lydekkerina huxleyi* (Tetrapoda, Temnospondyli) from the Lower Triassic of Australia. *Geological Magazine* 143:877–886.

Watson, D. M. S. 1912. The skeleton of *Lystrosaurus*. *Records of the Albany Museum* 2:287–299.

Watson, D. M. S. 1914. *Procolophon trigoniceps*, a cotylosaurian reptile from South Africa. *Proceedings of the Zoological Society of London* 1914:735–747.

Watson, D. M. S. 1919. The structure, evolution and origin of the Amphibia. The "orders" Rachitomi and Stereospondyli. *Philosophical Transactions of the Royal Society of London B* 209:1–73.

Weems, R. E. 1980. An unusual newly discovered archosaur from the Upper Triassic of Virginia, U.S.A. *Transactions of the American Philosophical Society* 70(7):1–53.

Weems, R. E. 1987. A Late Triassic footprint fauna from the Culpeper basin, northern Virginia (U.S.A.). *Transactions of the American Philosophical Society* 77(1):1–79.

Weems, R. E. 1992. The "terminal Triassic catastrophic extinction event" in perspective: A review of Carboniferous through Early Jurassic terrestrial vertebrate extinction patterns. *Palaeogeography, Palaeoclimatology, Palaeoecology* 94:1–29.

Wegener, A. 1915. *Die Entstehung der Kontinente und Ozeane*. Braunschweig: Friedrich Vieweg & Sohn.

Weinbaum, J. C., and A. Hungerbühler. 2007. A revision of *Poposaurus gracilis* (Archosauria: Suchia) based on two new specimens from the Late Triassic of the southwestern U.S.A. *Paläontologische Zeitschrift* 81:131–145.

Weishampel, D. B., P. Dodson, and H. Osmólska, eds. 2004. *The Dinosauria*. 2nd ed. Berkeley: University of California Press.

Welles, S. P. 1947. Vertebrates from the upper Moenkopi Formation of northern Arizona. *University of California Publications, Bulletin of the Department of Geological Sciences* 27:241–294.

Welles, S. P. 1984. *Dilophosaurus wetherilli* (Dinosauria, Theropoda). Osteology and comparisons. *Palaeontographica A* 185:85–180.

Welles, S. P. 1993. A review of lonchorhynchine trematosaurs (Labyrinthodontia), and the description of a new genus and species of the lower Moenkopi Formation. *PaleoBios* 14:1–24.

Welles, S. P., and J. W. Cosgriff. 1965. A revision of the labyrinthodont family Capitosauridae, and a description of *Parotosaurus peabodyi* n. sp. from the Wupatki Member of the Moenkopi Formation of northern Arizona. *University of California Publications in Geological Sciences* 54:1–148.

Welles, S. P., and R. Estes. 1969. *Hadrokkosaurus bradyi* from the upper Moenkopi Formation of Arizona, with a review of the brachyopid labyrinthodonts. *University of California Publications in Geological Sciences* 84:1–56.

Welman, J. 1998. The taxonomy of the South African proterosuchids (Reptilia, Archosauromorpha). *Journal of Vertebrate Paleontology* 18:340–347.

Whalley, P. 1986. Insects from the Italian Upper Trias. *Rivista del Museo Civico di Scienze Naturali "Enrico Caffi," Bergamo* 10:51–60.

Whiteside, D. I. 1986. The head skeleton of the Rhaetian sphenodontid *Diphydontosaurus avonis* gen. et sp. nov. and the modernizing of a living fossil. *Philosophical Transactions of the Royal Society of London B* 312:379–430.

Whiteside, D. I., and J. E. A. Marshall. 2008. The age, fauna and palaeoenvironment of the Late Triassic fissure deposits of Tytherington, South Gloucestershire, UK. *Geological Magazine* 145:105–147.

Whiteside, D. I., and D. Robinson. 1983. A glauconitic clay-mineral from a speleological deposit of Late Triassic age. *Palaeogeography, Palaeoclimatology, Palaeoecology* 41:81–85.

Wild, R. 1973. Die Triasfauna der Tessiner Kalkalpen. XXIII. *Tanystropheus longobardicus* (Bassani) (Neue Ergebnisse). *Abhandlungen der Schweizerischen Paläontologischen Gesellschaft* 95:1–162.

Wild, R. 1978. Die Flugsaurier (Reptilia, Pterosauria) aus der Oberen Trias von Cene bei Bergamo, Italien. *Bolletino della Società Paleontologica Italiana* 17:176–256.

Wild, R. 1984. A new pterosaur (Reptilia: Pterosauria) from the Upper Triassic of Friuli, Italy. *Gortania* 5:45–62.

Wild, R. 1989. *Aetosaurus* (Reptilia: Thecodontia) from the Upper Triassic (Norian) of Cene near Bergamo, Italy, with a revision of the genus. *Rivista del Museo Civico di Scienze Naturali "Enrico Caffi", Bergamo* 14:1–24.

Wilson, K. M., D. Pollard, W. W. Hay, S. L. Thompson, and C. N. Wold. 1994. General circulation model simulations of Triassic climates: Preliminary results. In G. de V. Klein, ed., *Pangea: Paleoclimate, Tectonics, and Sedimentation during Accretion, Zenith, and Breakup of a Supercontinent*, pp. 91–116. Geological Society of America Special Paper 288. Boulder, CO: Geological Society of America.

Winkler, T. C. 1886. Histoire de l'ichnologie. Étude ichnologique sur les empreintes de pas d'animaux fossiles suivie de la description des plaques à impressions d'animaux qui se trouvent au Musée Teyler. *Archives du Musée Teyler*, sér. 2, 2:241–440.

Witzmann, F., R. R. Schoch, and M. W. Maisch. 2008. A relict basal tetrapod from Germany: First evidence of a Triassic chroniosuchian outside Russia. *Naturwissenschaften* 95:67–72.

Woodward, A. S. 1907. On a new dinosaurian reptile (*Scleromochlus Taylori*, gen. et sp. nov.) from the Trias of Lossiemouth, Elgin. *Quarterly Journal of the Geological Society of London* 63:140–146.

Wopfner, H. 2002. Tectonic and climatic events controlling deposition in Tanzanian Karoo basins. *Journal of African Earth Sciences* 34:167–177.

Wu, X.-c. 1981. [The discovery of a new thecodont from north-east Shanxi.] *Vertebrata PalAsiatica* 19:122–132. [Chinese with English summary.]

Wu, X.-c. 1994. Late Triassic–Early Jurassic sphenodontians from China and the phylogeny of Sphenodontia. In N. C. Fraser and H.-D. Sues, eds., *In the Shadow of the Dinosaurs: Early Mesozoic Tetrapods*, pp. 38–69. New York: Cambridge University Press.

Wu, X.-c., and A. P. Russell. 2001. Redescription of *Turfanosuchus dabanensis* (Archosauriformes) and new information on its phylogenetic relationships. *Journal of Vertebrate Paleontology* 21:40–50.

Yates, A. M. 2003a. A new species of the primitive dinosaur *Thecodontosaurus* (Saurischia: Sauropodomorpha) and its implications for the systematics of early dinosaurs. *Journal of Systematic Palaeontology* 1:1–42.

Yates, A. M. 2003b. The species taxonomy of the sauropodomorph dinosaurs from the Löwenstein Formation (Norian, Late Triassic) of Germany. *Palaeontology* 46:317–337.

Yates, A. M., and J. W. Kitching. 2003. The earliest known sauropod dinosaur and the first steps towards sauropod locomotion. *Proceedings of the Royal Society of London B* 270:1753–1758.

Yates, A. M., and A. A. Warren. 2000. The phylogeny of 'higher' temnospondyls (Vertebrata, Choanata) and its implications for the monophyly and origins of the Stereospondyli. *Zoological Journal of the Linnean Society* 128:77–121.

Young, C. C. 1935. On two skeletons of Dicynodontia from Sinkiang. *Bulletin of the Geological Society of China* 14:483–517.

Young, C. C. 1936. On a new *Chasmatosaurus* from Sinkiang. *Bulletin of the Geological Society of China* 15:291–320.

Young, C. C. 1964. The pseudosuchians in China. *Palaeontologia Sinica*, n.s., C, 19:1–205. [Chinese with English summary.]

Zan, S., B. J. Axsmith, N. C. Fraser, F. Liu, and D. Xing. 2008. New evidence for Laurasian corystosperms: *Umkomasia* from the Upper Triassic of northern China. *Review of Palaeobotany and Palynology* 149:202–207.

Zeigler, K. E., and J. W. Geissman. 2008. Magnetostratigraphy of the Upper Triassic Chinle Group and implications for the age and correlation of Upper Triassic strata in North America. *Geological Society of America Abstracts with Program.* http://a-c-s.confex.com/crops/2008am/webprogram/Paper47897.html.

Zeigler, K. E., A. B. Heckert, and S. G. Lucas. 2003. Paleontology and geology of the Upper Triassic (Revueltian) Snyder Quarry, New Mexico. *New Mexico Museum of Natural History & Science Bulletin* 24:1–132.

Zeigler, K. E., S. Kelley, and J. W. Geissman. 2008. Revisions to the stratigraphic nomenclature of the Upper Triassic Chinle Group in New Mexico: New insights from geologic mapping, sedimentology, and magnetostratigraphic/paleomagnetic data. *Rocky Mountain Geology* 43:121–141.

Zerfass, H., F. Chemale Jr., C. L. Schultz, and E. Lavina. 2004. Tectonics and sedimentation in Southern South America during Triassic. *Sedimentary Geology* 166:265–292.

Zerfass, H., E. L. Lavina, C. L. Schultz, A. J. V. Garcia, U. F. Faccini, and F. Chemale Jr. 2003. Sequence stratigraphy of continental Triassic strata of Southernmost Brazil: A contribution to Southwestern Gondwana palaeogeography and palaeoclimate. *Sedimentary Geology* 161:85–105.

Ziegler, A. M., G. Eshel, P. M. Rees, T. A. Rothfus, D. B. Rowley, and D. Sunderlin. 2003. Tracing the tropics across land and sea: Permian to present. *Lethaia* 36:227–254.

Ziegler, A. M., J. M. Parrish, J. Yao, E. D. Gyllenhaal, D. B. Rowley, J. T. Parrish, S. Nie, A. Bekker, and M. L. Hulver. 1993. Early Mesozoic phytogeography and climate. *Philosophical Transactions of the Royal Society of London B* 341:297–305.

INDEX

Printed in the USA
CPSIA information can be obtained
at www.ICGtesting.com
JSHW051425221024
72173JS00006B/1401